Dielectric Properties of Agricultural Materials and Their Applications

Dielectric Properties of Agricultural Materials and Their Applications

Stuart O. Nelson

U.S. Department of Agriculture
Richard B. Russell Agricultural Research Center
Athens, GA
USA

AMSTERDAM • BOSTON • HEIDELBERG • LONDON
NEW YORK • OXFORD • PARIS • SAN DIEGO
SAN FRANCISCO • SINGAPORE • SYDNEY • TOKYO

Academic Press is an imprint of Elsevier

Academic Press is an imprint of Elsevier
32 Jamestown Road, London NW1 7BY, UK
525 B Street, Suite 1800, San Diego, CA 92101-4495, USA
225 Wyman Street, Waltham, MA 02451, USA
The Boulevard, Langford Lane, Kidlington, Oxford OX5 1GB, UK

Notices
Knowledge and best practice in this field are constantly changing. As new research and experience broaden
our understanding, changes in research methods, professional practices, or medical treatment may become
necessary.

Practitioners and researchers must always rely on their own experience and knowledge in evaluating and
using any information, methods, compounds, or experiments described herein. In using such information
or methods they should be mindful of their own safety and the safety of others, including parties for whom
they have a professional responsibility.

To the fullest extent of the law, neither the Publisher nor the authors, contributors, or editors, assume any
liability for any injury and/or damage to persons or property as a matter of products liability, negligence or
otherwise, or from any use or operation of any methods, products, instructions, or ideas contained in the
material herein.

ISBN: 978-0-12-802305-1

British Library Cataloguing-in-Publication Data
A catalogue record for this book is available from the British Library.

Library of Congress Cataloging-in-Publication Data
A catalog record for this book is available from the Library of Congress.

For Information on all Academic Press publications
visit our website at http://store.elsevier.com/

Typeset by MPS Limited, Chennai, India
www.adi-mps.com

Printed and bound in the United States of America

Working together
to grow libraries in
developing countries

www.elsevier.com • www.bookaid.org

Dedicated to my parents,
Irvin A. and Agnes E. Nelson, and to two remarkable women,
Joye Fricke Nelson and Ellen White Nelson,
who shared my productive years.

Contents

Foreword xiii
Preface xv
Acknowledgments xix

CHAPTER 1 Theory and Fundamental Principles 1
    1.1 Dielectric Properties of Materials 1
    1.2 Variation of Dielectric Properties 3
        1.2.1 Frequency Dependence 3
        1.2.2 Temperature Dependence 6
        1.2.3 Density Dependence 6
        References 8

CHAPTER 2 Measurement of Dielectric Properties 11
    2.1 General Principles 11
        2.1.1 Audio Frequencies 14
        2.1.2 1 to 50 MHz 15
        2.1.3 50 to 250 MHz 20
        2.1.4 200 to 500 MHz 21
        2.1.5 Microwave Frequencies 22
        2.1.6 Broadband Measurements 26
        References 28

CHAPTER 3 General Agricultural Applications 33
    3.1 Dielectric Heating 33
    3.2 Microwave Heating 34
    3.3 Basic Differences 35
    3.4 Product Quality Sensing 36
    3.5 Treating Seed-Borne Pathogens 37
        3.5.1 Loose Smut in Barley 37
        3.5.2 Bacteria on Alfalfa Seed 38
        References 39

CHAPTER 4 Insect Control Applications 41
    4.1 Stored-Grain Insects 41
        4.1.1 Selective Dielectric Heating 41
        4.1.2 Experimental Findings 42
        4.1.3 Entomologic Factors 43
        4.1.4 Physical Factors 45
        4.1.5 Practical Aspects 50

4.2 Pecan Insects...51
4.3 Summary..53
References...53

CHAPTER 5   Seed Treatment Applications.................................57
5.1 Background Information...57
5.2 Alfalfa Seed Studies..58
5.2.1 Basic Factors...59
5.2.2 Experimental Findings...60
5.2.3 Aspects of Practical Application...63
5.3 Sweetclover Seed...64
5.4 Other Small-Seeded Legumes and Some Cereals 65
5.5 Vegetable Seed 65
5.6 Tree Seed 66
5.6.1 Pine Seed 66
5.6.2 Other Tree and Woody Plant Seeds...................................67
5.7 Summary 68
References...69

CHAPTER 6   Product Conditioning Applications 73
6.1 Background Information...73
6.2 Drying of Chopped Alfalfa 73
6.3 Improving Nutritional Value of Soybeans.....................................74
6.4 Quality Maintenance in Pecans..74
References...76

CHAPTER 7   Grain and Seed Moisture Sensing Applications..........77
7.1 Background Information...77
7.2 Early History...77
7.3 Dielectric Properties..78
7.4 Moisture Content Sensing 83
7.5 Summary of Grain Moisture Sensing Development.........................88
7.6 Single Kernel or Seed Moisture Sensing 89
7.6.1 Single-Kernel Grain Moisture Measurements on Corn 90
7.6.2 Single Soybean Seed Moisture Measurements 93
7.6.3 Measuring Moisture Content in Single Kernels of Peanuts 94
7.6.4 Comparison of Four Single-Kernel Moisture Sensing
Techniques for Corn..95
7.6.5 Single Nut and Kernel Pecan Moisture Sensing...................97
7.6.6 Single-Kernel Microwave-Resonator Moisture Sensing 99
References...103

CHAPTER 8   Assessment of Soil Treatment for Pest Control........................... 109
    8.1 Soil Microorganisms and Nematodes 109
    8.2 Soil Treatment for Weed Control 110
    8.3 Initial Assessment...111
    8.4 Basic Principles..112
    8.5 Further Assessment...113
        8.5.1 Attenuation...113
        8.5.2 Selective Heating..115
        8.5.3 Soil Insect and Nematode Treatment 117
    8.6 Discussion 118
    8.7 Conclusions..120
    References...120

CHAPTER 9   Quality Sensing in Fruits and Vegetables 123
    9.1 Background Information...123
    9.2 Studies on the Use of Dielectric Properties..123
        9.2.1 Melon Studies 124
        9.2.2 Apple Studies...125
        9.2.3 Onion Studies...126
        9.2.4 Sensing the Moisture Content of Dates 127
    References...129

CHAPTER 10   Mining Applications 131
    10.1 Background Information—Coal..131
    10.2 Dielectric Properties Measurements on Coal 131
    10.3 Dielectric Heating of Coal—Pyrite Mixtures..132
    10.4 Background Information—Minerals..134
    10.5 Measurements of the Dielectric Properties of Minerals...................................134
    10.6 Coal and Limestone Measurements 136
    10.7 Sensing Pulverized Material Mixture Proportions...142
        10.7.1 Principles of Resonant Cavity Measurement...143
        10.7.2 Measuring Mixture Proportions 143
    References...145

CHAPTER 11   Dielectric Properties of Selected Food Materials 147
    11.1 Measurement of the Dielectric Properties of Some
         Food Materials...147
    11.2 Measurements on Hydrocolloid Food Ingredients..159
    11.3 Dielectric Properties of Chicken Meat for Quality Sensing..............................159
    References...164

CHAPTER 12   Sensing Moisture and Density of Solid Biofuels 167
   12.1 Pine Pellets (Pelleted Sawdust) 167
       12.1.1 Bulk Density Determination for Pine Pellets from
           Complex-Plane Representation 169
       12.1.2 Moisture Content Determination with Density-Independent
           Calibration Function 170
   12.2 Peanut-Hull Pellets 171
       References 173

CHAPTER 13   Dielectric Properties Models for Grain and Seed 175
   13.1 Model Development for Wheat 175
   13.2 Models for Corn 181
   13.3 Models for Soybeans 182
   13.4 Models for Barley 182
   13.5 Models for Rice 183
   13.6 Composite Model for Cereal Grain 185
       13.6.1 Density Dependence 185
       13.6.2 Frequency Dependence 186
       13.6.3 Moisture Dependence 186
       13.6.4 Model Development 187
   13.7 Models at Microwave Frequencies 188
       References 192

**CHAPTER 14   Development of Microwave Moisture-Sensing
                 Instrumentation 195**
   14.1 Peanut Kernel Moisture Meter 195
       14.1.1 Background 195
       14.1.2 Dielectric Properties 196
       14.1.3 Practical Meter Development 196
   14.2 Peanut Drier Control by Monitoring Kernel Moisture Content 206
       14.2.1 Background Information 206
       14.2.2 Peanut Drier Control System 207
       14.2.3 Drier Control System Tests 207
       References 208

CHAPTER 15   Dielectric Properties Data 211
   15.1 Earlier Tabulations 211
   15.2 Grain and Seed Data 212
       15.2.1 Hard Red Winter Wheat Data 212
       15.2.2 Shelled Hybrid Yellow-Dent Field Corn Data 216
       15.2.3 Rice Data 216

15.2.4 Other Grain and Seed Data 224
15.2.5 Individual Grain Kernels and Seeds Data 232
15.2.6 Microwave Data for Cereal Grains and Oilseeds 232
15.2.7 Fresh Fruit and Vegetable Data 238
15.2.8 Pecan Nut Data 238
15.3 Insect data 244
References 244

**CHAPTER 16 Closely Related Physical Properties Data for Grain
and Seed.. 247**
16.1 Moisture Dependence of Kernel and Bulk Densities for Wheat
and Corn 247
16.1.1 Wheat Lots 247
16.1.2 Corn Lots 248
16.1.3 Moisture Determinations 248
16.1.4 Density Measurements 248
16.1.5 Results for Hard Red Winter Wheat 249
16.1.6 Results for Yellow-Dent Field Corn 251
16.1.7 Porosity of Wheat and Corn 253
16.1.8 Summary 254
16.2 Grain Kernel and Seed Dimensions and Densities for Agricultural Crops 254
16.2.1 Grain and Seed Lots 255
16.2.2 Dimensional and Weight Measurements 256
16.2.3 Density Measurements 256
16.2.4 Resulting Data 256
16.2.5 Summary 262
References 263

Index 265

Foreword

From measuring, to modeling, and applying the dielectric properties, the author's scientific journey in *Dielectrics* has been unique, at times challenging, but always rewarding. And this book is a vivid illustration of such a journey.

Dielectric properties are intrinsic properties that describe the electromagnetic wave—material interaction. They are often referred to as the electrical signature of a given material. It is the intrinsic nature of these properties that enables useful development of sensing applications and dielectric heating applications. The dielectric properties are dependent on frequency, temperature, and composition. Their dependence on composition is the basis for developing methods and sensors for rapid and nondestructive assessment of physical properties of materials. For materials containing water, the field of application of such methods and sensors is broad and comprehensive, covering fields such as food, agriculture, pharmacy, and mining.

At radio frequencies, water strongly influences the dielectric properties, and therefore radio-frequency methods and sensors are most suitable for sensing moisture and other water-related parameters. However, the bound nature of water in most materials, and their dielectric properties, often described as being somewhere between those of ice and those of liquid water, make it difficult to derive analytical models that explicitly correlate the measured dielectric properties with the physical properties of interest. From an engineering standpoint, a full understanding of the phenomenon at hand is not always needed as long as one can develop a practical and useful solution for a particular problem.

For decades, since the discovery of the potential for using radio-frequency techniques for non-destructive, rapid assessment of physical properties of materials, correlations between the dielectric properties and physical properties of interest were established empirically and were successfully used in many industrial applications. Notable among these are the simultaneous and nondestructive determination of moisture content and bulk density of granular materials from measurement of the dielectric properties at a single microwave frequency. Moreover, in recent years, the use of advanced statistical methods and measurement of dielectric properties over a broad frequency range opened new possibilities for developing a new generation of multiparameter radio-frequency sensors for in-line and off-line measurement applications. As for all indirect methods, reliability of such sensors relies primarily on the accuracy with which the dielectric properties can be measured and the robustness of the calibration against the standard methods.

For over six decades, the author of this book dedicated his scientific career to developing radio-frequency techniques for accurate measurement of the dielectric properties, and he meticulously collected data over broad ranges of frequency for various agricultural, food, and mining materials. These methods and data have been widely used worldwide by scientists and engineers in various sensing and heating applications and are here compiled in one single volume. Therefore, I believe this book constitutes an important and unique reference for the scientist and engineer and all those interested in the field of dielectric measurements and their applications. I know that I have been privileged to work in this field with Dr. Nelson as a mentor and a co-worker. And it all started with reading some of his papers while I was working on my PhD in France. In some way, by encouraging Dr. Nelson to write this book, I wanted to share that privilege with everyone interested in this field. I remember Dr. Nelson telling me about a dream he had occasionally. In that dream, he conceived a simple and elegant theory valid for all dielectrics. However, upon awakening, it did not make sense, or he could not remember it all. Well, this book, in some way, realizes that dream.

Samir Trabelsi, B.Sc., M.Sc., Ph.D.
Research Electronics Engineer,
US Department of Agriculture,
Agricultural Research Service,
Athens, GA, USA

Preface

This book was prepared as a reference for engineers, scientists, and students interested in dielectric properties of materials and their use in solving problems related to agricultural production, marketing, storage, and product processing and distribution operations. It summarizes the findings of 65 years of research by the US Department of Agriculture on dielectric properties of agricultural products, methods of measurement, and various applications for this kind of information. It draws together materials from more than 50 different scientific and engineering journals published over that period of time, making that body of information available in a single source for reference by researchers and practicing engineers in the agricultural and related industries.

The book is not intended to be a comprehensive review of available literature relating to dielectric properties of foods and other agricultural materials, although extensive reviews of some topics are included. It will hopefully be helpful to research engineers and scientists in university, government, and private industry sectors, and graduate students and undergraduate students interested in new applications for the solution of agricultural and related problems. Agricultural and biological engineers, electrical engineers, and physicists with industrial companies developing sensors, instruments and equipment for farm production, marketing, and product processing operations should find the presented information useful. The book should also be helpful to anyone seeking a better understanding of dielectric properties of materials and the application of radio-frequency (RF) and microwave electromagnetic energy for solutions of problems in agricultural and related fields.

The book will provide a basic understanding of dielectric properties of agricultural products and materials and the variables that influence these properties. Discussions of a number of applications of information on dielectric properties to dielectric and microwave heating and quality sensing applications provide the reader with information on applications that have been studied and developed. Assessments of the potential for some of the more highly studied applications are provided, and these discussions may stimulate new ideas for solving problems associated with production, handling, and processing of agricultural products. The presentation of fundamental principles and historical reviews of applications should be helpful to readers.

In Chapter 1, the theory and fundamental principles of dielectric properties are presented. Dielectrics and dielectric properties of materials are defined, and the fundamental relationships of dielectric properties with electromagnetic energy are discussed. The variation of the dielectric properties of materials with respect to frequency of the fields to which they are subjected, the temperature of the materials, and their density, are discussed. Since water is an important component of agricultural materials, the dielectric behavior of water is included in the discussion.

General principles for measuring dielectric properties are discussed in Chapter 2, and techniques and instruments used for measurements on agricultural materials are briefly presented. Equipment and techniques are described for measurements ranging from audio frequencies through the radio frequencies and into the microwave region.

In Chapter 3, general agricultural applications are discussed. The use of information on dielectric properties for applications in agriculture can be divided into two general categories: dielectric heating and product quality sensing. Fundamentals of dielectric and microwave heating are discussed, and results of studies on controlling loose smut in barley and human pathogens on alfalfa seed are presented. Other applications that have been studied in more detail are covered in separate chapters.

Results of extensive experimental work on controlling stored-grain insects with RF (high-frequency and microwave) electric field exposures are summarized in Chapter 4. The basis for selectively heating the insects by RF dielectric heating is explained, and the frequency range from 10 to 100 MHz is identified for the best selective heating of insects in grain. Results of experimental work on controlling insects that infest pecans are also summarized. Major findings from research on controlling stored-product insects are presented in graphical and tabular forms.

An historical background on electrical treatment to improve seed germination and seedling performance is presented in Chapter 5. Extensive studies on treatment of alfalfa seed with RF dielectric heating exposures to increase germination through hard seed reduction, and examination of seeds to explain effects of treatment, are summarized. Experiments with RF treatment of seeds of sweetclover, other small-seeded legumes, vegetables, pine, and other woody plant and tree seeds are also summarized.

Experiments conducted to determine the influence of RF dielectric heating and microwave heating on quality attributes of a few agricultural products are summarized in Chapter 6. RF treatment of alfalfa forage resulted in significantly improved retention of carotene. Dielectric heating treatments, at 42 and 2450 MHz, of intact soybeans, containing only their innate moisture, inactivated the trypsin inhibitor of raw soybeans and improved their nutritional value. RF dielectric heating treatments of pecan nut kernels showed promise for maintaining quality of pecans in storage, and the dielectric heating treatments offer the advantage of maintaining the desirable lighter color of the pecan kernels.

Grain and seed moisture sensing applications, the most widely used practical agricultural application of information on dielectric properties, are discussed in Chapter 7. Basic principles of moisture sensing through dielectric properties are presented, and the dependence of dielectric properties on frequency, moisture content, temperature, and bulk density are discussed. The density-independence advantage of microwave moisture sensing is included. The sensing of peanut kernel moisture content in unshelled peanuts by microwave measurements is also presented. The use of RF impedance and microwave resonance techniques for sensing moisture in single kernels of grain, seeds, and pecan nuts is also discussed.

In Chapter 8, an assessment of microwave heating for soil pest control is presented. The use of microwave energy has been proposed frequently as an alternative method for controlling pests in the soil, such as weed seeds, insects, nematodes, and soil-borne plant pathogens. Basic principles governing absorption of microwave energy and microwave heating are presented, and a review of pertinent literature is included. Upon considering the basic principles of microwave energy absorption by dielectric materials and the experimental work that has been reported, there appears to be little probability for the practical application of microwave power for field use in controlling agricultural pests in the soil.

Quality sensing in fruits and vegetables through their dielectric properties is discussed in Chapter 9. New, nondestructive techniques for sensing quality of fresh fruits and vegetables would be helpful to producers, handlers, and consumers. Dielectric properties of fresh fruits and vegetables that are available in the literature are reviewed, and studies are described in which efforts were made to find correlations between qualities such as sweetness in melons and apples and their RF dielectric properties. Dielectric properties of onions have also been studied for use in measuring moisture content, which is important in onion curing. Impedance measurements on dates at frequencies of 1 and 5 MHz were found useful in sensing moisture content for the sorting of fresh dates.

The ability to measure dielectric properties developed for agricultural applications, spanning wide ranges of frequency, attracted interest from the mining industry, so efforts were made to provide assistance in solving some of the problems facing that industry, and the findings are summarized in Chapter 10. Dielectric properties of coal and coal—pyrite mixtures were measured to identify frequencies for selective heating of pyrite to enhance magnetic separation for reducing pollution from burning coal. Also, dielectric properties of 10 pulverized and purified mineral samples were measured from 1 to 22 GHz to provide information useful in research on rock fragmentation by microwave heating for recovery of minerals. Microwave dielectric properties of pulverized coal and limestone samples were measured in support of mine safety research for explosion prevention, and a resonant cavity technique was developed for measuring coal—limestone dust mixture proportions.

Dielectric properties of selected food materials are discussed in Chapter 11. Their dielectric properties are important in understanding and modeling their behavior in RF and microwave processing, heating, and cooking of food materials. They are important because they influence the absorption of energy and conversion to heat. Thus, they are also important in the design of RF and microwave processing equipment and in the design of foods and meals intended for microwave preparation. Some food materials were selected to illustrate the behavior of their dielectric properties with respect to frequency, temperature, and moisture content. The importance of dielectric relaxation and ionic conduction is discussed in explaining the behavior of food dielectric properties with respect to frequency and temperature.

In Chapter 12, the sensing of moisture and density of solid biofuels is considered. Biofuels are an alternative source of energy with the advantage of renewability, compared to fossil fuels. Moisture content is an important factor in pricing, optimizing combustion, and storing of solid biofuels. Because of the correlation between dielectric properties and moisture content, these properties of solid biofuels can be used for sensing or measuring moisture content. As with grain and seed, microwave measurement techniques can provide both moisture content and bulk density values through measurement of the dielectric properties of these materials. This application is described and illustrated for pine pellets, made from sawdust, and peanut-hull pellets made from that byproduct of the peanut industry.

Mathematical models for the dielectric properties of grain and seed are discussed in Chapter 13. These properties of grain and seed are of particular importance for their use in rapid measurement of moisture content. Development of mathematical models providing the dielectric constant as a function of frequency, moisture content, and temperature is explained in detail, and models are presented for corn, barley, and rice. A composite model for cereal grains provides both the dielectric constant and loss factor values at 24°C. Models for the microwave dielectric properties at 23°C of several cereal grains and oilseeds are presented for calculation as functions of frequency and moisture content.

Development of practical microwave moisture sensing instrumentation is discussed in Chapter 14. Two practical applications have been addressed, one for kernel moisture sensing in unshelled peanuts, or peanut pods—a microwave peanut kernel moisture meter—and another for control of peanut driers by monitoring kernel moisture in unshelled peanut pods during the drying process. Both are practical applications in the peanut industry, but extension to the grain and seed and other industries is only natural for the potential benefits to be derived.

An attempt is made in Chapter 15 to accumulate, for reference purposes, data on the dielectric properties of agricultural materials, and to identify other sources of such data that have already been tabulated. Data on dielectric properties over wide frequency ranges, and for ranges of other variables influencing the dielectric properties, are presented graphically or in tabular form for wheat, corn, rice, grain sorghum, soybeans, oats, barley, and winter rye. Microwave dielectric properties of individual kernels or seeds are presented for wheat, corn, soybeans, and rice. Data on microwave dielectric properties, at several frequencies from 5 to 15 GHz, are tabulated for wheat, corn, barley, oats, soybeans, and canola. Tabular data are also included for 11 fruits and vegetables, pecan nuts, and for insects.

Data on physical properties, closely related to the dielectric properties of grain and seed, are presented in Chapter 16. Some physical properties of grain have been studied in detail because of their influence on the dielectric properties of those materials. The moisture dependence of density, both kernel density and bulk density, are prominent among those. Also, kernel dimensions and kernel and seed densities are of interest, and they have been determined for a range of agricultural grain and crop seeds. Graphical relationships between bulk density and moisture content, and between kernel density and moisture content, are presented for both wheat and corn. Porosity and moisture content relationships for wheat and corn are shown graphically. Also, for a reasonable selection of seeds of grain and other agricultural crops, physical properties, including moisture content, test weight, bulk density, seed weight, seed volume, seed density, and seed dimensions, are tabulated for reference along with volume coefficients for relating kernel or seed dimensions and volume.

Acknowledgments

The author expresses his sincere appreciation to many colleagues in research and others whose efforts contributed to the development of information presented in this book. Truman E. Hienton, Chief of the Farm Electrification Research Branch, Bureau of Plant Industry, Soils and Agricultural Engineering, US Department of Agriculture, is recognized for establishing the initial research project on radio-frequency heating for agricultural applications in 1949 at the Nebraska Agricultural Experiment Station, University of Nebraska, Lincoln, NE, in cooperation with the Department of Agricultural Engineering. Leo H. Soderholm is recognized for his initial project leadership and recommending the measurement of dielectric properties of grain as a thesis project for the author, who joined the research effort as a graduate student in 1950 and assumed the project leadership in 1954 under the Agricultural Research Service (ARS), which came into being within the US Department of Agriculture in 1953. Special thanks are due Wayne W. Wolf, and LaVerne E. Stetson, long-term dedicated co-worker, for conduct of research at Lincoln, NE. Research cooperators there providing essential support included Elda R. Walker, Associate Professor of Botany, Emeritus, Harold J. Ball, Professor of Entomology (insect physiologist) and his students, Donald L. Silhacek, Ahmed M. Kadoum, and Patte S. Rai, Entomologists John J. Rhine and Benjamin H. Kantack, William R. Kehr, Research Agronomist, and Allen R. Edison, Professor of Electrical Engineering. Other physics and engineering students contributing significantly to the research effort included Carl. W. Schlaphoff, Paul T. Corcoran, James L. Jorgenson, W. Clint Jurgens, and Berlin P. Kwok.

Co-researchers at the Manhattan, KS, Stored-Grain Insects Laboratory, Agricultural Marketing Service, US Department of Agriculture, providing essential entomological cooperation were Norman M, Dennis, and W. Keith Whitney. Gratitude is also expressed to the following scientists from other locations who provided essential research cooperation: Robert M. Heckert, Eastern States Farmers Exchange, Buffalo, NY, Gabriel E. Nutile, Asgrow Seed Co., Twin Falls, ID, Leslie A. T. Ballard, Division of Plant Industry, CSIRO, Canberra, Australia, Rodney W. Bovey, College Station, TX, and Earl W. Belcher, Jr., Eastern Tree Seed Laboratory, US Forest Service, Macon, GA. Engineers contributing to the electrical seed treatment research included Russell B. Stone, and J, C. Webb, Knoxville, TN, C. Alan Pettibone, Pullman, Washington, and D. W. Works, University of Idaho, Moscow, ID. Biochemists providing cooperative research included Raymond Borchers, University of Nebraska, and Akiva Pour-El, PEACO, St. Paul, MN.

Following the relocation of the research project by ARS to the Richard B. Russell Agricultural Research Center, Athens, GA, in 1976, cooperation was initiated with the Department of Agricultural Engineering at The University of Georgia. Special thanks are due students and research colleagues, Kurt C. Lawrence, Chari V. K. Kandala, Wanda L. Bellamy, Wensheng Kuang, Philip G. Bartley, Jr., Micah A. Lewis and Murat Sean McKeown. Special gratitude is also expressed to Andrzej W. Kraszewski and Samir Trabelsi, who both joined the research effort as visiting scientists and became long-term co-researchers providing continuous essential contributions to the progress of the research. Significant contributions of other visiting scientists, including Anuradha Prakash, Shahab Sokhansanj, Sang Ha Noh, Tian-su You, and Wenchuan Guo are also acknowledged.

The cooperative research efforts of several ARS scientists, including Roy W. Forbus, Jr., Samuel D. Senter, and Jerry A. Payne, and the cooperation of Stanley J. Kays, Professor of Horticulture, The University of Georgia are gratefully acknowledged.

Finally, personal thanks are due Samir Trabelsi, without whose continuing encouragement, this book would never have been compiled, and the helpful advice of Micah A. Lewis, on software use in preparation of the manuscript and files for the illustrations, is gratefully acknowledged.

Stuart O. Nelson, B.Sc., M.Sc., M.A., Ph.D., D.Sc.
Collaborator, U.S. Department of Agriculture,
Agricultural Research Service, Athens, GA, USA

THEORY AND FUNDAMENTAL PRINCIPLES

1

The dielectric properties of materials are those electrical characteristics that determine the interaction of materials with electric fields. In radio-frequency (RF) and microwave heating of foods, agricultural products, and other dielectric materials, it is the interaction of the materials with the electric field component of the electromagnetic waves that produces the desired heating effects (Nelson and Trabelsi, 2014). Strictly speaking, radio frequencies range from about 10 kHz to about 100 GHz. These are the frequencies practicable for radio transmission; they span that portion of the electromagnetic spectrum between the audio frequencies and the infrared region (IEEE, 1990). Thus, the RF range includes those frequencies used for microwave heating. However, because RF dielectric heating applications were developed first in the frequency range of about 3−40 MHz, and microwave heating applications came later, there is a tendency—particularly in the food industry—to refer to the lower frequency applications as RF dielectric heating, and to refer to dielectric heating at microwave frequencies, about 1 GHz and higher, as microwave heating. Frequencies for RF communication are further designated as high frequency (HF, 3−30 MHz), very-high frequency (VHF, 30−300 MHz), ultra-high frequency (UHF, 300−3000 MHz), super-high frequency (SHF, 3−30 GHz), and extremely-high frequency (EHF, 30−300 GHz).

1.1 DIELECTRIC PROPERTIES OF MATERIALS

Dielectrics are a class of materials that are poor conductors of electricity, in contrast to materials such as metals that are generally good electrical conductors. Many materials, including foods, living organisms, and most agricultural products, conduct electric currents to some degree, but they are still classified as dielectrics. The electrical nature of these materials can be described by their dielectric properties, which influence the distribution of electromagnetic fields and currents in the region occupied by the materials, and which determine the behavior of the materials in electric fields. Thus, the dielectric properties determine how rapidly a material will warm up in RF or microwave dielectric heating applications. Their influence on electric fields also provides a means for sensing certain other properties of materials, which may be correlated with the dielectric

Dielectric Properties of Agricultural Materials and Their Applications. DOI: http://dx.doi.org/10.1016/B978-0-12-802305-1.00001-4

properties, by nondestructive electrical measurements. Therefore, dielectric properties of agricultural products may be important for quality-sensing applications in the agricultural industry as well as in dielectric heating applications.

A few simplified definitions of dielectric properties are useful in discussing their applications. A fundamental characteristic of all forms of electromagnetic energy is their propagation through free space at the velocity of light, c. The velocity of propagation v of electromagnetic energy in a material depends on the electromagnetic characteristics of that material and is given as:

$$v = \frac{1}{\sqrt{\mu\varepsilon}} \tag{1.1}$$

where μ is the magnetic permeability of the material and ε is the electric permittivity. For free space, this becomes:

$$c = \frac{1}{\sqrt{\mu_0\varepsilon_0}} \tag{1.2}$$

where μ_0 and ε_0 are the permeability and permittivity of free space. Most food and agricultural products are nonmagnetic, so their magnetic permeability has the same value as μ_0. These materials, however, have different permittivities when compared to free space. The absolute permittivity ε_a can be represented as a complex quantity,

$$\varepsilon_a = \varepsilon_a' - j\varepsilon_a'' \tag{1.3}$$

where $j = \sqrt{-1}$. The complex permittivity relative to free space is then given as:

$$\varepsilon_r = \frac{\varepsilon_a}{\varepsilon_0} = \varepsilon_r' - j\varepsilon_r'' \tag{1.4}$$

where ε_0 is the permittivity of free space (8.854×10^{-12} farad/m); the real part ε_r' is called the dielectric constant, and the imaginary part ε_r'' is called the dielectric loss factor. These latter two quantities are the dielectric properties of practical interest, and the subscript r will be dropped for simplification in the remainder of this book. The dielectric constant ε' is associated with the ability of a material to store energy in the electric field in the material, and the loss factor ε'' is associated with the ability of the material to absorb or dissipate energy, that is, to convert electric energy into heat energy. The dielectric loss factor, for example, is an index of the tendency of the material to warm up in a microwave oven. The dielectric constant is also important because of its influence on the distribution of electric fields. For example, the electric capacitance of two parallel conducting plates separated by free space or air will be multiplied by the value of the dielectric constant of a material if the space between the plates is filled with that material.

It should also be noted that $\varepsilon = \varepsilon' - j\varepsilon'' = |\varepsilon|e^{-j\delta}$ where δ is the loss angle of the dielectric. Often, the loss tangent, $\tan\delta = \varepsilon''/\varepsilon'$, or dissipation factor, is also used as a descriptive dielectric parameter, and sometimes the power factor, $\tan\delta/\sqrt{1 + \tan^2\delta}$, is used. The ac conductivity of the dielectric σ in S/m is $\sigma = \omega\varepsilon_0\varepsilon''$, where $\omega = 2\pi f$ is the angular frequency, with frequency f in hertz (Hz). In this book, ε'' is interpreted to include the energy losses in the dielectric due to all operating dielectric relaxation mechanisms and ionic conduction.

1.2 VARIATION OF DIELECTRIC PROPERTIES

The dielectric properties of most materials vary with several influencing factors (Nelson, 1981, 1991; Nelson and Datta, 2001). In hygroscopic materials such as agricultural products, the amount of water in the material is generally a dominant factor. The dielectric properties also depend on the frequency of the applied alternating electric field, on the temperature of the material, and on the density, composition, and structure of the material. In granular or particulate materials, the bulk density of the air–particle mixture is another factor that influences the dielectric properties. Of course, the dielectric properties of materials are dependent on their chemical composition and especially on the presence of mobile ions and the permanent dipole moments associated with water and any other molecules making up the material of interest.

1.2.1 FREQUENCY DEPENDENCE

With the exception of some extremely low-loss materials, that is, materials that absorb essentially no energy from RF and microwave fields, the dielectric properties of most materials vary considerably with the frequency of the applied electric fields. This frequency dependence has been discussed previously (Nelson and Datta, 2001; Nelson, 1973, 1991). An important phenomenon contributing to the frequency dependence of the dielectric properties is the polarization, arising from the orientation with the imposed electric field, of molecules which have permanent dipole moments. The mathematical formulation developed by Debye to describe this process for pure polar materials (Debye, 1929) can be expressed as:

$$\varepsilon = \varepsilon_\infty + \frac{\varepsilon_s - \varepsilon_\infty}{1 + j\omega\tau} \tag{1.5}$$

where ε_∞ represents the dielectric constant at frequencies so high that molecular orientation does not have time to contribute to the polarization, ε_s represents the static dielectric constant, that is, the value at zero frequency (dc value), and τ is the relaxation time, the period associated with the time for the dipoles to revert to random orientation when the electric field is removed. Separation of Eq. (1.5) into its real and imaginary parts yields:

$$\varepsilon' = \varepsilon_\infty + \frac{\varepsilon_s - \varepsilon_\infty}{1 + (\omega\tau)^2} \tag{1.6}$$

$$\varepsilon'' = \frac{(\varepsilon_s - \varepsilon_\infty)\omega\tau}{1 + (\omega\tau)^2} \tag{1.7}$$

The relationships defined by these equations are illustrated in Figure 1.1.

Thus, at frequencies very low and very high with respect to the molecular relaxation process, the dielectric constant has constant values, ε_s and ε_∞, respectively, and the losses are zero. At intermediate frequencies, the dielectric constant undergoes a dispersion, and dielectric losses occur with the peak loss at the relaxation frequency, $\omega = 1/\tau$.

The Debye equation can be represented graphically in the complex ε''-versus-ε' plane as a semicircle with locus of points ranging from $(\varepsilon' = \varepsilon_s, \varepsilon'' = 0)$ at the low-frequency limit to $(\varepsilon' = \varepsilon_\infty, \varepsilon'' = 0)$ at the high-frequency limit (Figure 1.2).

Such a representation is known as a Cole–Cole diagram (Cole and Cole, 1941).

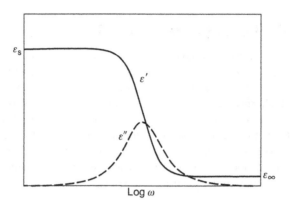

FIGURE 1.1

Dielectric constant and loss factor for a material following the Debye relaxation (Nelson, 1973).

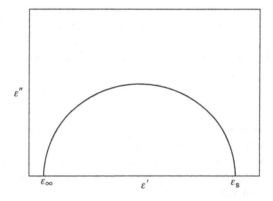

FIGURE 1.2

Cole–Cole diagram for a material following the Debye relaxation (Nelson, 1973).

Because few materials of interest here consist of pure polar materials with a single relaxation time, many other equations have been developed to better describe the frequency-dependent behavior of materials with more relaxation times or a distribution of relaxation times (Cole and Cole, 1941; Bottcher and Bordewijk, 1978; Davidson and Cole, 1951; Havriliak and Negami, 1967; Nigmatullin et al., 2006, 2008; Nigmatullin and Nelson, 2006).

Water, in its liquid state, is a good example of a polar dielectric. The microwave dielectric properties of liquid water are listed in Table 1.1 for several frequencies at temperatures of 20°C and 50°C as selected from data in the literature (Hasted, 1973; Kaatze, 1989).

Kaatze has shown that the dielectric spectra for pure water can be well represented by the Debye equation when using the relaxation parameters given in Table 1.2.

Table 1.1 Microwave dielectric constant ε' and dielectric loss factor ε'' of water at indicated temperatures (Hasted, 1973; Kaatze, 1989)

Frequency, GHz	20°C		50°C	
	ε'	ε''	ε'	ε''
0.6	80.3	2.75	69.9	1.25
1.7	79.2	7.9	69.7	3.6
3.0	77.4	13.0	68.4	5.8
4.6	74.0	18.8	68.5	9.4
7.7	67.4	28.2	67.2	14.5
9.1	63.0	31.5	65.5	16.5
12.5	53.6	35.5	61.5	21.4
17.4	42.0	37.1	56.3	27.2
26.8	26.5	33.9	44.2	32.0
36.4	17.6	28.8	34.3	32.6

Table1.2 Debye dielectric relaxation parameters for water (Kaatze, 1989)

Temperature, °C	ε_s	ε_∞	τ, ps	Relaxation Frequency, GHz
0	87.9	5.7	17.67	9.007
10	83.9	5.5	12.68	12.552
20	80.2	5.6	9.36	17.004
30	76.6	5.2	7.28	21.862
40	73.2	3.9	5.82	27.346
50	69.9	4.0	4.75	33.506
60	66.7	4.2	4.01	39.690

ε_s = static dielectric constant.
ε_∞ = high-frequency dielectric constant.
τ = relaxation time.

The relaxation frequency, $(2\pi\tau)^{-1}$, is provided in Table 1.2 along with the static and HF values of the dielectric constant, ε_s and ε_∞, for water at temperatures between 0°C and 60°C. Thus, Eqs (1.6) and (1.7), together with relaxation parameters listed in Table 1.2, can be used to provide close estimates for the dielectric properties of water over a wide range of frequencies and temperatures. However, water in its pure liquid state appears in food products very rarely (Hasted, 1973). Most often it has dissolved constituents, is physically absorbed in material capillaries or cavities, or is chemically bound to other molecules of the material. Dielectric relaxations of absorbed water take place at lower frequencies than the relaxation of free water (Hasted, 1973), which occurs at about 19.5 GHz for water at 25°C. Depending upon the material structure, there may be various forms of bound water, differing in energy of binding and in dielectric properties. Moist material, in practice, is usually an inhomogeneous mixture, often containing more than one substance with unknown dielectric properties. Thus, it is difficult to understand and predict the dielectric behavior of such materials at different frequencies, temperatures, and hydration levels.

1.2.2 TEMPERATURE DEPENDENCE

The dielectric properties of materials are also temperature dependent, and the nature of that dependence is a function of the dielectric relaxation processes operating under the particular conditions existing and the frequency being used. As temperature increases, the relaxation time decreases, and the loss-factor peak, and the accompanying dispersion noted for ε', illustrated in Figure 1.1, will shift to higher frequencies. Thus, in a region of dispersion, the dielectric constant will tend to increase with increasing temperature as a result of dielectric relaxation, whereas the loss factor may either increase or decrease, depending on whether the operating frequency is higher or lower than the relaxation frequency. However, for complex dielectrics such as agricultural materials, other mechanisms may mask or dominate the dielectric relaxation effects. The temperature dependence of ε_∞ is relatively small (Hasted, 1973), and while that of ε_s is larger, its influence is minor in a region of dispersion. Below and above the dispersion region, the dielectric constant tends to decrease with increasing temperature. Distribution functions can be useful in expressing the temperature dependence of dielectric properties (Bottcher and Bordewijk, 1978), but the frequency- and temperature-dependent behavior of the dielectric properties of most materials is complicated and can perhaps best be determined by measurement at the frequencies and under the other conditions of interest.

1.2.3 DENSITY DEPENDENCE

Because the influence of a dielectric depends on the amount of mass interacting with the electromagnetic fields, the mass per unit volume, or density, will have an effect on the dielectric properties. This is especially notable with particulate dielectrics such as pulverized or granular materials. In understanding the nature of the density-dependence of the dielectric properties of particulate materials, relationships between the dielectric properties of solid materials and those of air—particle mixtures, such as granular or pulverized samples of such solids, are useful.

In some instances, the dielectric properties of a solid may be needed when particulate samples are the only available form of the material. This was true for cereal grains, where kernels were too small for the dielectric sample holders used for measurements (You and Nelson, 1988; Nelson and You, 1989), and in the case of pure minerals that had to be pulverized for purification (Nelson et al., 1989). For some materials, fabrication of samples to exact dimensions required for the measurement of dielectric properties is difficult, and measurements on pulverized materials are more easily performed. In such instances, proven relationships for converting dielectric properties of particulate samples to those for the solid material are important. Several well-known dielectric mixture equations have been considered for this purpose (Nelson and You, 1990; Nelson, 1990, 1992).

The notation used here applies to two-component mixtures, where ε represents the effective permittivity of the mixture, ε_1 is the permittivity of the medium in which particles of permittivity ε_2 are dispersed, and v_1 and v_2 are the volume fractions of the respective components, where $v_1 + v_2 = 1$. Two of the mixture equations found particularly useful for cereal grains were the Complex Refractive Index mixture equation:

$$(\varepsilon)^{1/2} = v_1(\varepsilon_1)^{1/2} + v_2(\varepsilon_2)^{1/2} \tag{1.8}$$

and the Landau and Lifshitz, Looyenga equation:

$$(\varepsilon)^{1/3} = v_1(\varepsilon_1)^{1/3} + v_2(\varepsilon_2)^{1/3} \tag{1.9}$$

To use these equations to determine ε_2, one needs to know the dielectric properties (permittivity) of the pulverized sample at its bulk density (air–particle mixture density), ρ, and the specific gravity or density of the solid material, ρ_2. The fractional part of the total volume of the mixture occupied by the particles (volume fraction), v_2, is then given by ρ/ρ_2. Solving Eqs (1.8) and (1.9), respectively, for the complex permittivity of the solid material and substituting $1 - j0$ for ε_1 (the permittivity of air), the permittivity of the solid materials can be calculated as:

$$\varepsilon_2 = \left(\frac{\varepsilon^{1/2} + v_2 - 1}{v_2}\right)^2 \tag{1.10}$$

$$\varepsilon_2 = \left(\frac{\varepsilon^{1/3} + v_2 - 1}{v_2}\right)^3 \tag{1.11}$$

It has been noted that Eqs (1.8) and (1.9) imply the linearity of $\varepsilon^{1/2}$ and $\varepsilon^{1/3}$, respectively, with the bulk density of the mixture (Nelson, 1992). The Complex Refractive Index and the Landau and Lifshitz, Looyenga relationships thus provide a relatively reliable method for adjusting the dielectric properties of granular and powdered materials with characteristics like grain products from known values at one bulk density to corresponding values for a different bulk density. It follows from Eq. (1.8) that for an air–particle mixture, where $\varepsilon_1 = 1 - j0$, and because $v_1 = 1 - v_2$ and $v_2 = \rho/\rho_2$, that:

$$(\varepsilon_x)^{1/2} = \left[\left(\frac{(\varepsilon_2)^{1/2} - 1}{\rho_2}\right)\right]\rho_x + 1 \tag{1.12}$$

for a mixture of density ρ_x. Similarly,

$$(\varepsilon_y)^{1/2} = \left[\left(\frac{(\varepsilon_2)^{1/2} - 1}{\rho_2}\right)\right]\rho_y + 1 \tag{1.13}$$

for the same mixture of density ρ_y. Equating the slopes of these two lines (the terms in brackets in Eqs (1.12) and (1.13)) and solving for ε_x gives the following:

$$\varepsilon_x = \left[(\varepsilon_y^{1/2} - 1)\frac{\rho_x}{\rho_y} + 1\right]^2 \tag{1.14}$$

which provides an expression for the permittivity of the mixture at any given density ρ_x when the permittivity ε_y is known at density ρ_y. In an analogous way, it follows from Eq. (1.9) that:

$$\varepsilon_x = \left[(\varepsilon_y^{1/3} - 1)\frac{\rho_x}{\rho_y} + 1\right]^3 \tag{1.15}$$

Either Eq. (1.14) or (1.15) should provide reliable conversions of permittivity from one mixture density to another, but the Landau and Lifshitz, Looyenga relationship (Eq. (1.9)) provided somewhat closer estimates within the range of measured densities in work with whole kernel wheat, ground wheat, and finely pulverized coal (Nelson, 1983); so Eq. (1.15) is preferred.

REFERENCES

Bottcher, C.J.F., Bordewijk, P., 1978. Theory of Electric Polarization, Vol. II, Dielectrics in Time-Dependent Fields. Elsevier Scientific Publishing Company, Amsterdam, Oxford, New York, NY.

Cole, K.S., Cole, R.H., 1941. Dispersion and absorption in dielectrics. I. Alternating current characteristics. J. Chem. Phys. 9, 341–351.

Davidson, D.W., Cole, R.H., 1951. Dielectric relaxation in glycerol, propylene glycol, and n-propanol. J. Chem. Phys. 19 (12), 1484–1490.

Debye, P., 1929. Polar Molecules. The Chemical Catalog Co., New York, NY.

Hasted, J.B., 1973. Aqueous Dielectrics. Chapman and Hall, London.

Havriliak, S., Negami, S., 1967. A complex plane representation of dielectric and mechanical relaxation processes in some polymers. Polymer 8, 161–210.

IEEE, 1990. IEEE Standard Dictionary of Electrical and Electronics Terms. The Institute of Electrical and Electronics Engineers, Inc., New York, NY.

Kaatze, U., 1989. Complex permittivity of water as a function of frequency and temperature. J. Chem. Eng. Data 34, 371–374.

Nelson, S.O., 1973. Electrical properties of agricultural products—a critical review. Trans. ASAE 16 (2), 384–400.

Nelson, S.O., 1981. Review of factors influencing the dielectric properties of cereal grains. Cereal Chem. 58 (6), 487–492.

Nelson, S.O., 1983. Density dependence of the dielectric properties of particulate materials. Trans. ASAE 26 (6), 1823–1825, 1829.

Nelson, S.O., 1990. Use of dielectric mixture equations for estimating permittivities of solids from data on pulverized samples. In: Cody, G.D., Geballe, T.H., Sheng, P. (Eds.), Physical Phenomena in Granular Materials, vol. 195. Materials Research Society, Pittsburgh, PA, pp. 295–300.

Nelson, S.O., 1991. Dielectric properties of agricultural products—measurements and applications. IEEE Trans. Electr. Insul. 26 (5), 845–869.

Nelson, S.O., 1992. Estimation of permittivities of solids from measurements on pulverized or granular materials. In: Priou, A. (Ed.), Dielectric Properties of Heterogeneous Materials, vol. PIER 6. Elsevier, New York, Amsterdam, London, Tokyo, pp. 231–271.

Nelson, S.O., Datta, A.K., 2001. Dielectric properties of food materials and electric field interactions. In: Datta, A.K., Anantheswaran, R.C. (Eds.), Handbook of Microwave Technology for Food Applications. Marcel Dekker, Inc., New York, NY.

Nelson, S.O., Trabelsi, S., 2014. Dielectric properties of agricultural products: fundamental principles, influencing factors, and measurement techniques. In: Awuah, G., Ramaswamy, H.S., Tang, J. (Eds.),

Radio-Frequency Heating in Food Processing: Principles and Applications. CRC Press, Taylor & Francis Group, Boca Raton, FL.

Nelson, S.O., You, T.-S., 1989. Microwave dielectric properties of corn and wheat kernels and soybeans. Trans. ASAE 32 (1), 242–249.

Nelson, S.O., You, T.-S., 1990. Relationships between microwave permittivities of solid and pulverised plastics. J. Phys. D Appl. Phys. 23, 346–353.

Nelson, S.O., Lindroth, D.P., Blake, R.L., 1989. Dielectric properties of selected minerals at 1 to 22 GHz. Geophysics 54 (10), 1344–1349.

Nigmatullin, R.R., Nelson, S.O., 2006. Recognition of the "fractional" kinetics in complex systems: dielectric properties of fresh fruits and vegetables from 0.01 to 1.8 GHz. Signal Process. 86, 2744–2759.

Nigmatullin, R.R., Arbuzov, A.A., Nelson, S.O., Trabelsi, S., 2006. Dielectric relaxation in complex systems: quality sensing and dielectric properties of honeydew melons from 10 MHz to 1.8 GHz. J. Instrum. 1, P10002.

Nigmatullin, R.R., Osokin, S.I., Nelson, S.O., 2008. Application of fractional-moments statistics to data for two-phase dielectric mixtures. IEEE Trans. Dielectr. Electr. Insul. 15 (5), 1385–1392.

You, T.-S., Nelson, S.O., 1988. Microwave dielectric properties of rice kernels. J. Microw. Power Electromagn. Energy 23 (3), 150–159.

MEASUREMENT OF DIELECTRIC PROPERTIES

Because the dielectric properties of agricultural materials were unknown when they were needed for various applications, it was necessary to determine the required data by measurement. The measurement techniques appropriate for any particular application depend on the frequency of interest; the nature of the dielectric material to be measured, both physical and electrical; and the degree of accuracy required. Many different kinds of instrument can be used for measurement, but any instrument that provides reliable determinations of the required electrical parameters involving the unknown material in the frequency range of interest can be considered.

2.1 GENERAL PRINCIPLES

For the radio frequencies, a material can be modeled electrically at any given frequency as a series or parallel equivalent lumped-element circuit. Therefore, if one can measure the radio-frequency circuit parameters appropriately—the impedance or admittance, for example—the dielectric properties of that material, at that particular frequency, can be determined from equations that properly relate the way in which the permittivity of the material affects those circuit parameters. The challenge in making accurate measurements of dielectric properties or permittivity is in designing the material sample holder for those measurements and adequately modeling the circuit for reliable calculation of the permittivity from the electrical measurements.

Techniques for the measurement of permittivity, or dielectric properties, in the low-frequency, medium-frequency, and high-frequency ranges were reviewed by Field (1954); these measurements included the use of various bridges and resonant circuits. Determination of the dielectric properties of grain samples at audio frequencies from 250 Hz to 20 kHz were determined from measurements using a precision bridge with samples confined in a coaxial sample holder (Corcoran et al., 1970). At very low frequencies, attention must also be paid to electrode polarization, which can invalidate measurement data, and the frequency below which this affects measurements depends on the nature and conductivity of the materials being measured (Foster and Schwan, 1989; Kuang and Nelson, 1998).

A large quantity of data on dielectric properties was obtained in the 1- to 50-MHz range on grain and seed samples with a Q-Meter based on a series resonant circuit (Nelson et al., 1953; Nelson, 1965, 1973a, 1979a). Techniques were developed for higher-frequency ranges with coaxial sample holders modeled as transmission-line sections with lumped parameters and measured with

Dielectric Properties of Agricultural Materials and Their Applications. DOI: http://dx.doi.org/10.1016/B978-0-12-802305-1.00002-6

an RX-Meter for the 50- to 250-MHz range (Jorgensen et al., 1970), and measured with an Admittance Meter for the 200- to 500-MHz range (Stetson and Nelson, 1970).

Components of the sample holders used in the earlier studies with the RX Meter and Admittance Meter were assembled to provide a shielded open-circuit coaxial sample holder for grain, and a technique was developed for measurements from 100 kHz to 1 GHz with two imped-ance analyzers, the use of proper calibrations, and the invariance-of-the-cross-ratio technique (Lawrence et al., 1989). The shielded open-circuit coaxial sample holder was also used by Bussey (1980) and by Jones et al. (1978) for determining dielectric properties of grain with high-frequency bridge measurements from 1 to 200 MHz.

A coaxial sample holder, designed to accommodate flowing grain, was modeled and character-ized by full two-port scattering parameter measurements, with the use of several alcohols of known permittivities, and signal-flow-graph analysis to provide dielectric properties of grain over the range from 25 to 350 MHz (Lawrence et al., 1998a).

For measurements at frequencies above those just mentioned, a number of microwave measure-ment techniques are available. At microwave frequencies, generally about 1 GHz and higher, trans-mission-line, resonant cavity, and free-space techniques have been useful. Principles and techniques for the measurement of microwave dielectric properties have been discussed in several reviews (Westphal, 1954; Altschuler, 1963; Redheffer, 1964; Bussey, 1967; Franceschetti, 1967; Baker-Jarvis, 1990; Kraszewski and Nelson, 2004).

Techniques for the measurement of microwave dielectric properties can be classified as reflec-tion or transmission measurements using resonant or nonresonant systems, with open or closed structures for the sensing of the properties of material samples (Kraszewski, 1980). Closed-structure methods include waveguide and coaxial-line transmission measurements and short-circuited waveguide or coaxial-line reflection measurements. Open-structure techniques include free-space transmission measurements and open-ended coaxial-line or open-ended waveguide mea-surements. Resonant structures can include either closed resonant cavities or open resonant struc-tures operated as two-port devices for transmission measurements or as one-port devices for reflection measurements.

With the development of suitable equipment for time-domain measurements, techniques were developed for measuring dielectric properties of materials over wide ranges of frequency (Fellner-Feldegg, 1969; Nicolson and Ross, 1970; van Gemert, 1973; Kent, 1975; Kwok et al., 1979; Bellamy et al., 1985). Since modern microwave network analyzers have become available, the methods of obtaining dielectric properties over wide frequency ranges have become even more effi-cient. Extensive reviews have included methods for both frequency-domain and time-domain tech-niques (Kaatze and Giese, 1980; Afsar et al., 1986).

The design of the dielectric sample holder for specific materials is an important aspect of the measurement technique. The Roberts and von Hippel (1946) short-circuited line technique for the measurement of dielectric properties provides a suitable method for many materials. For this method, the sample holder can be simply a short section of coaxial-line or rectangular or circular waveguide with a shorting plate or other short-circuit termination at the end of the line against which the sample rests. This is convenient for particulate samples because the sample holder, and also the slotted line or slotted section to which the sample holder is connected, can be mounted in a vertical orientation so that the top surface of the sample can be maintained perpendicular to the axis of wave propagation, as required for the measurement. The vertical orientation of the sample holder is also convenient for liquid or particulate materials when the measurements are taken with a network analyzer instead of a slotted line.

Dielectric properties of cereal grains, seed, and powdered or pulverized material have been determined with various microwave measurement systems assembled for such measurements. Twenty-one-mm, 50-ohm coaxial-line systems were used for these measurements at frequencies from 1 to 5.5 GHz (Nelson, 1973b, 1980; Nelson and You, 1989). A 54-mm, 50-ohm coaxial sample holder, designed for minimal reflections from the transition, was used with this same system for measurements on larger-kernel cereals such as corn (Nelson, 1978b, 1979b). A rectangular-waveguide X-band system was used to determine dielectric properties of grain and seed samples at 8−12 GHz (Nelson, 1972b). A rectangular-waveguide K-band system was used for measurements on fruit and vegetable samples (Nelson, 1983) and on ground and pulverized materials for measurements at 22 GHz (Nelson and You, 1989, 1990; You and Nelson, 1988; Nelson et al., 1989a).

The Roberts and von Hippel (1946) method requires measurements to determine the standing-wave ratios (SWRs) in the line with and without the sample inserted. From the shift of the standing-wave node and changes in node widths related to SWRs, sample length, and waveguide dimensions, ε' and ε'' can be calculated with suitable computer programs (Nelson et al., 1972, 1974). Similar determinations can be made with a network analyzer or other instrumentation by measurement of the complex reflection coefficient of the empty and filled sample holder.

Computer control of impedance analyzers and network analyzers (Lawrence et al., 1989; Waters and Brodwin, 1988) has facilitated the automatic measurement of dielectric properties over wide frequency ranges. Special calibration methods have also been developed to eliminate errors caused by unknown reflections in the coaxial-line systems (Lawrence et al., 1989; Kraszewski et al., 1983).

Microwave dielectric properties of wheat and corn have been measured at several frequencies by free-space measurements with a network analyzer and dielectric sample holders with rectangular cross-sections between horn antennas and other types of radiating elements. Measurement of the complex transmission coefficient, the components of which are attenuation and phase shift, permits the calculation of ε' and ε''. For free-space permittivity measurements, it is important that an attenuation of at least 10 dB through the sample layer be maintained to avoid disturbances resulting from multiple reflections within the sample and between the sample and the antennas; the sample size, laterally, must be sufficiently large to avoid problems caused by diffraction at the edges of the sample (Trabelsi et al., 1998a). Other practices and techniques have been developed to improve the accuracy and reliability of free-space measurements of the microwave dielectric properties of such granular materials (Trabelsi and Nelson, 2003).

Open-ended coaxial-line probes have been used successfully for convenient broadband permittivity measurements (Blackham and Pollard, 1997; Grant et al., 1989) on liquid and semisolid materials of relatively high loss, which include most agricultural materials. This technique has been used for permittivity measurements on fresh fruits and vegetables (Tran et al., 1984; Nelson et al., 1994a, 1994b, 1995; Nelson, 2003). The technique is subject to errors if there are significant density variations in the material, or if there are air gaps or air bubbles between the end of the coaxial probe and the sample. Moreover, the technique is not suitable for determining permittivities of very low-loss materials (Nelson and Bartley, 1998), but it has been used to provide broad-band information on granular and pulverized materials when sample bulk densities were established by auxiliary permittivity measurements (Nelson and Bartley, 1998; Nelson et al., 1997, 1998).

The choices of measurement technique, equipment, and sample holder design depend upon the dielectric materials to be measured and the frequency range of interest. Vector network analyzers and impedance analyzers are very versatile and useful if studies are extensive. For limited studies, more commonly available microwave laboratory measurement equipment can suffice if suitable sample holders are constructed. When data are required at only one microwave frequency, or a limited number of frequencies, a resonant cavity technique may be the logical choice (Bussey, 1967; Rzepecka, 1973). Such cavities can be constructed with rectangular waveguide sections or from waveguide flanges and waveguide stock (Kraszewski and Nelson, 1996). Construction of a cylindrical cavity (Risman and Bengtsson, 1971; Li et al., 1981) may be advantageous, depending on the needs. For temperature-dependent studies, a cavity with provision for the measurement of alternate dielectric properties and microwave heating of the sample for temperature control may be advantageous (Bosisio et al., 1986). Resonant cavities can also be used to measure other permittivity-related characteristics of materials such as moisture content, mass, volume, and mixture proportions (Kraszewski and Nelson, 1992, 1994; Kraszewski, 1992; Kraszewski et al., 1990; Nelson and Kraszewski, 1998).

In this chapter, a few measurement techniques that have been used successfully for the measurement of dielectric properties of agricultural and other materials will be briefly described. Although the instruments and equipment used may no longer be available—or even appropriate for new work in view of modern instruments that have become available—the fundamental principles employed are still valid and instructive. The techniques will be presented in order of ascending frequency.

2.1.1 AUDIO FREQUENCIES

For measurements to determine the dielectric properties of grain and seed at these low frequencies, a coaxial sample holder providing a large surface area for sensitivity in determining dielectric loss (see Figure 2.1) was used with a General Radio[1] 1608-A precision impedance bridge and an external audio-frequency oscillator (Corcoran et al., 1970).

The outside diameter of the inner conductor was 11.43 cm, and the inside diameter of the outer conductor was 14.61 cm, thus providing a sample thickness between the electrodes of about 1.6 cm. The length of the sample holder provided a sample height of about 17 cm. Measurements with the impedance bridge at a given frequency for capacitance C and dissipation factor $D = \tan \delta$ (loss tangent) for the empty sample holder and for the sample holder filled with grain, provided the necessary information for calculation of the dielectric properties. Measurements were taken in both series and parallel-equivalent modes, and results were averaged for better accuracy. Data for dielectric properties over the frequency range 250 Hz to 20 kHz were reported for corn, wheat, oats, grain sorghum, soybeans, cottonseed, and seed of alfalfa, Kentucky bluegrass, switchgrass, and Western wheatgrass over ranges of moisture content (Stetson and Nelson, 1972).

---

[1]Mention of trade names or commercial products in this publication is solely for the purpose of providing specific information and does not imply recommendation or endorsement by the US Department of Agriculture.

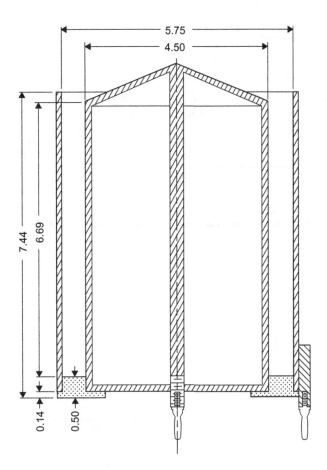

FIGURE 2.1

Sectional view of coaxial sample holder for the measurement of audio-frequency dielectric properties on grain and seed (Corcoran et al., 1970). Dimensions are shown in inches.

2.1.2 **1 TO 50 MHz**

For measurements on grain and seed samples in the 1- to 50-MHz range, coaxial sample holders were designed and built for use with the Boonton 160-A Q-Meter, which was a versatile instrument based on the principle of series resonance (Nelson et al., 1953; Nelson, 1979a). As shown in the equivalent circuit diagram of Figure 2.2, the instrument had a variable frequency source, terminals for serial connection of a coil and an unknown capacitor, two internal variable capacitors, a main capacitor and a vernier capacitor, and a voltmeter reading the potential across the parallel capacitors.

A parallel-equivalent circuit and corresponding phasor diagram for the unknown capacitor is shown in Figure 2.3 illustrating the in-phase current I_R through the resistive branch of the circuit,

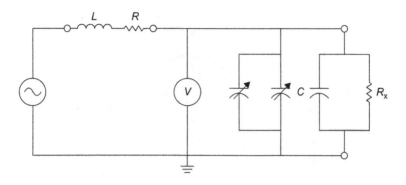

FIGURE 2.2

Simplified diagram for the Q-Meter series-resonant measuring circuit with sample holder connected (Nelson, 1979a).

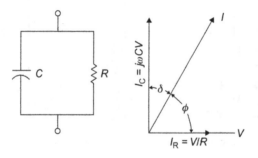

FIGURE 2.3

Parallel-equivalent circuit representing a lossy dielectric and the corresponding phasor diagram (Nelson, 1979a).

the quadrature current I_C through the capacitive branch leading the voltage phasor V by 90°, the phase angle ϕ, and the loss angle δ.

A coaxial sample holder (Figure 2.4), 22.9 cm high, with inner conductor diameter of 1.9 cm and outer conductor diameter of 7.62 cm had an insulating disk at its midsection upon which samples could be supported.

A variable air capacitor was mounted in the section below the sample-supporting disk, connected between the inner and outer conductors and calibrated with a range of 16 pF.

The sample holder was equipped with banana plugs for convenient connection to the Q-Meter by plugging it into the jacks of the capacitor terminals. When the sample holder was connected to the Q-Meter capacitor terminals and a suitable coil was connected to the inductor terminals, the circuit could be tuned to resonance with the main variable capacitor in the Q-Meter, as indicated by a peak on the voltmeter reading the potential across the capacitors. The sample holder with a grain sample was first connected to the Q-Meter, and the circuit was tuned to resonance. Then, the circuit

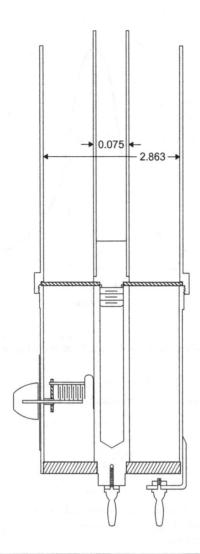

FIGURE 2.4

Sectional view of coaxial sample holder for the measurement of dielectric properties of grain and seed with the Q-Meter (Nelson et al., 1953). Dimensions are shown in inches (Nelson et al., 1953).

was detuned from V_m to V (Figure 2.5) on both sides of resonance with the variable-air vernier capacitor to determine ΔC_s for the sample.

Then it was retuned to resonance, the sample was poured out, and the sample holder was reconnected to the Q-Meter. The capacitance lost when the sample was removed was then restored by adjusting the calibrated variable capacitor built into the sample holder to restore resonance and thus measure C_x. With the empty sample holder still connected, the circuit was again detuned on either

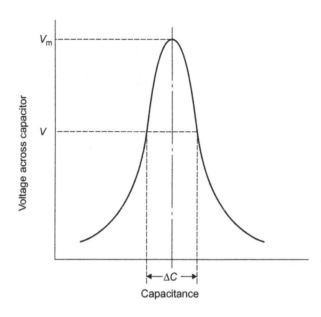

FIGURE 2.5

Relationship between voltage across the capacitor and capacitance in the region of resonance (Nelson, 1979a).

side of resonance to determine V_m, V, and ΔC_a for the air-filled (empty) sample holder. From these measurements, the dielectric properties can be calculated as follows:

$$\varepsilon' = \frac{1.412\, C_x(b^2 - a^2)\, \log_e b/a}{v} + 1 \tag{2.1}$$

$$\tan\delta = \frac{\left(\Delta C_s/\sqrt{(V_m/V)_s^2 - 1}\right) - \left(\Delta C_a/\sqrt{(V_m/V)_a^2 - 1}\right)}{2C_x} \tag{2.2}$$

Detailed information on the method and sample holders is provided in previous publications (Nelson et al., 1953; Nelson, 1979a); extensive data on dielectric properties for grain and seed were reported (Nelson, 1965; ASAE, 2000).

The use of Q-Meters for measurements of the dielectric properties of grain by several investigators was identified in a publication describing an improved sample holder for Q-Meter measurements of dielectric properties (Nelson, 1979a). The improved coaxial sample holder (Figure 2.6) allowed control of the sample temperature through circulating liquids for temperatures below or above room temperature.

The new design also included a precision vernier capacitor for loss measurements that eliminated the potential influence of lead inductance between the Q-Meter measuring circuit and the sample, and featured additional improvements. The detail of the vernier variable capacitor is shown in Figure 2.7.

Small differences in measured dielectric properties attributable to different interelectrode spacing of sample holders for granular samples were also noted and explained (Nelson, 1979a).

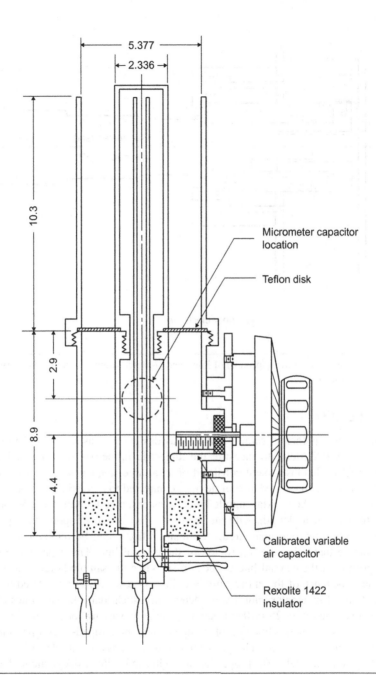

FIGURE 2.6

Improved coaxial sample holder for use with Q-Meter for the measurement of dielectric properties (Nelson, 1979a). Dimensions are shown in inches.

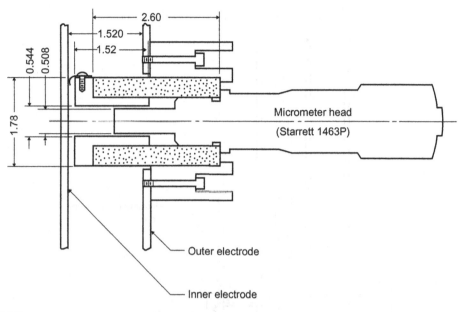

FIGURE 2.7

Micrometer capacitor (precision vernier capacitor) of improved Q-Meter sample holder (Nelson, 1979a). Dimensions are shown in inches.

2.1.3 50 TO 250 MHz

For this frequency range, a coaxial sample holder was designed, constructed and modeled for use with a Boonton 250-A RX Meter to measure the dielectric properties of grain and seed samples (Jorgensen et al., 1970). This instrument consisted of a variable-frequency oscillator and a Schering bridge designed to measure the impedance of devices connected to a coaxial N-Type terminal on the instrument. In this instance, the coaxial sample holder was designed to maintain a 50-ohm characteristic impedance throughout its length, with an insulating disk at the midpoint to support the grain sample in the top portion and an open-circuit termination that could be removed for insertion and removal of the sample that filled the top region of the coaxial line. The capacitance of the empty sample-holding portion of the coaxial line C became εC when the sample holder was filled with the grain sample. The impedance of the entire sample holder, or test cell, was modeled as transmission line sections and lumped parameter values were determined such that reliable values of $\varepsilon = \varepsilon' - j\varepsilon''$ could be obtained over the 50- to 250-MHz frequency range by solving for ε' and ε'' in terms of four cell constants, each of which was a function of frequency and the lumped-circuit parameter values of the test cell (Jorgensen et al., 1970). The performance of the coaxial sample holder with the RX-Meter in providing reliable dielectric properties was checked with n-decyl alcohol and provided values very close to the known values for the alcohol over the 50- to 250-MHz frequency range. The measurement system was used successfully for measurements on many materials, including grain and seed (Nelson and Stetson, 1975; Nelson, 1978a; Noh and Nelson, 1989), pecans (Nelson, 1981), insects (Nelson and Charity, 1972), and coal (Nelson et al., 1980, 1981).

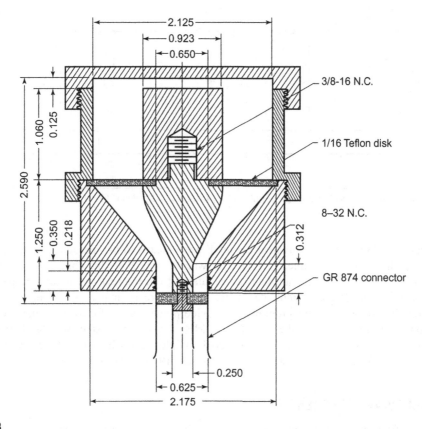

FIGURE 2.8

Sectional view of coaxial dielectric sample holder (Stetson and Nelson, 1970). Dimensions are shown in inches.

2.1.4 **200 TO 500 MHz**

For measurements in this frequency range, a coaxial sample holder was designed, constructed, and modeled for use with the General Radio 1602-B U-H-F Admittance Meter and auxiliary oscillators, power supply, detector, and 50-ohm GR-874 line components (Stetson and Nelson, 1970). This system was used to measure the admittance of the coaxial sample holder designed as a 50-ohm coaxial line (Figure 2.8).

The sample to be measured occupied the region between the supporting Teflon disc and the top of the inner conductor. The removable open-circuit termination permitted easy insertion and removal of the sample material. The capacitance of that section of the empty sample holder is multiplied by the complex permittivity of the sample when that portion is filled with the sample. The sample holder was modeled as three transmission line sections with lumped-circuit parameters and open-circuit termination as illustrated in Figure 2.9.

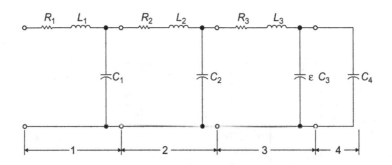

FIGURE 2.9

Lumped-circuit parameter model of coaxial sample holder (Stetson and Nelson, 1970).

The input admittance of the equivalent circuit Y was equated to the measured admittance of the sample holder filled with the sample, $Y_m = G + jB$, where G and B are the conductance and susceptance, respectively, and ε' and ε'' can then be calculated from the real and imaginary parts of the resulting equation. The lumped-circuit parameter values were estimated by calculation and then adjusted through various computer programs to provide proper values for measurements on dielectrics of known permittivities over the 200- to 500-MHz range. The system was used successfully for measurements on grain, seed, insects and coal samples, for which references were cited in Section 2.1.3, and also on yellow-dent field corn (Nelson, 1978a, 1978b, 1979b).

2.1.5 MICROWAVE FREQUENCIES

At microwave frequencies, different techniques are required for measuring dielectric properties of materials. This chapter describes two methods that were found suitable for measurements on grain and seed samples. A waveguide measurement technique, in which the material sample is placed inside the waveguide in contact with a short-circuit termination, was very useful for granular and powdered materials (Roberts and von Hippel, 1946). For larger samples, a free-space transmission technique was found very helpful (Trabelsi and Nelson, 2003).

2.1.5.1 Short-circuited line technique

An X-band system was developed for measurements of the dielectric properties of grain and seed samples (Nelson, 1972b), and a computer program for computation of dielectric properties from short-circuited waveguide measurements was developed for use with this system and with coaxial-line and cylindrical-waveguide systems as well (Nelson et al., 1972, 1974). This program provided precise calculation of dielectric properties of high- or low-loss materials and was very general with respect to the measurement system used and the character of input data.

When using this method, determination of the real and imaginary parts of the relative complex permittivity of the material requires the measurement of the standing-wave ratio S and the position of the voltage or current standing-wave node locations, both with and without the material sample in the short-circuited end of the waveguide (Figures 2.10 and 2.11).

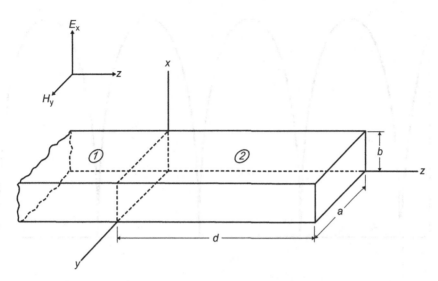

FIGURE 2.10

Short-circuited rectangular waveguide with sample of length d at the shorted end (Nelson et al., 1974).

Values for the dielectric constant and loss factor are obtained from the following relationships:

$$\varepsilon' - j\varepsilon'' = (\lambda_o/\lambda_c)^2 - (\gamma_2 d\lambda_o/(2\pi d))^2 \tag{2.3}$$

$$\frac{\tanh(\gamma_2 d)}{\gamma_2 d} = -j\frac{\lambda_g}{2\pi d}\left[\frac{S - j\tan(2\pi z_o/\lambda_o)}{1 - jS\tan(2\pi z_o/\lambda_g)}\right] \tag{2.4}$$

where λ_o and λ_c are, respectively, the free-space wavelength at the frequency of measurement and the cutoff wavelength of the waveguide used, $\gamma_2 = \alpha_2 + j\beta_2$ is the complex propagation constant in the sample in the waveguide, where α and β represent the attenuation and phase constants, respectively, d is the sample length, z_o is the distance from the air–sample interface to the first standing wave node outside the sample (Figure 2.11), $\lambda_g = \lambda_1$, is the guide wavelength in the air space in the waveguide, and S is the standing-wave ratio with the sample in the waveguide. The value of $\lambda_2 d$ must first be extracted from Eq. (2.4) before the dielectric constant and loss factor can be determined from Eq. (2.3).

This technique and the associated computer program were used with several systems that were used for the measurement of microwave dielectric properties, including the X-band system mentioned for measurements at 8.2–12.4 GHz (Nelson, 1972), two Rohde & Schwarz 21-mm coaxial line systems, one with a slotted line at 1 and 2.44 GHz and another with a nonslotted line at about 6 GHz (Nelson, 1973b), a Central Research Laboratories Microwave Dielectrometer at 1, 3, and 8.6 GHz (Nelson, 1972a), and a K-band rectangular waveguide system at 22 GHz (Nelson, 1983).

Microwave dielectric properties measured by these techniques were reported for grain and seed (Nelson, 1973b, 1978b, 1979b; Nelson and You, 1989; You and Nelson, 1988; Nelson and

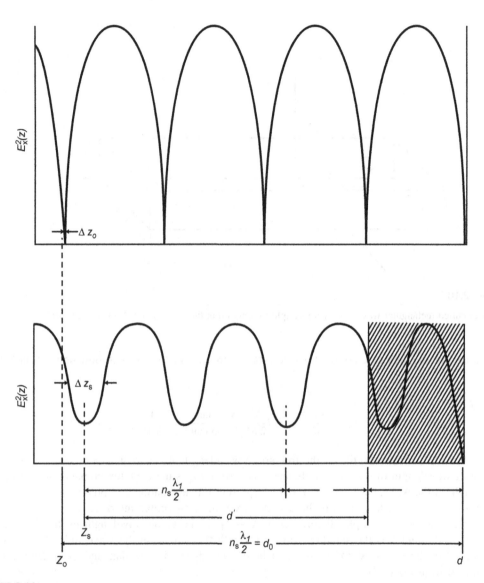

FIGURE 2.11

Voltage standing-wave pattern relationships in an empty waveguide and in a waveguide with dielectric sample in place (Nelson et al., 1974).

Stetson, 1975, 1976; Noh and Nelson, 1989), insects (Nelson, 1976), fruits and vegetables (Nelson, 1980, 1983, 1987, 1992), nuts (Nelson, 1981), food materials (Nelson et al., 1991), insects (Nelson et al., 1997; Nelson, 1972a), coal (Nelson et al., 1980, 1981), coal and limestone (Nelson, 1996), and minerals (Nelson et al., 1989a, 1989b).

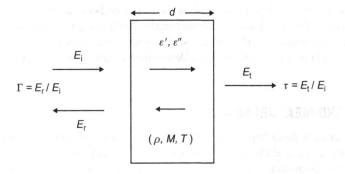

FIGURE 2.12

Diagram of wave−dielectric-material interaction (Trabelsi et al., 1998b).

2.1.5.2 Free-space techniques

Free-space transmission measurements with a network analyzer have been found very useful for larger grain samples and for measurements at higher frequencies. Important principles for such measurements have been discussed previously (Trabelsi and Nelson, 2003; Kraszewski and Nelson, 1990; Nelson and Trabelsi, 2004).

From a dielectrics viewpoint, a mass of grain can be characterized by the effective complex permittivity of the material, relative to free space, $\varepsilon = \varepsilon' - j\varepsilon''$, where ε' is the dielectric constant, and ε'' is the dielectric loss factor. For cereal grains, the permittivity is not only a function of the moisture content M, but also of the frequency f of the applied electric field, the temperature T of the grain, and the bulk density ρ of the grain. Thus, a plane wave traversing a layer of grain of thickness d will interact with the granular material as depicted in Figure 2.12, where E_i represents the incident wave electric field, E_r is the reflected wave electric field, E_t is the electric field of the transmitted wave, Γ is the reflection coefficient, and τ is the transmission coefficient.

The components of the permittivity, ε' and ε'', can be obtained by measurement of the attenuation A and phase shift ϕ of the wave as it traverses the dielectric layer, because for a plane wave, when $\varepsilon'' \ll \varepsilon'$, $\varepsilon' = (\beta/\beta_0)^2$ and $\varepsilon'' = 2\alpha\beta/\beta_0$, where $\alpha = A/d$ is the attenuation constant, $\beta = \varphi/d + \beta_0$ is the phase constant, and $\beta_0 = 2\pi/\lambda_0$ is the phase constant for free-space wavelength λ_0. The disturbance caused by any reflected waves within the grain layer can be rendered negligible by using a layer thickness that provides at least 10 dB of attenuation for waves traveling one way through the layer. In terms of the measured attenuation and phase shift as the wave traverses the dielectric layer of thickness d, the dielectric properties of a relatively low-loss material are given by:

$$\varepsilon' = \left(1 + \frac{\phi\lambda_0}{2\pi d}\right)^2 \tag{2.5}$$

$$\varepsilon'' = \frac{A\lambda_0\sqrt{\varepsilon'}}{8.686\pi d} \tag{2.6}$$

For many of these measurements, a pair of horn/lens antennas providing a plane wave a short distance from the antenna were used, and sample holders constructed of Styrofoam sheet material, which has dielectric properties close to those of air, were placed between the antennas for the network analyzer measurements. This provided the measurements for the transmission scattering parameter S_{21} from which the attenuation and phase shift were determined as $A = 20 \log|S_{21}|$ and $\phi = \mathrm{Arg}(S_{21}) - 2n\pi$, where n in an integer to be determined (Trabelsi et al., 2000).

2.1.6 BROADBAND MEASUREMENTS

All of the measurements described so far, except for the network analyzer measurements, were made at single frequencies, and the sources had to be tuned or adjusted to new frequencies for measurements at different frequencies. With the advent of synthesized signal sources that can be swept over wide frequency ranges, as applied in the network analyzers, measurements over ranges of frequency are possible in short time periods.

2.1.6.1 Time-domain measurements

In addition to such frequency-domain measurements, time-domain techniques are also available, whereby a fast rise-time step waveform is launched into in a low-loss coaxial transmission line containing a section with the dielectric sample. At the air–dielectric interface, partial reflectance and partial transmission of the incident wave occur. The resultant transmitted or reflected waveforms carry the necessary information from which the frequency-dependent behavior of the dielectric sample that is responsible for the change in the characteristic impedance of the line can be determined. Either the reflected or transmitted signals may be used for time-domain spectroscopy measurements (Fellner-Feldegg, 1969; Nicolson and Ross, 1970; van Gemert, 1973). Both single reflection from long samples or multiple or total reflection from thin samples can be used for the measurements of dielectric properties (Kwok et al., 1979; Clark et al., 1974). A technique for measurements on small samples between the end of the center conductor and a shorting cap on a coaxial line section (Figure 2.13) was also developed (Bellamy et al., 1985; Iskander and Stuchly, 1972).

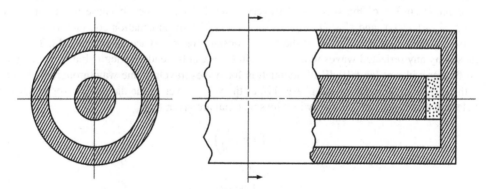

FIGURE 2.13

Coaxial-line shunt capacitance sample holder (Bellamy et al., 1985).

Dielectric properties of wheat, alfalfa seed, and adult rice weevils from 30 MHz to 1 GHz were measured by time-domain techniques (Kwok et al., 1979). Dielectric properties of frozen cod have also been measured from 3 to 100 MHz by time-domain methods (Kent, 1975).

2.1.6.2 Impedance and network analyzer measurements

Dielectric properties of materials over the frequency range 100 kHz to 1 GHz were made utilizing a Hewlett-Packard 4192A LF Impedance Analyzer and a Hewlett-Packard 4191A RF Impedance Analyzer with a 50-ohm shielded open-circuit coaxial sample holder of 5.4-cm outer electrode diameter designed such that an open, short and known liquid (*n*-butanol) could be used for calibration in the plane of the central conductor truncation (Lawrence et al., 1989). The invariance-of-the-cross-ratio technique was employed to transform the complex impedance of the sample into its complex permittivity components. The performance of the measurement system was checked with several other alcohols, and dielectric properties of wheat were measured as a function of temperature and moisture content with automatic stepping of frequency over the range 0.1 to 100 MHz (Lawrence et al., 1989, 1990). Frequency- and temperature-dependent dielectric properties of pecans were also measured with this system (Lawrence et al., 1992). Another coaxial sample holder, designed for flowing grain, was used with a Hewlett-Packard 8753C network analyzer, and dielectric properties from 1 to 350 MHz were obtained with a signal-flow-graph model from full two-port S-parameter measurements (Lawrence et al., 1998b). Measurements were made on several alcohols to verify accuracy, and data on dielectric properties were reported for four hard red winter wheat cultivars (Lawrence et al., 1998b).

2.1.6.3 Open-ended coaxial-line measurements

The open-ended coaxial-line technique for the measurement of dielectric properties of materials, which was commercialized by Hewlett-Packard, with continuing development by Agilent Technologies, has been used for measurements on a number of agricultural products. The method, using an open-ended probe with network and impedance analyzers, measures the reflection coefficient with a dielectric sample in contact with the probe and calculates the permittivity. It is well suited for broadband measurements on liquid and semisolid high-loss materials with which problematic air voids between the tip of the probe and the sample can generally be avoided. However, as with most measurement techniques for dielectric properties, check measurements on samples of known permittivities are necessary to confirm that the resulting data are reliable. And, because the calibration generally utilizes an open air measurement, and measurements on a short-circuit termination and pure water at 25°C, care must be taken to assess the reliability of measurements at temperatures too far above or below that temperature.

Dielectric properties of 23 kinds of fresh fruit and vegetables were measured over the 0.2- to 20-GHz frequency range with an open-ended probe made from a 3.6-mm (0.141-in) diameter, coaxial, semirigid, Teflon-insulated copper transmission line and a Hewlett-Packard 8510A network analyzer (Nelson et al., 1994a, 1994b). Many more measurements were obtained with Hewlett-Packard and Agilent Technologies 85070 open-ended probes used with different network and impedance analyzers and a sample temperature control system for grain (Nelson and Trabelsi, 2006), fruits and vegetables (Nelson, 2003, 2005; Guo et al., 2007a), insects (Nelson et al., 1997, 1998), coal and limestone (Nelson and Bartley, 1998), food materials (Nelson and Bartley, 2000, 2002), melons (Nelson et al., 2006, 2007, 2008), and poultry products (Guo et al., 2007b; Zhuang et al., 2007).

REFERENCES

Afsar, M.N., Birch, J.R., Clarke, R.N., Chantry, G.W., 1986. The measurement of the properties of materials. Proc. IEEE 74 (1), 183−199.

Altschuler, H.M., 1963. Dielectric constant. In: Sucher, M., Fox, J. (Eds.), Handbook of Microwave Measurements (Chapter IX). Polytechnic Press of the Polytechnic Inst., Brooklyn, New York, NY.

ASAE, 2000. ASAE D293.2 Dielectric properties of grain and seed. ASAE Standards 2000. American Society of Agricultural Engineers, St. Joseph, MI, pp. 549−558.

Baker-Jarvis J., 1990. Transmission/Reflection and Short-Circuit Line Permittivity Measurements. U. S. Department of Commerce, Nat'l Inst. Standards and Technol., NIST Technical Note 1341.

Bellamy, W.L., Nelson, S.O., Leffler, R.G., 1985. Development of a time-domain reflectrometry system for dielectric properties measurement. Trans. ASAE 28 (4), 1313−1318.

Blackham, D.V., Pollard, R.D., 1997. An improved technique for permittivity measurements using a coaxial probe. IEEE Trans. Instrum. Meas. 46 (5), 1093−1099.

Bosisio, R., Akyel, C., Labelle, R.C., Wang, W., 1986. Computer-aided permittivity measurements in strong RF Fields (Part II). IEEE Trans. Instrum. Meas. 35 (4), 606−611.

Bussey, H.E., 1967. Measurement of RF properties of materials—a survey. Proc. IEEE 55 (6), 1046−1053.

Bussey, H.E., 1980. Dielectric measurements in a shielded open circuit line. IEEE Trans. Instrum. Meas. 29 (2), 120−124.

Clark, A.H., Quickenden, P.A., Suggett, A., 1974. Multiple reflection time domain specroscopy. J. Chem. Soc. Faraday Trans. II 70, 1847−1962.

Corcoran, P.T., Nelson, S.O., Stetson, L.E., Schlaphoff, C.W., 1970. Determining dielectric properties of grain and seed in the audiofrequency range. Trans. ASAE 13 (3), 348−351.

Fellner-Feldegg, H., 1969. The measurement of dielectrics in the time domain. J. Phys. Chem. 73, 616−623.

Field, R.F., 1954. Dielectric measuring techniques. In: von Hippel, A. (Ed.), Dielectric Materials and Applications (Sec. 1, Lumped circuits, A. Permittivity, Chapter II). John Wiley & Sons., New York, NY.

Foster, K.R., Schwan, H.P., 1989. Dielectric properties of tissues and biological materials: a critical review. In: Bourne, J.R. (Ed.), Critical Reviews in Biomedical Engineering. CRC Press, Inc., Boca Raton, FL.

Franceschetti, G., 1967. A complete analysis of the reflection and transmission methods for measuring the complex permittivity of materials at microwaves. Alta Frequenza 36 (8), 757−764.

Grant, J.P., Clarke, R.N., Symm, G.T., Spyrou, N.M., 1989. A critical study of the open-ended coaxial line sensor technique for RF and microwave complex permittivity measurements. J. Phys. E 22, 757−770.

Guo, W., Nelson, S.O., Trabelsi, S., Kays, S.J., 2007a. 10−1800-MHz dielectric properties of fresh apples during storage. J. Food Eng. 83, 562−569.

Guo, W., Trabelsi, S., Nelson, S.O., Jones, D.R., 2007b. Storage effects on dielectric properties of eggs from 10 to 1800 MHz. J. Food Sci. 72 (5), E335−E340.

Iskander, M.F., Stuchly, S.S., 1972. A time-domain technique for measurement of the dielectric properties of biological substances. IEEE Trans. Instrum. Meas. IM-21 (4), 425−429.

Jones R.N., Bussey H.E., Little W.E. and Metzker R.F., 1978. Electrical Characteristics of Corn, Wheat, and Soya in the 1−200 MHz Range Report No. NBSIR 78-897.

Jorgensen, J.L., Edison, A.R., Nelson, S.O., Stetson, L.E., 1970. A bridge method for dielectric measurements of grain and seed in the 50- to 250-MHz range. Trans. ASAE 13 (1), 18−20, 24.

Kaatze, U., Giese, K., 1980. Dielectric relaxation spectroscopy: frequency domain and time domain experimental methods. J. Phys. E 13, 133−141.

Kent, M., 1975. Time domain measurements of the dielectric properties of frozen fish. J. Microw. Power 10 (1), 37−48.

Kraszewski, A., 1980. Microwave aquametry—a review. J. Microw. Power 15 (4), 209−220.

Kraszewski, A., 1992. Microwave resonator technique for sorting dielectric objects—advantages and limitations. J. Wave Mater. Interact. 7 (1), 39–55.

Kraszewski, A., Nelson, S.O., 1990. Study on grain permittivity measurments in free space. J. Microw. Power Electromagn. Energy 25 (4), 202–210.

Kraszewski, A., Nelson, S.O., 1994. Microwave resonator for sensing moisture content and mass of single wheat kernels. Can. Agric. Eng. 36 (4), 231–238.

Kraszewski, A., Nelson, S.O., 2004. Microwave permittivity determination in agricultural products. J. Microw. Power Electromagn. Energy 39 (1), 41–52.

Kraszewski, A., Stuchly, S.S., Stuchly, M.A., 1983. ANA calibration method for measurement of dielectric properties. IEEE Trans. Instrum. Meas. 32 (2), 385–386.

Kraszewski, A., Nelson, S.O., You, T.-S., 1990. Use of a microwave cavity for sensing dielectric properties of arbitratily shaped biological objects. IEEE Trans. Microw. Theory Tech. 38 (7), 858–863.

Kraszewski, A.W., Nelson, S.O., 1992. Observations on resonant cavity perturbtion by dielectric objects. IEEE Trans. Microw. Theory Tech. 40 (1), 151–155.

Kraszewski, A.W., Nelson, S.O., 1994. Resonant-cavity perturbation measurement for mass determination of the perturbing object. IEEE Instrum. Meas. Technol. Conf. Proc. 3, 1261–1264.

Kraszewski, A.W., Nelson, S.O., 1996. Resonant cavity perturbation—some new applications of an old measuring technique. J. Microw. Power Electromagn. Energy 31 (3), 178–187.

Kuang, W., Nelson, S.O., 1998. Low-frequency dielectric properties of biological tissues: a review with some new insights. Trans. ASAE 41 (1), 173–184.

Kwok, B.P., Nelson, S.O., Bahar, E., 1979. Time-domain measurements for determination of dielectric properties of agricultural materials. IEEE Trans. Instrum. Meas. 28 (2), 109–112.

Lawrence, K.C., Nelson, S.O., Kraszewski, A.W., 1989. Automatic system for dielectric properties measurements from 100 kHz to 1 GHz. Trans. ASAE 32 (1), 304–308.

Lawrence, K.C., Nelson, S.O., Kraszewski, A.W., 1990. Temperature dependence of the dielectric properties of wheat. Trans. ASAE 33 (2), 535–540.

Lawrence, K.C., Nelson, S.O., Kraszewski, A., 1992. Temperature dependence of the dielectric properties of pecans. Trans. ASAE 35 (1), 251–255.

Lawrence, K.C., Nelson, S.O., Bartley Jr., P.G., 1998a. Flow-through coaxial sample holder design for dielectric properties measurements from 1 to 350 MHz. IEEE Trans. Instrum. Meas. 47 (3), 613–622.

Lawrence, K.C., Nelson, S.O., Bartley Jr., P.G., 1998b. Measuring dielectric properties of hard red winter wheat from 1 to 350 MHz with a flow-through coaxial samle holder. Trans. ASAE 41 (1), 143–149.

Li, S., Akyel, C., Bosisio, R.G., 1981. Precise calculations and measurements on the complex dielectric constant of lossy materials using TM_{010} cavity perturbation techniques. IEEE Trans. Microw. Theory Tech. 29 (10), 1041–1048.

Nelson, S.O., 1965. Dielectric properties of grain and seed in the 1 to 50-mc range. Trans. ASAE 8 (1), 38–48.

Nelson S.O., 1972a. Frequency Dependence of the Dielectric Properties of Wheat and the Rice Weevil. (Ph.D. dissertation), Iowa State University.

Nelson, S.O., 1972b. A system for measuring dielectric properties at frequencies from 8.2 to 12.4 GHz. Trans. ASAE 15 (6), 1094–1098.

Nelson, S.O., 1973a. Electrical properties of agricultural products—a critical review. Trans. ASAE 16 (2), 384–400.

Nelson, S.O., 1973b. Microwave dielectric properties of grain and seed. Trans. ASAE 16 (5), 902–905.

Nelson, S.O., 1976. Microwave dielectric properties of insects and grain kernels. J. Microw. Power 11 (4), 299–303.

Nelson, S.O., 1978a. Frequency and moisture dependence of the dielectric properties of high-moisture corn. J. Microw. Power 13 (2), 213–218.

Nelson S.O., 1978b. Radiofrequency and Microwave Dielectric Properties of Shelled Field Corn. Agricultural Research Service, U. S. D. A., ARS-S-184.

Nelson, S.O., 1979a. Improved sample holder for Q-meter dielectric measurements. Trans. ASAE 22 (4), 950−954.

Nelson, S.O., 1979b. RF and microwave dielectric properties of shelled, yellow-dent field corn. Trans. ASAE 22 (6), 1451−1457.

Nelson, S.O., 1980. Microwave dielectric properties of fresh fruits and vegetables. Trans. ASAE 23 (5), 1314−1317.

Nelson, S.O., 1981. Frequency and moisture dependence of the dielectric properties of chopped pecans. Trans. ASAE 24 (6), 1573−1576.

Nelson, S.O., 1983. Dielectric properties of some fresh fruits and vegetables at frequencies of 2.45 to 22 GHz. Trans. ASAE 26 (2), 613−616.

Nelson, S.O., 1987. Frequency, moisture, and density dependence of the dielectric properties of small grains and soybeans. Trans. ASAE 30 (5), 1538−1541.

Nelson, S.O., 1992. Microwave dielectric properties of fresh onions. Trans. ASAE 35 (3), 963−966.

Nelson, S.O., 1996. Determining dielectric properties of coal and limestone by measurements on pulverized samples. J. Microw. Power Electromagn. Energy 31 (4), 215−220.

Nelson, S.O., 2003. Frequency- and temperature-dependent permittivities of fresh fruits and vegetables from 0.0l to 1.8 GHz. Trans. ASAE 46 (2), 567−574.

Nelson, S.O., 2005. Dielectric spectroscopy of fresh fruit and vegetable tissues from 10 to 1800 MHz. J. Microw. Power Electromagn. Energy 40 (1), 31−47.

Nelson, S.O., Bartley Jr., P.G., 1998. Open-ended coaxial-line permittivity measurements on pulverized materials. IEEE Trans. Instrum. Meas. 47 (1), 133−137.

Nelson, S.O., Bartley Jr., P.G., 2000. Measuring frequency- and temperature-dependent dielectric properties of food materials. Trans. ASAE 43 (6), 1733−1736.

Nelson, S.O., Bartley Jr., P.G., 2002. Frequency and temperature dependence of the dielectric properties of food materials. Trans. ASAE 45 (4), 1223−1227.

Nelson, S.O., Charity, L.F., 1972. Frequency dependence of energy absorption by insects and grain in electric fields. Trans. ASAE 15 (6), 1099−1102.

Nelson, S.O., Kraszewski, A., 1998. Sensing pulverized material mixture proportions by resonant cavity measurements. IEEE Trans. Instrum. Meas. 47 (5), 1201−1204.

Nelson, S.O., Stetson, L.E., 1975. 250-Hz to 12-GHz dielectric properties of grain and seed. Trans. ASAE 18 (4), 714,715,718.

Nelson, S.O., Stetson, L.E., 1976. Frequency and moisture dependence of the dielectric properties of hard red winter wheat. J. Agric. Eng. Res. 21, 181−192.

Nelson, S.O., Trabelsi, S., 2004. Principles for microwave moisture and density measurement in grain and seed. J. Microw. Power Electromagn. Energy 39 (2), 107−117.

Nelson, S.O., Trabelsi, S., 2006. Dielectric spectroscopy of wheat from 10 MHz to 1.8 GHz. Meas. Sci. Technol. 17, 2294−2298.

Nelson, S.O., You, T.-S., 1989. Microwave dielectric properties of corn and wheat kernels and soybeans. Trans. ASAE 32 (1), 242−249.

Nelson, S.O., You, T.-S., 1990. Relationships between microwave permittivities of solid and pulverised plastics. J. Phys. D Appl. Phys. 23, 346−353.

Nelson, S.O., Soderholm, L.H., Yung, F.D., 1953. Determining the dielectric properties of grain. Agric. Eng. 34 (9), 608−610.

Nelson S.O., Stetson L.E. and Schlaphoff C.W., 1972. Computer Program for Calculating Dielectric Properties of Low- or High-Loss Materials from Short-Circuited Waveguide Measurements. ARS-NC-4, Agricultural Research Service, U. S. Department of Agriculture.

Nelson, S.O., Stetson, L.E., Schlaphoff, C.W., 1974. A general computer program for precise calculation of dielectric properties from short-circuited waveguide measurements. IEEE Trans. Instrum. Meas. 23 (4), 455−460.

Nelson, S.O., Fanslow, G.E., Bluhm, D.D., 1980. Frequency dependence of the dielectric properties of coal. J. Microw. Power 15 (4), 277−282.

Nelson, S.O., Beck-Montgomery, S.R., Fanslow, G.E., Bluhm, D.D., 1981. Frequency dependence of the dielectric properties of coal—part II. J. Microw. Power 16 (3−4), 319−326.

Nelson, S.O., Lindroth, D.P., Blake, R.L., 1989a. Dielectric properties of selected and purified minerals at 1 to 22 GHz. J. Microw. Power Electromagn. Energy 24 (4), 213−220.

Nelson, S.O., Lindroth, D.P., Blake, R.L., 1989b. Dielectric properties of selected minerals at 1 to 22 GHz. Geophysics 54 (10), 1344−1349.

Nelson, S.O., Prakash, A., Lawrence, K.C., 1991. Moisture and temperature dependence of the permittivities of some hydrocolloids at 2.45 GHz. J. Microw. Power Electromagn. Energy 26 (3), 178−185.

Nelson, S.O., Forbus Jr., W.R., Lawrence, K.C., 1994a. Microwave permittivities of fresh fruits and vegetables from 0.2 to 20 GHz. Trans. ASAE 37 (1), 181−189.

Nelson, S.O., Forbus Jr., W.R., Lawrence, K.C., 1994b. Permittivities of fresh fruits and vegetables at 0.2 to 20 GHz. J. Microw. Power Electromagn. Energy 29 (2), 81−93.

Nelson, S.O., Forbus Jr., W.R., Lawrence, K.C., 1995. Assessment of microwave permittivity for sensing peach maturity. Trans. ASAE 38 (2), 579−585.

Nelson, S.O., Bartley Jr., P.G., Lawrence, K.C., 1997. Measuring RF and microwave permittivities of adult rice weevils. IEEE Trans. Instrum. Meas. 46 (4), 941−946.

Nelson, S.O., Bartley Jr., P.G., Lawrence, K.C., 1998. RF and microwave dielectric properties of stored-grain insects and their implications for potential insect control. Trans. ASAE 41 (3), 685−692.

Nelson, S.O., Trabelsi, S., Kays, S.J., 2006. Dielectric spectroscopy of honeydew melons from 10 MHz to 1.8 GHz for quality sensing. Trans. ASABE 49 (6), 1977−1981.

Nelson, S.O., Guo, W., Trabelsi, S., Kays, S.J., 2007. Dielectric spectroscopy of watermelons for quality sensing. Meas. Sci. Technol. 18, 1887−1892.

Nelson, S.O., Trabelsi, S., Kays, S.J., 2008. Dielectric spectroscopy of melons for potential quality sensing. Trans. ASABE 51 (6), 2209−2214.

Nicolson, A.M., Ross, G.R., 1970. Measurement of the intrinsic properties of materials by time-domain techniques. IEEE Trans. Instrum. Meas. 19 (4), 377−382.

Noh, S.H., Nelson, S.O., 1989. Dielectric properties of rice at frequencies from 50 Hz to 12 GHz. Trans. ASAE 32 (3), 991−998.

Redheffer, R.M., 1964. The measurement of dielectric constants. In: Montgomery, C.G. (Ed.), Technique of Microwave Measurements, Vol. 11. MIT Radiation Laboratory Series. Boston Technical Publishers, Inc., Boston, MA, pp. 561−676.

Risman, P.O., Bengtsson, N.E., 1971. Dielectric properties of foods at 3 GHz as determined by cavity perturbation technique. J. Microw. Power 6 (2), 101−106.

Roberts, S., von Hippel, A., 1946. A new method for measuring dielectric constant and loss in the range of centimeter waves. J. Appl. Phys. 17 (7), 610−616.

Rzepecka, M.A., 1973. A cavity perturbation method for routine permittivity measurements. J. Microw. Power 8 (1), 3−11.

Stetson, L.E., Nelson, S.O., 1970. A method for determining dielectric properties of grain and seed in the 200- to 500-MHz Range. Trans. ASAE 13 (4), 491−495.

Stetson, L.E., Nelson, S.O., 1972. Audiofrequency dielectric properties of grain and seed. Trans. ASAE 15 (1), 180−184, 188.

Trabelsi, S., Nelson, S.O., 2003. Free-space measurement of dielectric properties of cereal grain and oilseed at microwave frequencies. Meas. Sci. Technol.

Trabelsi, S., Kraszewski, A., Nelson, S.O., 1998a. Nondestructive microwave characterization for determining the bulk density and moisture content of shelled corn. Meas. Sci. Technol. 9, 1548–1556.

Trabelsi, S., Kraszewski, A., Nelson, S.O., 1998b. A microwave method for on-line determination of bulk density and moisture content of particulate materials. IEEE Trans. Instrum. Meas. 47 (1), 127–132.

Trabelsi, S., Kraszewski, A.W., Nelson, S.O., 2000. Phase-shift ambiguity in microwave dielectric properties measurements. IEEE Trans. Instrum. Meas. 49 (1), 56–60.

Tran, V.N., Stuchly, S.S., Kraszewski, A., 1984. Dielectric properties of selected vegetables and fruits. J. Microw. Power 19 (4), 251–258.

van Gemert, M.J.C., 1973. High frequency time domain methods in dielectric spectroscopy. Philips Res. Rep. 28, 530–572.

Waters, D.G., Brodwin, M.E., 1988. Automatic material characterization at microwave frequencies. IEEE Trans. Instrum. Meas. 37 (2), 280–284.

Westphal, W.B., 1954. In: von Hippel, A. (Ed.), Dielectric Materials and Applications (Sec. 2. Distributed circuits, A. Permittivity, Chapter II). John Wiley & Sons, New York, NY.

You, T.-S., Nelson, S.O., 1988. Microwave dielectric properties of rice kernels. J. Microw. Power Electromagn. Energy 23 (3), 150–159.

Zhuang, H., Nelson, S.O., Trabelsi, S., Savage, E.M., 2007. Dielectric properties of uncooked chicken breast muscles from ten to one thousand eight hundred megahertz. Poult. Sci. 86, 2433–2440.

GENERAL AGRICULTURAL APPLICATIONS

3

The use of information on dielectric properties for applications in agriculture can be divided into two general categories: dielectric heating, and the sensing of product quality. Dielectric heating of agricultural materials through the application of high-frequency electric fields and electric fields associated with microwave energy has been considered for a number of applications. These include drying of products such as grain and seed, controlling stored-product insects, seed treatment to improve germination, improving nutritional value of products, controlling product-borne plant and animal pathogens, and inactivating weed seed and plant-infecting organisms in soils. The sensing of product quality has included the use of dielectric properties of agricultural products for sensing moisture content and other quality attributes of some products.

3.1 DIELECTRIC HEATING

When dielectric materials with relatively high dielectric loss factors are subjected to radio-frequency (RF) electric fields (high-frequency and microwave frequencies) of sufficient intensity, these materials will absorb energy from the electric fields by conversion of electric field energy to heat energy in the materials. The phenomenon is known as RF dielectric heating, high-frequency dielectric heating, or microwave heating, depending on the frequencies employed. The degree of heating depends on the power absorbed and on the characteristics of the material.

The power dissipated per unit volume in a nonmagnetic, uniform material exposed to RF or microwave fields can be expressed as:

$$P = E^2 \sigma = 55.63 \times 10^{-12} f E^2 \varepsilon'' \tag{3.1}$$

where P is in W/cm^3, E is the rms electric field intensity in V/m, σ is the conductivity in S/m, f is the frequency in hertz (Hz), and ε'' is the dielectric loss factor, the imaginary part of the relative complex permittivity $\varepsilon = \varepsilon' - j\varepsilon''$. Power dissipated over a period of time provides energy to raise the temperature of the material, and this time rate of temperature increase (°C/s) is given by:

$$\frac{dT}{dt} = P/(c\rho) \tag{3.2}$$

where c is the specific heat of the material in kJ/(kg°C) and ρ is density in kg/m^3. If water is evaporated in the heating process, the energy required for the vaporization and release of the water must be taken into account, and the temperature rise of the material is reduced accordingly.

Dielectric Properties of Agricultural Materials and Their Applications. DOI: http://dx.doi.org/10.1016/B978-0-12-802305-1.00003-8

At dielectric heating frequencies between about 1 and 100 MHz, where materials are often exposed between conducting parallel-plate electrodes, the electric field intensity E between the electrodes is determined by the RF voltage across the electrodes, V, and their separation, d, as $E = V/d$ V/m. If two parallel layers of different materials fill the space between the electrodes with surfaces parallel to the electrodes, the electric field intensity in the material, E_1, can be calculated as

$$E_1 = \frac{V}{d_1 + d_2(\varepsilon_1/\varepsilon_2)} \tag{3.3}$$

where d_1 and d_2 are the thicknesses of the two materials and ε_1 and ε_2 are the respective permittivities of the two materials. This equation is frequently useful when layers of material are treated with an air gap between the material and the top electrode, in which case $\varepsilon_2 = 1$. Equations (3.1) and (3.2) can then be used to estimate the power absorption and heating rate.

3.2 MICROWAVE HEATING

When microwave radiation is directed at a material layer, the absorption of microwave energy propagating through the material also depends on the variables of Eq. (3.1), but the absorption of the energy as the wave travels through the material must also be taken into account. Thus, the dielectric loss factor of the material is important. The frequency of the wave is also a factor, and the power absorption depends on the square of the electric field intensity. For a plane wave, the electric field intensity E which has $e^{j\omega t}$ dependence, can be expressed as (von Hippel, 1954a):

$$E(z) = E_0 e^{j\omega t - \gamma z} \tag{3.4}$$

where E_0 is the rms electric field intensity at a point of reference, z is the distance in the direction of travel, and γ is the propagation constant for the medium in which the wave is traveling. The propagation constant is a complex quantity,

$$\gamma = \alpha + j\beta = j\frac{2\pi}{\lambda_o}\sqrt{\varepsilon} \tag{3.5}$$

where α is the attenuation constant, β is the phase constant, λ_o is the free-space wavelength and ε is the relative complex permittivity.

The attenuation constant α and the phase constant β are related to the dielectric properties of the medium as follows:

$$\alpha = \frac{2\pi}{\lambda_o}\sqrt{\frac{\varepsilon'}{2}\left(\sqrt{1 + \tan^2 \delta} - 1\right)} \text{ nepers/m} \tag{3.6}$$

$$\beta = \frac{2\pi}{\lambda_o}\sqrt{\frac{\varepsilon'}{2}\left(\sqrt{1 + \tan^2 \delta} + 1\right)} \text{ rad/m} \tag{3.7}$$

As the wave travels through a material that has a significant dielectric loss, its energy will be attenuated. For a plane wave traversing a dielectric material, the electric field intensity at the site of interest can be obtained by combining Eqs. (3.4) and (3.5) as follows:

$$E(z) = E_0 e^{-\alpha z} e^{j(\omega t - \beta z)} \tag{3.8}$$

where the first exponential term controls the magnitude of the electric field intensity at the point of interest; it should be noted that this magnitude decreases as the wave advances into the material. Since the power dissipated is proportional to E^2, $P \propto e^{-2\alpha z}$. The penetration depth, D_p, is defined as the distance at which the power drops to $e^{-1} = 1/2.718$ of its value at the surface of the material. Thus, $D_p = 1/2\alpha$. If attenuation is high in the material, the dielectric heating will taper off rapidly as the wave penetrates the material. Attenuation is often expressed in decibels per meter (dB/m). In terms of power densities and electric field intensity values, this can be expressed as (von Hippel, 1954b):

$$10 \log_{10}\left[\frac{P_0}{P(z)}\right] = 20 \log_{10}\left[\frac{E_0}{E(z)}\right] = 8.686\alpha z \tag{3.9}$$

The attenuation in decibels, combining Eqs. (3.6) and (3.9), can be expressed in terms of the dielectric properties, when $(\varepsilon'')^2 \ll (\varepsilon')^2$, as follows:

$$\alpha = \frac{8.686\pi\varepsilon''}{\lambda_0\sqrt{\varepsilon'}} \, \text{dB/m} \tag{3.10}$$

A plane wave incident upon a material surface will have some of the power reflected, and the rest, P_t, will be transmitted into the material. The relationship is given by the following expression:

$$P_t = P_0(1 - |\Gamma_0|^2) \tag{3.11}$$

where P_0 is the incident power at the material surface and Γ is the reflection coefficient. For an air—material interface, the reflection coefficient can be expressed in terms of the complex relative permittivity of the material as (Stratton, 1941):

$$\Gamma = \frac{1 - \sqrt{\varepsilon}}{1 + \sqrt{\varepsilon}} \tag{3.12}$$

The power density diminishes as an exponential function of the attenuation and distance traveled (Eq. (3.8)) as the wave propagates through the material:

$$P = P_t \, e^{-2\alpha z} \tag{3.13}$$

with α expressed in nepers/m. For attenuation in decibels, dB/cm = 0.08686 × (nepers/m).

3.3 BASIC DIFFERENCES

The term RF dielectric heating is generally understood to involve frequencies between about 1 and 100 MHz. This field of application was developed earlier than microwave heating (Brown et al., 1947). Commonly used frequencies have been 13.56, 27.12, and 40.68 MHz, although other allocations exist. Microwave heating (Metaxas and Meredith, 1983), which is also dielectric heating, evolved after World War II, as magnetrons and other microwave sources became available. Microwave heating generally involves frequencies above 500 MHz, and the principal frequencies used for application have been 896, 915, and 2450 MHz.

As already mentioned, RF dielectric heating is often accomplished with the load between parallel-plate electrodes, although many other configurations are available (Brown et al., 1947). In this case, rapid heating is produced by using very high RF electrode voltages to achieve high electric field intensities in the material. The maximum field intensities are usually limited by dielectric breakdown, arcing, and consequent damage to the product being heated. For cereal grains such as wheat, field intensities of 1.4–1.5 kV/cm were used with little arcing difficulty (Nelson and Whitney, 1960). With parallel-plate geometry, the field intensity is quite uniform if the load is relatively homogeneous in character, and then Eqs. (3.1)–(3.3) are sufficient for general description.

For microwave heating, Eqs. (3.1) and (3.2) still apply, but attenuation and consequent penetration depth become important. Because frequency is much greater, rapid heating can be achieved with much lower field intensities (Eq. (3.1)), and the problems of arcing in the product are diminished. However, it becomes much more complicated to estimate electric field intensities because of attenuation and the need to irradiate materials from more than one direction to obtain required interior heating. Often, the product is moved through the microwave fields or rotated during exposure to achieve better heating uniformity.

With RF dielectric heating, penetration is not so much the problem, but the dimensions of the material to be heated are limited by high-voltage RF insulation problems in producing desired high field intensities in the material. However, all material between the parallel-plate electrodes will be subjected to high field intensities, whereas the field intensity decreases with penetration in microwave heating.

Microwave attenuation at 9.4 GHz in cereal grains varies from about 0.3 dB/cm for moisture contents of 7% to 2–4 dB/cm when the grain is at 21% moisture (Kraszewski, 1988). Attenuation values computed by Eq. (3.10) from permittivity data from similar frequencies (Kraszewski et al., 1995; Kraszewski and Nelson, 1990) agree, showing about 3.5 dB/cm for wheat at 20% moisture. At 4.8 GHz, the attenuation in the same wheat was about 2 dB/cm. These attenuations are significant, because for 3 dB the power densities are reduced by one half, and this occurs only a centimeter or so deep in the grain. Thus only a small fraction of the power available for heating grain at the surface is available after a few centimeters of penetration. The attenuation is somewhat lower at the common microwave heating frequencies. Eq. (3.10) indicates the inverse relationship between wavelength and attenuation. Consequently, 915-MHz microwave heating is known to provide deeper penetration than heating at 2450 MHz (2.45 GHz). Attenuation is also lower for grain with a lower moisture content. For example, the attenuation at 2.45 GHz for wheat of 12% moisture is about 0.3 dB/cm (Kraszewski et al., 1995), so the power density at a depth of 10 cm would drop to half its value at the surface.

Dielectric heating has been explored for several agricultural applications in considerable detail, and those areas are summarized in separate chapters. These include applications for insect control (Chapter 4), seed treatment (Chapter 5), and product conditioning (Chapter 6).

3.4 PRODUCT QUALITY SENSING

Sensing the quality of agricultural products through the application of electric fields requires correlations between the dielectric properties of materials and those quality attributes of interest.

The moisture content of agricultural products is one of the most important characteristics determining quality and storability. The correlation of dielectric properties with moisture content is high, and therefore sensing of moisture content has received much attention; that work is summarized in Chapters 7 and 12.

Sensing of quality factors such as sugars in fruits has also been explored to some extent, and that information is summarized in Chapter 9.

3.5 TREATING SEED-BORNE PATHOGENS
3.5.1 LOOSE SMUT IN BARLEY

Loose smut in barley, caused by the fungus *Ustilago nuda* (Jens.) Rostr., is a disease that can cause a significant loss in yield. The fungus infects the growing barley plant during the flowering stage. The infected seed heads become sources of spores that infect other plants during flowering. Loose smut survives to the next growing season as dormant mycelium inside the seed embryos. As internal seed-borne pathogens, the smut fungi have been difficult to control with surface fungicide treatments, and tedious hot-water treatments—with critical temperature control to avoid damaging seed viability—were used to rid barley seed lots of the disease. Therefore, RF dielectric heating has been studied for potential control of seed-borne pathogens. The use of infrared and RF heating to control loose smut in wheat was attempted, but no satisfactory control was achieved (Kleis et al., 1951). Experiments were also conducted with loose smut in barley (Nelson and Walker, 1961).

A small amount of "Mars" barley, *Hordeum vulgare* L. with about 25% smut-infected seeds, was obtained, and the degree of infection was determined according to procedures reported by Popp (1951). Small samples of this seed were treated at 8% and 11% moisture content, and also some seeds were treated after being soaked in distilled water for 6 h at 40°F (4.4°C) (Nelson and Walker, 1961). Soaked samples were allowed to dry on paper for about 30 min prior to treatment. Frequencies of 10 and 39 MHz and field intensities ranging from 1.7 to 4.8 kV/in (0.67−1.89 kV/cm) were employed with exposure times chosen to produce a range of grain mass temperatures from below 110°F (43.3°C) to those high enough to seriously lower seed germination. After RF treatment, soaked seeds were allowed to dry thoroughly before being treated with a fungicide to kill any superficial fungal spores which might be present. While mycelium from these germinating spores is usually very different in appearance and staining reaction, this precaution was deemed advisable. Seeds were thoroughly dried for a few days following the application of fungicide, and were then planted 2.5 cm deep in soil in pots in the greenhouse. To produce greater elongation of the subcrown internodes to facilitate removal of the crown nodes, the pots were kept in dim light until the seedlings were 2.5−5 cm high. When the second leaf appeared (about 2 weeks after planting), the crown node was removed with about 3 mm of the subcrown internode and 1 cm of the shoot (plumule). In a few cases the scutellar node was also removed for examination. The excised portions of the seedlings were then macerated, stained with cotton blue, and mounted on slides according to the method used by Popp (1951). In this way the smut mycelium was easily recognized under the microscope by its thick, deeply stained walls and short, irregularly swollen cells. For one experiment, seed samples were divided, one set being planted in the greenhouse and examined by the laboratory method, and the other planted in a field plot providing three 10-ft (3 m) rows (about 100 seeds per row) for each treated sample and the controls. The number of smut-infected heads and the total number of heads for each sample were counted at maturity, and their ratio taken for the degree of smut infection.

Microscopic examination of seedlings grown from samples of smut-infected barley, treated at 8% and 11% moisture, showed infection still present for treatments that raised the temperature so high that germination of the barley was seriously reduced. The percentage of infection found by microscopic examination and by head counts in the field test agreed. Results for the seed samples that were soaked in water prior to treatment suggested that some control of the fungi may have been achieved, but that germination was seriously lowered by such treatments (Nelson and Walker, 1961). Subsequent trials with treatment of presoaked seed were no more successful. In some of the treated seedlings, examination showed traces of smut infection in the scutellar node which had failed to develop and infect the crown node. These observations suggest that the treatment may have affected the smut in some way to prevent its normal development. However, no likelihood for successful use of RF heating to control the fungi without damaging the viability of the barley seeds was indicated.

3.5.2 BACTERIA ON ALFALFA SEED

The production of sprouts from alfalfa, *Medicago sativa* L., and other seed for human consumption is a substantial industry, However, there have been several outbreaks of illness associated with sprouts, and contamination by *Salmonella* and *Escherichia coli* O157:H7 has been identified as the cause (Beuchat, 1996; Taormina et al., 1999). *Listeria monocytogenes* is another organism that poses a potential threat. Because most of the outbreaks of infection were attributed to contaminated sprouting seed, several methods have been studied for decontaminating seed. Treatment of alfalfa seed in hot water at 54°C for 5−10 min resulted in moderate reduction of *Salmonella* populations, but temperatures above 54°C significantly reduced seed viability (Jaquette et al., 1996). Several aqueous solutions of chemicals—including chlorine, chlorine dioxide, hydrogen peroxide, trisodium phosphate, ethanol, peracetic acid, and some commercial fruit and vegetable produce wash solutions—have been studied for decontaminating alfalfa seed. None of these treatments eliminated *E. coli* O157:H7 or *Salmonella* from alfalfa seed intended for sprouting (Jaquette et al., 1996; Beuchat, 1997; Taormina and Beuchat, 1999; Weissinger and Beuchat, 2000).

Earlier research has shown that RF and microwave dielectric heating treatments are effective for increasing the germination percentage of alfalfa seed lots containing high percentages of hard seed (Nelson and Walker, 1961; Nelson and Wolf, 1964; Nelson et al., 1968; Stetson and Nelson, 1972). Hard seeds occur naturally and are viable seeds with seed coats that are impermeable to water. Therefore, they will not germinate promptly when planted, but they may germinate several weeks, months, or years later when the seed coat becomes permeable through natural processes. Mechanical scarification of such seed lots to increase germination is a common practice for alfalfa, but the abrasive process scratches the seed coat, thus providing a favorable environment for bacterial attachment, which may make sanitization with liquids more difficult (Holliday et al., 2001). Thus, it appeared reasonable to explore the possible use of dielectric heating for reduction of bacterial populations on alfalfa seed, especially since the improvement of germination and subsequent sprout yield can be achieved without mechanical abrasion of the seed coat. Similar consistent increases in alfalfa seed germination through hard seed reduction were achieved by dielectric heating at frequencies of 5, 10, 39, and 2450 MHz (Nelson and Wolf, 1964; Stetson and Nelson, 1972). Treatment at 39 MHz was selected for study because treating seed samples between parallel-plate electrodes provides a more uniform electric field for exposure of the samples than is commonly available in microwave ovens.

The potential for controlling human bacterial pathogens on alfalfa seed by dielectric heating was studied by experimental exposure of alfalfa seed—artificially contaminated with *Salmonella, E. coli* O157:H7, and *L. monocytogenes*—to RF dielectric heating treatments at 39 MHz and different electric field intensities for varying times of exposure (Nelson et al., 2002). Moisture content of alfalfa seed and final temperatures produced by the RF exposures were determined, and control and treated seed samples were analyzed in the laboratory for reduction of bacterial populations and effects on seed germination (Lu, 2001).

Significant reductions in populations of bacteria (*Salmonella, E. coli* O157:H7, and *L. monocytogenes*) on artificially inoculated alfalfa seeds were achieved by exposures of several seconds to RF dielectric heating treatments at 39 MHz and different electric field intensities, which did not lower seed germination or affect sprout appearance (Nelson et al., 2002). However, the desired levels of reduction in bacterial populations (5log CFU/g) were not achieved without severe reductions in germination of the alfalfa seed. Although not clearly delineated, bacterial reductions appeared better when seed moisture content at the time of treatment was in the 5−7% (wet basis) range than at higher moisture levels. Final seed temperatures at the termination of RF exposures, along with moisture content, provide guidance as to the level of exposure that can be used without reducing rates of seed germination. Relationships among final temperature, seed moisture content, and germination percentage confirmed earlier reports that seeds can tolerate higher temperatures without damage to viability when their moisture content at the time of dielectric heating is lower (Nelson and Wolf, 1964). Because RF dielectric heating treatments can increase the germination of alfalfa seed lots with substantial hard-seed content, thus increasing sprouting yield, without abrasion of the seed coat, the process might be considered for this purpose as well as for reduction of human pathogen populations in sprouting seed (Nelson et al., 2002).

REFERENCES

Beuchat, L.R., 1996. Pathogenic microorganisms associated with fresh produce. J. Food Prot. 59, 204−216.

Beuchat, L.R., 1997. Comparison of chemical treatments to kill *Salmonella* on alfalfa seeds destined for sprout production. Int. J. Food Microbiol. 34, 329−333.

Brown, G.H., Hoyler, C.N., Bierwirth, R.A., 1947. Radio-Frequency Heating. D. Van Nostrand Co., New York, NY.

Holliday, S.L., Scouten, A.J., Beuchat, L.R., 2001. Efficacy of chemical treatments in eliminating *Salmonella* and *Escherichia coli* O157:H7 on scarified and polished alfalfa seeds. J. Food Prot. 64, 1489−1495.

Jaquette, C.B., Beuchat, L.R., Mahon, B.E., 1996. Efficacy of chlorine and heat treatment in killing *Salmonella* Stanley inoculated onto alfalfa seeds and growth and survival of the pathogen during sprouting and storage. Appl. Environ. Microbiol. 62, 2212−2215.

Kleis, R.W., Lucas, E.H., Brown, H.M. and de Aeeuw, D.J., 1951. Dry Treatment of Wheat for Loose Smut. Quarterly Bulletin, Michigan Agricultural Experiment Station, Michigan State University, East Lansing, MI, 34(2), 162−165.

Kraszewski, A., Nelson, S.O., 1990. Study on grain permittivity measurments in free space. J. Microw. Power Electromagn. Energy 25 (4), 202–210.

Kraszewski, A.W., 1988. Microwave monitoring of moisture content in grain—further considerations. J. Microw. Power 23 (4), 236–246.

Kraszewski, A.W., Trabelsi, S., Nelson, S.O., 1995. Microwave dielectric properties of wheat. In: Proceedings of the 30th Microwave Power Symposium (IMPI), Denver, CO, 90–93.

Lu, C.-Y., 2001. Application of Dielectric Heating and Ultrasound in Combination with Heat and Chemical Solutions to Kill Human Pathogenic Bacteria on Alfalfa Seeds (M.S. thesis), The University of Georgia, Athens, GA.

Metaxas, R.C., Meredith, R.J., 1983. Industrial Microwave Heating. Peter Peregrinus Ltd, London.

Nelson, S.O., Walker, E.R., 1961. Effects of radio-frequency electrical seed treatment. Agric. Eng. 42 (12), 688–691.

Nelson, S.O., Whitney, W.K., 1960. Radio-frequency electric fields for stored-grain insect control. Trans. ASAE 3 (2), 133–137.

Nelson, S.O., Wolf, W.W., 1964. Reducing hard seed in alfalfa by radio-frequency electrical seed treatment. Trans. ASAE 7 (2), 116–119, 122.

Nelson, S.O., Stetson, L.E., Works, D.W., 1968. Hard-seed reduction in alfalfa by infrared and radio-frequency electrical treatments. Trans. ASAE 11 (5), 728–730.

Nelson, S.O., Lu, C.-Y., Beuchat, L.R., Harrison, M.A., 2002. Radio-frequency heating of alfalfa seed for reduction of human pathogens. Trans. ASAE 45 (6), 1937–1942.

Popp, W., 1951. Infection in seeds and seedlings of wheat and barley in relation to development of loose smut. Phytopathology 41, 261–275.

Stetson, L.E., Nelson, S.O., 1972. Effectiveness of hot-air, 39-MHz dielectric, and 2450-MHz microwave heating for hard-seed reduction in alfalfa. Trans. ASAE 15 (3), 530–535.

Stratton, J.A., 1941. Electromagnetic Theory. McGraw Hill Book Co., New York, NY.

Taormina, P.J., Beuchat, L.R., 1999. Comparison of chemical treatments to eliminate enterohemorrhagic *Escherichia coli* O157:H7 on alfalfa seeds. J. Food Prot. 62, 318–324.

Taormina, P.J., Beuchat, L.R., Slutsker, L., 1999. Infections associated with eating seed sprouts. Emerg. Infect. Dis. 5, 626–634.

von Hippel, A.R., 1954a. Dielectric Materials and Applications. The Technology Press of M.I.T. and John Wiley & Sons, New York, NY.

von Hippel, A.R., 1954b. Dielectrics and Waves. John Wiley & Sons, New York, NY.

Weissinger, W.R., Beuchat, L.R., 2000. Comparison of aqueous chemical treatments to eliminate *Salmonella* on alfalfa seeds. J. Food Prot. 63, 1475–1482.

INSECT CONTROL APPLICATIONS

4

Interest in the possibility of controlling insects with high-frequency electric energy dates back more than 80 years. In 1929, Headlee and Burdette reported results of experiments determining lethal exposures for several insect species subjected to 12-MHz electric fields and the determination of body temperatures produced in honey bees by the consequent dielectric heating. These and later studies have been summarized in numerous reviews (Thomas, 1952; Frings, 1952; Peredel'skii, 1956; Watters, 1962; Nelson and Seubert, 1966; Nelson, 1967, 1973). A patent (Davis, 1934) was issued in 1934 on the high-frequency method and equipment for exterminating insect life in seed, grain, or other materials. Concern about the health hazards of chemical pesticides in the 1950s through the 1970s stimulated further studies on the possible use of radio-frequency (RF) and microwave energy for controlling stored-grain and other stored-product insects. Still today, even though the concerns about the use of chemical pesticides remain, the techniques have not found their way into practical use. This chapter is written to summarize the findings and to examine the reasons for the lack of acceptance for practical use as a method for insect control in stored products.

4.1 STORED-GRAIN INSECTS

Insects that infest grain after harvest and in storage account for tremendous losses of cereal grains and other similar crops around the world, especially in tropical zones but also in subtropical and temperate zones. Grain storage sanitation practices and chemical controls have been relied upon for many years to limit such losses, but the losses of such food materials are still substantial, and health hazards of chemical residues are still of concern. Thus, nonchemical methods for controlling stored-grain insects are still of interest.

4.1.1 SELECTIVE DIELECTRIC HEATING

Raising the body temperatures of insects to lethal levels is one method for controlling them. However, temperatures at these levels generally produce damage to the host product. If selective heating of the insects in relation to the grain they infest were possible, dielectric heating would offer an advantage over conventional heating for the control of stored-grain insects. The general

Dielectric Properties of Agricultural Materials and Their Applications. DOI: http://dx.doi.org/10.1016/B978-0-12-802305-1.00004-X

principles of dielectric heating are presented in Chapter 3, Section 3.1. From Eqs (3.1) and (3.2) we can write the following:

$$\frac{dT}{dt} \propto \frac{fE^2 \varepsilon''}{c\rho} \tag{4.1}$$

which states that the time rate of temperature increase is directly proportional to the frequency, the square of the electric field intensity E, and the dielectric loss factor ε'' of the material, and inversely proportional to the specific heat c and the density of the material ρ. In considering the effects of the variables of Eq. (4.1) on the relative heating rates for insects and grain, the frequency for the two materials will be the same. However, the dielectric loss factor for the insects and the grain may be different, and in that instance, the electric field intensities in the insects and the grain might also differ.

To examine the relative electric field intensities in the insect, E_i, and that in the host-grain medium, E_m, we can consider that, for a plane wave interacting with a spherical insect in a uniform, infinite medium, the electric field in the insect is (Stratton, 1941):

$$E_i = E_m \left(\frac{3\varepsilon_m}{2\varepsilon_m + \varepsilon_i} \right) \tag{4.2}$$

where ε_i and ε_m represent the complex relative permittivities of the insect and the host medium, respectively. If the necessary permittivity values are available, $(E_i/E_m)^2$ can be calculated, which gives us the ratio of the E-field's contribution to the power dissipation per unit volume in the insect relative to that in the host medium. If the permittivity values are known, the $\varepsilon_i''/\varepsilon_m''$ ratio can also be obtained. The product of the $(E_i/E_m)^2$ and $\varepsilon_i''/\varepsilon_m''$ ratios gives the estimated power absorption ratio for the insect in relation to the host medium.

For selective dielectric heating of the insects, one must also consider the other two variables (Thomas, 1952) that, along with power absorbed, affect the heating rate in the two different materials (Eq. (4.1)). The specific heat and the density have an inverse influence, because c and ρ appear in the denominator of the right-hand side of Eq. (4.1). Therefore, the reciprocal of the $(c_i/c_m) \times (\rho_i/\rho_m)$ value should be multiplied by the power absorption ratio to determine the differential heating factor for the insects in relation to the host-grain medium.

4.1.2 EXPERIMENTAL FINDINGS

The findings of many early experiments in which insects were exposed to RF electric fields can be explained by considering the physical principles already outlined. The heating rate of materials exposed in RF electric fields increases with increasing field intensity and with increasing frequency (Thomas, 1952; Frings, 1952). Since the loss factor of hygroscopic materials, such as grain, generally increases with increasing moisture content, their heating rates also are higher when moisture content is greater.

Experiments have shown that many insect species that infest grain and cereal products can be controlled by short exposures to RF fields that do not damage the host material. Generally, for successful RF insect-control treatments, resulting temperatures in host materials of this kind range between 40°C and 90°C, depending upon the characteristics of the host material, the insect species and developmental stage, and the nature of the RF or microwave treatment. RF treatments at 13.6 and 39 MHz necessary for insect control have not been damaging to wheat germination or milling and baking qualities (Nelson and Walker, 1961; Munzel, 1975). So far, however, RF and microwave methods have not come into practical use because they have been considered too costly compared to conventional chemical and other physical control methods.

4.1.3 ENTOMOLOGIC FACTORS

Differences among various stored-grain insect species in their susceptibility to control by exposure to RF dielectric heating have been noted when they were treated in common host grains under similar conditions (Nelson, 1973; Nelson and Whitney, 1960; Whitney et al., 1961; Nelson and Kantack, 1966; Nelson et al., 1966). Some differences are attributable to interspecific characteristics of a biologic or physiologic nature, but some can be explained by variations in size and in geometric relationships. Based on adult insect mortality, comparison of several species—treated in wheat at a frequency of 39 MHz and electric field intensity of 1.2 kV/cm in the grain—showed the following rank in order of decreasing susceptibility: rice weevil, *Sitophilus oryzae* (L.), and granary weevil, *Sitophilus granarius* (L.); sawtoothed grain beetle, *Oryzaephilus surinamensis* (L.); confused flour beetle, *Triboleum confusum* Jacquelin duVal, and red flour beetle, *Triboleum castaneum* (Herbst); the deremstid, *Trogoderma varabile* Ballion; cadelle, *Tenebroides mauritanicus* (L.); and the lesser grain borer, *Rhyzopertha dominica* (F.) (Whitney et al., 1961; Nelson and Kantack, 1966). The greater susceptibility of the rice weevil compared to the confused flour beetle was also noted in studies at 2.45 GHz by Tateya and Takano (1977).

In these studies and in others (Webber et al., 1946; Baker et al., 1956; van den Bruel et al., 1960; van Dyck, 1965; Benz, 1975; Anglade et al., 1979), differences in susceptibility were also found between developmental stages within species. In general, the adult stages were more susceptible to control by RF treatment than were the immature stages. With rice weevils, granary weevils and lesser grain borers, the immature forms develop inside the grain kernel. The adults were more susceptible to control in each instance (Whitney et al., 1961; Nelson et al., 1966). The closely surrounding kernel material might provide some benefit to the immature forms in affecting the field intensities to which they are subjected or in conducting heat from the insects because of better physical contact. Differences in larval and adult susceptibility seem to vary among species. Little difference was noted in mortalities of the adult and larval stages of the confused flour beetle when treated at 39 MHz and 1.4 kV/cm in wheat shorts (Whitney et al., 1961). However, Webber et al. (1946) found in experiments at 11 MHz that larvae of the same species treated in flour were somewhat more susceptible than the adults. Baker et al. (1956) found that, when treated in whole-wheat flour at 2.45 GHz, the adult confused flour beetle was more susceptible than either the larval or egg stages. Granary weevil adults were also found to be more susceptible than the eggs in these experiments. However, other experiments at the same frequency with the confused flour beetle in wheat flour showed that the larval stage was significantly more susceptible than adult, pupal, or egg stages (Tateya and Takano, 1977). In the same studies, young (2—7 days postemergence) adults of the confused flour beetle had significantly higher survival rates than older (37—107 days) adults of the same species.

Larvae of the cadelle were more susceptible to 39-MHz exposures than were adults of this species (Nelson et al., 1966). With the cadelle, however, the larva is much larger than the adult, and it feeds outside the kernels. Thus, physical factors such as size and geometric relationships may well account for the difference noted in susceptibility to RF treatments.

Interspecific differences were also noted in the degree of delayed mortality following 39-MHz treatment of adult insects. A substantial increase in mortality was obtained with rice weevils and granary weevils between 1 day and 1 week after treatment—whereas, practically no change in mortality attributable to the RF treatment occurred thereafter. The lesser grain borer also exhibited a substantial delayed mortality, but it occurred during the second week after treatment rather than during the first week. Adults of the confused flour beetle and the red flour beetle exhibited less delayed mortality, most of which occurred during the second and third weeks after treatment (Whitney et al., 1961; Nelson et al., 1966). In other studies, mortality of the yellow meal worm, *Tenebrio molitor* L., larvae continued to increase during a 2-week period after treatment at 39 MHz (Kadoum et al., 1967a).

The age of yellow mealworm larvae at the time of sublethal exposures at 39 MHz influenced the degree of resulting morphologic abnormality observed in adults developing from the treated larvae (Kadoum et al., 1967c; Rai et al., 1971). One-day-old eggs of the yellow mealworm were also more susceptible than 3-day-old eggs to 39-MHz treatment (Rai et al., 1972).

Physiologic injuries in stored-grain insects following exposure to RF electric fields were noted in the appendages, particularly in the joints of the legs (Whitney et al., 1961). Heat injury to the histoblasts was suspected in RF-treated larvae of the yellow mealworm that developed into adults with badly deformed or missing legs (Kadoum et al., 1967c). The same type of injury was obtained by either RF treatment or heat injury from a hot needle, as observed by Rai et al. (1971), who described abnormal development in both cephalic and thoracic appendages of the same species after RF treatment in the larval and pupal stages. Other types of abnormal development, characterized by incomplete metamorphosis, were reported by Carpenter and Livstone (1971) who exposed pupae of the same species to 10-GHz microwave energy. Similar results were obtained by Lindauer et al. (1974) and Liu et al. (1975); however, the results of these microwave exposures were attributed to thermal causes by Olsen (1982), and thermal effects were also suggested to be likely by Ondracek and Brunnhofer (1984).

In studies with 39-MHz treatment of yellow mealworm larvae, Kadoum et al. (1967a) accounted for observed mortality of the insects as a result of internal body heating. Losses in body weight and increased oxygen uptake rates as a result of RF exposures were also observed (Kadoum et al., 1967b). Increased rates of oxygen uptake were similar to those in surgically injured larvae, and accompanying increases in the rate of protein synthesis were also observed (Kadoum, 1969a, 1969b).

Experiments with adult insects, treated at 39 MHz in wheat and wheat shorts, showed that rice weevils and confused flour beetles that survived exposures were capable of reproduction (Whitney et al., 1961). However, the more severe sublethal treatments greatly reduced the number of progeny. Studies with lesser grain borers showed that the reproduction rate was lowered when adults exposed to RF treatment suffered greater than about 50% mortality (Nelson et al., 1966). Earlier, Webber et al. (1946) reported reproductive capacity observed in confused flour beetles that survived exposures in flour at 11 MHz. Confused flour beetles surviving 2.45-GHz treatments in wheat flour were also capable of reproduction, but at rates somewhat lower than those of untreated insects (Tateya and Takano, 1977).

Studies on reproductive tissues of yellow mealworms exposed as adults to 39-MHz treatments indicated that lowered reproductive capacity resulted from probable heat damage to sperm cells and ovarian tissues (Rai et al., 1974). Reduced fecundity following exposure of adult insects appeared more dependent on treatment of the males than on treatment of the females, but treatment in the larval stage resulted in both reduced ovary size and in the number of eggs developing in females that emerged from treated larvae (Rai et al., 1975). Insemination was inhibited by exposure of either sex of adult mealworms (Rai et al., 1977). The numbers of active spermatozoa were reduced by treatment, both in the males and in those transferred to the females, with consequent reductions in egg hatch percentage. Reduced fecundity as a result of RF treatment probably resulted from the inactivation of spermatozoa and impaired ability of treated adults to mate successfully (Rai et al., 1977).

4.1.4 PHYSICAL FACTORS

The influence of various physical factors on the control of stored-grain insects by exposures to RF and microwave energy has been studied. The importance of some of these factors, such as frequency, electric-field intensity, and permittivities of the materials involved, is evident in Eqs. (4.1) and (4.2). Other factors include heating rate, modulation of the applied energy, and other characteristics of the insects and the host media. General conclusions are difficult to draw concerning some of these factors. High rates of heating are to be preferred, generally, to minimize thermal energy loss from the insects to the host material. Therefore, high rates of power dissipation are desired; power dissipation depends on frequency, electric-field intensity and the dielectric loss factor of the material (Eq. (3.1)).

The possibilities for selective heating of the insects, as discussed in Section 4.2, are important here. Thomas (1952) concluded that selective heating of insects in grain should be possible with high-frequency dielectric heating. Differences in the temperatures of host media necessary for control of the rice weevil and the confused flour beetle, when exposed to 39-MHz treatments, were explained by selective heating analysis (Nelson and Whitney, 1960). For adult rice weevils and confused flour beetles treated in wheat, the theory predicted selective heating of the insects, but for confused flour beetles in wheat shorts, it did not. When insects were treated in wheat, exposures producing grain temperatures below lethal temperatures for the insects achieved complete mortality.

The electric field intensity to which insects are subjected depends on geometric and spatial factors as well as on the dielectric properties of the insects and their host medium. Therefore, host-medium particle size in relation to insect dimensions might be expected to influence lethal exposure levels. Experiments with adult granary weevils in wheat, and in corn of the same moisture content, showed that insects suffered lower mortalities in wheat (smaller kernels) than in corn (larger kernels) at comparable grain temperatures (Nelson and Kantack, 1966). Rice weevil adults exposed to RF electric fields in host media of glass beads of different sizes survived better in small beads than in larger ones (Nelson et al., 1966).

Immature forms of the rice weevil, granary weevil, and lesser grain borer, all of which develop inside the grain kernels, are less susceptible to control than are the adults which are outside the kernels. The kernels have too low an electrical conductivity to provide an effective shield, but they may alter the electric field distribution in a way that is favorable to survival of the internal insects. Experiments with rice weevil adults and with adults of the lesser grain borer, treated inside and outside wheat kernels, indicated that insects treated outside the kernels suffered somewhat higher mortalities than did insects of the same age that were exposed while inside the kernels (Nelson et al., 1966).

Because the dielectric properties of grain vary with moisture content (Nelson, 1982), the moisture content of the host medium might be expected to have some influence on the effectiveness of RF insect treatment. However, mortalities of adult rice weevils treated at 39 MHz in wheat of 11.4% and 12.8% moisture content were the same (Whitney et al., 1961). Later experiments with insects in wheat of moisture contents between 12% and 16% indicated that 39 MHz treatments were slightly more effective as wheat moisture content increased. In treating adult rice weevils at 2.45 GHz in wheat ranging from 12.3% to 16.0% moisture, Tateya and Takano (1977) observed no significant difference in insect mortality attributable to wheat moisture level.

Because the electric-field intensity has an important influence on the heating rate, its effects on insect mortality have received attention. Webber et al. (1946) used field intensities in the 1.2- to 1.8-kV/cm range in treating the Mediterranean flour moth, *Anagasta kuehniella* (Zeller), and the confused flour beetle in flour at a frequency of 11 MHz. No differences were apparent in the mortality that might be attributable to field intensity differences. Effects of field intensity and frequency on insect mortality are difficult to distinguish, because both influence the heating rate. The three factors of frequency, electric-field intensity, and heating rate must therefore, be considered in relation to one another. Several experiments at 10 and 39 MHz, involving exposure times ranging from a few seconds to a minute or more, with different combinations of field intensities and heating rates, indicated that there are subtle frequency effects which depend upon the species and the developmental stages of the insects (Nelson et al., 1966). In comparing 10- and 39-MHz treatments with similar heating rates, the 10-MHz treatment was consistently more effective in producing mortality for some species and stages, whereas, the 39-MHz treatment was consistently better for others. For some, the two frequencies produced similar mortalities.

High field intensities were much more efficient than low intensities in killing adult rice weevils in wheat at both 10 and 39 MHz, but with immature stages of the same species, high and low intensities produced the same results (Nelson and Whitney, 1960; Whitney et al., 1961). Generally, differences in insect mortality attributable to field intensity diminished at field intensities greater than 1.2 kV/cm. For a given frequency, an increase in field intensity increases the heating rate. The more rapid elevation of temperature that accompanies high-field-intensity treatments appears to offer a possible explanation for the greater effectiveness of high-field-intensity treatments which might produce a higher degree of thermal shock. Loss of heat energy to surroundings during treatment can also be a factor when exposure times are long. Heating rate alone, however, does not appear to determine the effectiveness of treatment at different frequencies (Nelson et al., 1966). In studies on the control of stored-grain insects with 13.6-MHz exposures, Benz (1975) reported that end-point temperatures of the host medium and not thermal shock (or heating rate) was the important factor determining insect mortality.

Efforts to improve the efficiency of control of stored-grain insects through the use of higher-field-intensity, pulse-modulated RF electric fields were not successful. Treatment of adult rice weevils and confused flour beetles in wheat with pulse-modulated 39-MHz fields of 5- to 40-ms pulse widths and 10−40 pulses per second (pps) pulse repetition rates did not show any improvement compared to unmodulated (CW) treatments (Nelson et al., 1966). Even with pulses as short as 50 μs and field intensities of 4 kV/cm, no improvement over CW, 1.4-kV/cm treatments was obtained for treatment of adult rice weevils and granary weevils in wheat.

The opinion sometimes expressed that selective frequencies might exist for certain insects has not been demonstrated through experimental or theoretical work. Studies in which confused flour beetles in flour and granary weevils in wheat were exposed to 2.45-GHz microwave treatments indicated that no selective heating was obtained (Baker et al., 1956). Temperatures in the host media in excess of 82°C were required for control of immature stages of both species. Corresponding control of granary weevils, rice weevils, and lesser grain borers in wheat with 39-MHz treatments was achieved when grain temperatures were momentarily raised to the 60–66°C range (Whitney et al., 1961; Nelson and Stetson, 1974). Adult rice weevils in hard red winter wheat were all killed by 39-MHz dielectric heating treatments of a few seconds that raised the grain temperature to 39°C (Nelson and Whitney, 1960); whereas these insects are able to survive for many hours at this temperature in a hot-air oven. Thus, selective heating of the adult insects was indicated as the explanation. Holding infested cereal products at 60°C for 10 min effectively controls stored-grain insects (Cotton, 1963).

RF treatment of confused flour beetle adults in wheat at 39 MHz required grain temperatures of 47°C for complete mortality (Nelson and Whitney, 1960). In other studies with 39-MHz treatment, complete control of granary weevil adults in wheat was achieved at wheat temperatures of 41°C (Nelson and Kantack, 1966). When adults of the same species were treated in wheat at 2.45 GHz, grain temperatures above 57°C were required for complete insect morality (Baker et al., 1956).

Results of these studies and several others at different RF and microwave dielectric heating frequencies are summarized for comparison in Table 4.1.

Examination of data in Table 4.1 shows that considerably higher temperatures in the host medium are required for complete mortality of the insects when treated at 2.45 GHz than when treated at frequencies in the 10- to 100-MHz range. Therefore, it is likely that the degree of selective heating of the insects obtained in the lower frequency range is much better than that obtained in the microwave range at 2.45 GHz.

To help in assessing the degree of potential selective heating of insects in grain, the dielectric properties of adult rice weevils and of hard red winter wheat were measured throughout the frequency range from 250 Hz to 12 GHz (Nelson and Charity, 1972). Figures 4.1 and 4.2 show resulting values obtained for the dielectric constant ε' and loss factor ε'' for bulk samples of hard red wither wheat and bulk samples of adult rice weevils for the range 50 kHz to 12 GHz.

An analysis of the data, based on relationships presented in Eqs (4.1) and (4.2), revealed that the loss factor is the dominant factor influencing differential energy absorption from the RF and microwave fields. Therefore, the best selective heating of the adult rice weevil in wheat is expected to be obtained in the frequency range between about 5 and 100 MHz. On the basis of these data, little differential heating can be expected at frequencies between 1 and 12 GHz. It appears that predictions from these measurements are consistent with experimental data available on exposure of insects to RF electric fields at the different frequencies shown in Table 4.1. The predictions were also confirmed by experimental results obtained when exposing hard red winter wheat infested with rice weevils at 39 and 2450 MHz for sequences of time exposures ranging from one to several seconds, providing comparable wheat heating rates at both frequencies (Nelson and Stetson, 1974). Observed relationships between resulting temperatures in the wheat and insect moralities are illustrated in Figure 4.3.

Complete mortality was obtained with much lower grain temperatures when infested grain was treated at 39 MHz than when it was treated at 2.45 GHz. This result indicates that a much higher degree of differential heating was obtained in the lower frequency range than was obtained at microwave frequencies. The delayed mortality of the insects was also much more severe when they were treated at 39 MHz compared to treatment at 2.45 GHz.

Table 4.1 Reported Host-Media Temperatures Following RF and Microwave Dielectric Heating Exposures Necessary for 99–100% Insect Mortality (Nelson, 1996)

Insect Species	Developmental Stage	Frequency, MHz	Host Medium	Temperature, °C	Reference
Rice weevil	Mixed immature	27	Wheat	56	Anglade et al. (1979)
Sitophilus	Adult	39	Wheat	39	Nelson and Whitney (1960)
oryzae (L.)	Mixed immature	39	Wheat	61	Nelson and Whitney (1960)
	Adult	39	Wheat	40	Nelson and Stetson (1974)
	Adult	2450	Wheat	83	Nelson and Stetson (1974)
	Adult	2450	Wheat	>60	Tateya and Takano (1977)
	Pupal	2450	Wheat	>60	Tateya and Takano (1977)
	Larval	2450	Wheat	>58	Tateya and Takano (1977)
	Egg	2450	Wheat	>57	Tateya and Takano (1977)
Granary weevil	All	13.6	Wheat	62	Benz (1975)
Sitophilus	Egg	13.6	Wheat	61	Benz (1975)
granarius (L.)	Adult	27	Wheat	55	Anglade et al. (1979)
	Larval	27	Wheat	58	Anglade et al. (1979)
	Pupal	27	Wheat	61	Anglade et al. (1979)
	Adult	39	Wheat	41	Nelson and Kantack (1966)
	Adult	39	Wheat	42	Nelson et al. (1966)
	Adult	2450	Wheat	86	Anglade et al. (1979)
	Adult	2450	Wheat	>92	Hamid et al. (1968)
	Adult	2450	Wheat	>57	Baker et al. (1956)
	Larval	2450	Wheat	>82	Baker et al. (1956)
	Egg	2450	Wheat	72	Baker et al. (1956)
Confused flour	Adult	11	Flour	75	Webber et al. (1946)
beetle	Larval	11	Flour	65	Webber et al. (1946)
Tribolium	Adult	27	Flour	60	Anglade et al. (1979)
confusum	Larval	27	Four	65	Anglade et al. (1979)
Jacquelin	Adult	39	Wheat	47	Nelson and Whitney (1960)
duVal	Adult	39	Wheat shorts	>60	Nelson et al. (1966)
	Adult	90	Flour	59	van den Bruel et al. (1960)
	Larval	90	Flour	53	van den Bruel et al. (1960)
	Adult	2450	Flour	>68	Baker et al. (1956)
	Larval	2450	Flour	>82	Baker et al. (1956)
	Adult	2450	Wheat	65	Hamid and Boulanger (1969)
	Adult	2450	Flour	>66	Tateya and Takano (1977)
	Pupal	2450	Flour	>60	Tateya and Takano (1977)
	Larval	2450	Flour	>58	Tateya and Takano (1977)
	Egg	2450	Flour	>57	Tateya and Takano (1977)

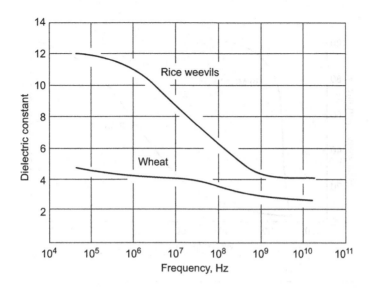

FIGURE 4.1

Frequency dependence of the dielectric constant of bulk samples of adult rice weevils and of hard red winter wheat of 10.6% moisture content at 24°C (Nelson and Charity, 1972).

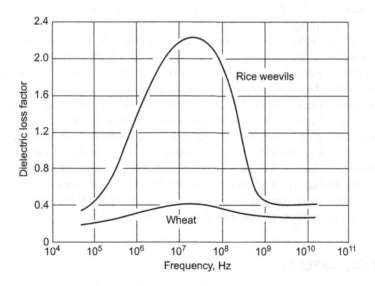

FIGURE 4.2

Frequency dependence of the dielectric loss factor of bulk samples of adult rice weevils and of hard red winter wheat of 10.6% moisture content at 24°C (Nelson and Charity, 1972).

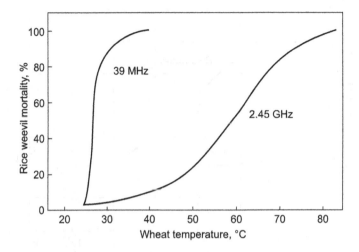

FIGURE 4.3

Comparison of mortalities of adult rice weevils treated in hard red winter wheat of 13.3% moisture content at frequencies of 39 and 2450 MHz 8 days after treatment (Nelson and Stetson, 1974).

Following the same approach as Nelson and Charity (1972) and Nelson (1972) to assess the selective heating of insects in grain, Shrestha and Baik (2013) recently confirmed theoretically the likely selective heating at 27.12 MHz of the rusty grain beetle, *Cryptolestes ferrugineus* (Stephens), in Canadian western red spring wheat.

In early tests with 2.45-GHz microwave energy radiated from a horn antenna into 10.5% moisture-content wheat infested with granary weevils and confused flour beetles, grain temperatures of 92°C were reported for 90% mortality of the insects (Hamid et al., 1968). It was also determined that 10 cm was the maximum grain thickness that could be treated because of attenuation of the energy in the grain. Later experiments with combined 2.45-GHz microwave energy and partial vacuum treatments of several species of stored-grain insect—in several kinds of grain in an installation designed for microwave-vacuum drying of grain—showed that exposures producing wheat temperatures of 62°C were necessary for complete mortality of all developmental stages of the granary weevil and lesser grain borer (Tilton and Vardell, 1982b). In other tests with the same equipment, control of rice weevil and lesser grain borer infestations in wheat and rye of 14.7% and 14.3% moisture content, respectively, was reported with final grain temperatures of 41°C, but the lack of temperature data precluded meaningful comparisons for other species (Tilton and Vardell, 1982a).

4.1.5 PRACTICAL ASPECTS

Some of the aspects of practical application of RF dielectric heating for insect control have been considered previously (Thomas, 1952; Nelson, 1973; Nelson and Whitney, 1960; Whitney et al., 1961; Nelson and Stetson, 1974; Hamid and Boulanger, 1969). Costs of RF treatment of grain to control stored-grain insects—based on heating wheat from 27°C to 66°C (80–150°F) and operating a 200-kW

RF generator with a capacity of 400 bu/h (10.89 MT/h) for 2000 h/year with a cost of $0.02 per kW-h in 1960—were estimated at $0.035/bu ($1.29/MT) (Nelson and Whitney, 1960; Whitney et al., 1961). This cost was several times the cost of chemical fumigants in use at that time. The cost estimates included a 50% efficiency in the conversion of 60-Hz electric energy to RF energy. The overall efficiency could be improved by use of heat exchangers that could capture waste heat for use in preheating the incoming grain; however, this would increase the capital equipment costs, which already constituted a significant portion (almost half) of the total estimated cost of treatment.

Although RF treatment of grain for insect control at frequencies between 10 and 40 MHz can be achieved without damaging germination if grain moisture content is below about 14% (Nelson and Walker, 1961) and without damaging milling and baking qualities (Munzel, 1975), exposures required for control of insects by microwave treatment (Table 4.1) would damage the grain. Adverse effects of dielectric heating depend very much on the moisture content of the grain, being more damaging at higher moisture contents. Hard red winter wheat of 18.3% moisture showed some reduction in germination when raised temporarily to 50°C; whereas, the same wheat at 6.7% moisture tolerated temperatures of 95°C without reduced germination when exposed to dielectric heating at 39 MHz (Nelson and Walker, 1961).

Other practical problems include the need to move the grain through RF or microwave heating equipment, because the penetration is limited for microwave irradiation, and the layer thickness is also limited by high-voltage design limitations for RF oscillators. It may be necessary to handle grain at much higher rates than the 400 bu/h rate considered in the above estimate because grain is transferred in commerce at rates of several thousand bushels per hour (hundreds of metric tons per hour). Thus, RF power equipment much larger than any in existence today might be required. Another practical problem would be the improved sanitation requirement to keep grain from becoming reinfested, because RF or microwave treatment would provide no residual protection. Dual use of dielectric heating equipment for possible accelerated drying, and seed treatment to improve germination, were suggestions that might improve the economics of the proposal (Nelson, 1973), but these uses also seem too expensive for practical consideration.

As noted earlier (Nelson and Stetson, 1974), the peak of the dielectric loss curve for the insects in Figure 4.2 will shift to higher frequencies as the temperature of the insects increases. However, in view of the shift for the relaxation frequency of liquid water from about 20 to 50 GHz as the temperature increases from 25°C to 75°C (Hasted, 1973) and the broad nature of the relaxation indicated for the insects in Figure 4.2, which is no doubt attributable to bound forms of water, the shift of relaxation frequency is not likely to be very significant. Thus, the suggested tracking of the relaxation frequency during the heating process, or the use of power oscillators operating at two different frequencies, starting the treatment at one frequency and finishing the treatment at a higher frequency (Nelson and Stetson, 1974), are not likely to provide a significant advantage.

4.2 PECAN INSECTS

The pecan weevil, *Curculio caryae* (Horn), is a serious insect pest in pecan-producing areas of the southeastern and south central United States. The adult weevils emerge from the soil in late summer or early autumn and feed on the developing nuts on the pecan trees. The female weevils also bore through the shucks and shells of the immature nuts and deposit their eggs in the developing pecan kernels. These eggs hatch in a few days, and the larvae feed on the kernels for several weeks. When they are fully grown (creamy-white grub about 1.5 cm long with reddish-brown head) in late autumn or early winter, they usually bore out of the pecan, fall to the ground, and enter the soil to diapause in the larval stage. One or two years later, they pupate and metamorphose into adults, which remain in the soil another year before emerging to start the cycle again (Payne et al., 1979).

A portion of the larval population, however, remains within the pecans until after harvest (Dutcher and Payne, 1981). Even when chemical spray programs are used to provide economic control of the insect, some harvested pecans are still infested with weevil larvae, and quarantine action is necessary to prevent shipment of in-shell pecans into areas where the pecan weevil has not become established. Fumigation with presently registered compounds is not always effective, especially at levels that do not leave harmful chemical residues in the product (Payne and Wells, 1974). Holding infested pecans at 43°C for 24 h can control pecan weevil larvae in in-shell pecans (Kirkpatrick et al., 1976). Hot water, steam, and low-temperature treatments can also provide control of larvae in in-shell pecans (Payne and Wells, 1974), but these treatments all add substantially to the costs of pecan processing.

An economic treatment that could safely provide complete control of pecan weevils in in-shell pecans would be very helpful. Therefore, experiments were conducted to learn whether pecan weevil larvae might be controlled by RF dielectric heating. Since earlier studies had identified the 10- to 100-MHz frequency range as the most effective frequencies for controlling all developmental stages of several stored-grain insect species, experimental treatment of infested pecans at 40 and 43 MHz was explored for this purpose (Nelson and Payne, 1982a). Some tests included treatments at 2450 MHz for comparison with the lower-frequency dielectric heating exposures.

The results of RF treatments at 43 MHz on in-shell pecans of 6.1% kernel moisture content, infested with pecan weevil larvae, showed increasing insect mortality with increasing time of exposure and increasing kernel temperature. Complete larval mortality was achieved with 25-s exposures resulting in 80°C kernel temperatures. For pecan weevil larvae treated in broken pecan kernel pieces of 2.6% moisture content at 40 MHz, insect mortality also increased with exposure time and pecan-kernel-piece temperature, reaching 100% mortality at pecan temperature of 50°C. Dielectric heating treatments at 40 and 2450 MHz of larvae in broken pecan kernels and pecan shells of 3.2% and 14.3% moisture, respectively, were compared. Complete control was achieved by 40-MHz treatment at 47°C and by 2450-MHz treatment at 55°C in the host medium. Temperatures of these levels in in-shell pecans result in reductions of viability of pecans for seed purposes (Nelson and Payne, 1982b), however, nearly all of the pecan crop is supplied for food, and these kernel temperatures do not adversely affect flavor or storage quality (Senter et al., 1984; Nelson et al., 1985).

The ratios of the dielectric properties of the insects and the host medium are important factors determining the degree to which the insects can be selectively heated by the high-frequency electric fields (Nelson and Charity, 1972). In general, a low insect-to-host dielectric constant ratio and a high insect-to-host dielectric loss-factor ratio are desirable for selective heating of the insect. At 40 MHz, the dielectric constant of pecan kernels at 6.1% moisture is about 1.3 times the dielectric constant of the kernels at 2.6% moisture (Nelson, 1981). The loss factor of the kernels at 6.1% moisture is about 3.6 times its value at 2.6% moisture. Because the dielectric properties of the insects should not vary much with this change in pecan moisture content, the selective heating of the insects and subsequent mortality should be greater when treated in pecans of lower moisture content. Also, the heat conductivity of the pecans should decrease with decreasing moisture content, so the transfer of heat from the insects to the surrounding host medium should be slower in low-moisture material, thus contributing to higher insect mortality.

Short exposures of less than half a minute of in-shell pecans with pecan weevil infestations to 40-MHz dielectric heating treatments can achieve 100% larval morality (Nelson and Payne, 1982b). Although pecans have a much higher unit value than grain, further study of moisture-related relationships, selective heating, and economic factors would be required for assessment of possibilities for practical use of dielectric heating for pecan weevil control.

4.3 SUMMARY

Experiments conducted during the past 70 years have shown that exposures of grain infested by stored-grain insects to RF and microwave energy can control the insect infestations by dielectric heating of the insects and the grain. Differences have been noted in the susceptibility to control by RF and microwave treatments among different stored-grain insect species and among the developmental stages of those species, depending also on the ages of the insects and on the characteristics of the host medium. In general, the immature stages of the insects are less susceptible to control by dielectric heating than are the adults of the species; however, there are exceptions.

Experimentally, RF treatments at frequencies between 10 and 90 MHz have achieved control of the insects treated in grain and grain products by exposures that raised grain temperatures to about 60–65°C. Microwave treatments at 2450 MHz have required exposures that raised temperatures of the host medium to much higher temperatures, often 80–90°C for complete control. Measurements of the RF and microwave dielectric properties of insects and grain, and consideration of the theoretical basis for differential dielectric heating of the insects, show that the frequency range from 10 to 100 MHz should be expected to provide selective heating of the insects; whereas, little differential heating can be expected at microwave frequencies. Thus, the theoretical predictions explain the experimental findings with respect to host-medium temperatures resulting from treatments that produced complete control of the insects.

Costs of electric energy and RF and microwave equipment of sufficient power capability to treat grain at practical rates are much too high to justify dielectric heating as a practical means of controlling stored-grain insects. Further research would be necessary to assess the practicality of RF dielectric heating for controlling other insects, such as pecan weevils.

REFERENCES

Anglade, P., Cangardel, H., Lessard, F.F., 1979. Application des O.E.M. de haute frequence et des micro-ondes a la desinsectisation des denrees stockees. Proc. Microw. Power Symp. 67–69.

Baker, V.H., Wiant, D.E., Taboada, O., 1956. Some effects of microwaves on certain insects which infest wheat and flour. J. Econ. Entomol. 49 (1), 33–37.

Benz, G., 1975. Entomologishe untersuchungen zur entwesung von getreide mittels hochfrequenz. Alimenta 14 (1), 11–15.

Carpenter, R.L., Livstone, E.M., 1971. Evidence for nonthermal effects of microwave radiation: abnormal development of irradiated insect pupae. IEEE Trans. Microw. Theory Tech. MTT-19 (2), 173−178.

Cotton, R.T., 1963. Pests of Stored Grain and Grain Products. Burgess Publishing Co., Minneapolis, M.

Davis J.H., 1934. High Frequency Method of and Apparatus for Exterminating Insect Life in Seed or Grain or Other Materials, U. S. Patent No. 1,972,050.

Dutcher, J.D., Payne, J.A., 1981. Pecan weevil (*Curculio caryae*) bionomics: a regional research problem. Misc. Publ. Entomol. Soc. Am 12 (2), 45−68.

Frings, H., 1952. Factors determining the effects of radio-frequency electromagnetic fields on insects and materials they infest. J. Econ. Entomol. 45 (3), 396−408.

Hamid, M.A.K., Boulanger, R.J., 1969. A new method for the control of moisture and insect infestations of grain by microwave power. J. Microw. Power 4 (11), 11−18.

Hamid, M.A.K., Kashyap, C.S., Van Cauwenberghe, R., 1968. Control of grain insects by microwave power. J. Microw. Power 3 (3), 126−135.

Hasted, J.B., 1973. Aqueous Dielectrics. Chapman and Hall, London.

Headlee, T.J., Burdette, R.C., 1929. Some facts relative to the effect of high frequency radio waves on insect activity. J. N.Y. Entomol. Soc. 37, 59−64.

Kadoum, A.M., 1969a. Electrophoretic studies of serum protein obtained from radiofrequency-treated yellow mealworm larvae. J. Econ. Entomol. 62 (3), 745−746.

Kadoum, A.M., 1969b. Effect of radiofrequency electric field treatment on protein metabolism in yellow mealworm larvae. J. Econ. Entomol. 62 (1), 220−223.

Kadoum, A.M., Nelson, S.O., Stetson, L.E., 1967a. Mortality and internal heating in radiofrequency-treated larvae of *Tenebrio molitor*. Ann. Entomol. Soc. Am. 60 (5), 885−889.

Kadoum, A.M., Ball, H.J., Nelson, S.O., 1967b. Metabolism in the yellow mealworm, *Tenebrio molitor* (Coleoptera: Tenebrionidae), following exposure to radiofrequency electric fields. Ann. Entomol. Soc. Am. 60 (6), 1195−1199.

Kadoum, A.M., Ball, H.J., Nelson, S.O., 1967c. Morphological abnormalities resulting from radiofrequency treatments of *Tenebrio molitor*. Ann. Entomol. Soc. Am. 60 (5), 889−892.

Kirkpatrick, R.L., Neel, W.W., Mody, N.V., 1976. High temperature as a method of controlling pecan weevil. J. Ga. Entomol. Soc. 11 (4), 293−296.

Lindauer, G.A., Liu, L.M., Skews, G.W., Rosenbaum, F.J., 1974. Further experiments seeking evidence of nonthermal biological effects of microwave radiation. IEEE Trans. Microw. Theory Tech. MTT-22 (8), 790−793.

Liu, L.M., Rosenbaum, F.J., Pickard, W.F., 1975. The relation of teratogenesis in *Tenebrio molitor* to the incidence of low level microwaves. IEEE Trans. Microw. Theory Tech. MTT. 23 (11), 929−931.

Munzel, F., 1975. Schadlingsbekampfung in Zeralienlagern mittels physikalischen Methoden. Alimenta 4 (1), 7−9.

Nelson, S.O., 1967. Electromagnetic energy. In: Kilgore, W.W., Doutt, R.L. (Eds.), Pest Control—Biological, Physical and Selected Chemical Methods. Academic Press, New York and London.

Nelson S.O., 1972. Frequency Dependence of the Dielectric Properties of Wheat and the Rice Weevil. (Ph.D. dissertation), Iowa State University.

Nelson, S.O., 1973. Insect-control studies with microwave and other radiofrequency energy. Bull. Entomol. Soc. Am. 19 (3), 157−163.

Nelson, S.O., 1981. Frequency and moisture dependence of the dielectric properties of chopped pecans. Trans. ASAE 24 (6), 1573−1576.

Nelson, S.O., 1982. Factors affecting the dielectric properties of grain. Trans. ASAE 25 (4), 1045−1049, 1056.

Nelson, S.O., 1996. Review and assessment of radio-frequency and microwave energy for stored-grain insect control. Trans. ASAE 39 (4), 1475−1484.

Nelson, S.O., Charity, L.F., 1972. Frequency dependence of energy absorption by insects and grain in electric fields. Trans. ASAE 15 (6), 1099–1102.

Nelson, S.O., Kantack, B.H., 1966. Stored-grain insect control sutdies with radio-frequency energy. J. Econ. Entomol. 59 (3), 588–594.

Nelson, S.O., Payne, J.A., 1982a. Pecan weevil control by dielectric heating. J. Microw. Power 17 (1), 51–55.

Nelson, S.O., Payne, J.A., 1982b. RF dielectric heating for pecan weevil control. Trans. ASAE 25 (2), 456–458, 464.

Nelson, S.O., Seubert, J.L., 1966. Electromagnetic and sonic energy for pest control, Publication No. 1402 . Scientific Aspects of Pest Control. National Academy of Sciences–National Research Council, Washington, DC

Nelson, S.O., Stetson, L.E., 1974. Comparative effectiveness of 39- and 2450-MHz electric fields for control of rice weevils in wheat. J. Econ. Entomol. 67 (5), 592–595.

Nelson, S.O., Walker, E.R., 1961. Effects of radio-frequency electrical seed treatment. Agric. Eng. 42 (12), 688–691.

Nelson, S.O., Whitney, W.K., 1960. Radio-frequency electric fields for stored-grain insect control. Trans. ASAE 3 (2), 133–137.

Nelson, S.O., Stetson, L.E., Rhine, J.J., 1966. Factors influencing effectiveness of radio-frequency electric fields for stored-grain insect control. Trans. ASAE 9 (6), 809–815.

Nelson, S.O., Senter, S.D., Forbus Jr., W.R., 1985. Dielectric and steam heating treatments for quality maintenance in stored pecans. J. Microw. Power 20 (2), 71–74.

Olsen, R.G., 1982. Constant dose microwave irradiation of insect pupae. Radio Sci. 17 (5S), 145.

Ondracek, J., Brunnhofer, V., 1984. Dielectric properties of insect tissues. Gen. Physiol. Biophys. 3, 251–257.

Payne, J.A., Wells, J.M., 1974. Postharvest control of the pecan weevil in in-shell pecans. J. Econ. Entomol. 67 (6), 789–790.

Payne, J.A., Ellis, J.M., Lockwood, D.W., 1979. Biology and distribution of the pecan weevil in Georgia and Tennesse. Pecan South 6 (1), 30–33.

Peredel'skii, A.A., 1956. The problem of electrotechnical measures for combatting harmful insects. Usp. Sovrem Biol. 41, 221–245 (in Russian)

Rai, P.S., Ball, H.J., Nelson, S.O., Stetson, L.E., 1971. Morphological changes in adult *Tenebrio molitor* (Coleoptera: Tenebrionidae) resulting from radiofrequency or heat treatment of larvae or pupae. Ann. Entomol. Soc. Am. 64 (5), 1116–1121.

Rai, P.S., Ball, H.J., Nelson, S.O., Stetson, L.E., 1972. Lethal effects of radiofrequency energy on eggs of *Tenebrio molitor* (Coleoptera: Tenebrionidae). Ann. Entomol. Soc. Am. 65 (4), 807–810.

Rai, P.S., Ball, H.J., Nelson, S.O., Stetson, L.E., 1974. Cytopathological effects of radiofrequency electric fields on reproductive tissue of adult *Tenebrio molitor* (Coleoptera: Tenebrionidae). Ann. Entomol. Soc. Am. 67 (4), 687–690.

Rai, P.S., Ball, H.J., Nelson, S.O., Stetson, L.E., 1975. Effects of radiofrequency electrical treatment on fecundity of *Tenebrio molitor* L. (Coleoptera: Tenebrionidae). Ann. Entomol. Soc. Am. 68 (3), 542–544.

Rai, P.S., Ball, H.J., Nelson, S.O., Stetson, L.E., 1977. Spermatozoan activity and insemination in *Tenebrio molitor* following radiofrequency electrical treatment (Coleoptera: Tenebrionidae). Ann. Entomol. Soc. Am. 70 (2), 282–284.

Senter, S.D., Forbus Jr., W.R., Nelson, S.O., Wilson Jr., R.L., Horvat, R.J., 1984. Effects of dielectric and steam heating treatments on the storage stability of pecan kernels. J. Food Sci. 49 (3), 893–895.

Shrestha, B., Baik, O.-D., 2013. Radio frequency selective heating of stored-grain insects at 27.12 MHz: a feasibility study. Biosyst. Eng. 114, 195–204.

Stratton, J.A., 1941. Electromagnetic Theory. McGraw Hill Book Co., New York, NY.

Tateya, M.A., Takano, T., 1977. Effects of microwave radiation on two species of stored-product insects. Res. Bull. Pl. Prot. Japan 14, 52–59.

Thomas, A.M., 1952. Pest Control by High-Frequency Electric Fields—Critical Resume Technical Report W/T23, British Electrical and Allied Industries Research Association, Lealtherhead, Surrey, UK.

Tilton, E.W., Vardell, H.H., 1982. Combination of microwaves and partial vacuum for control of four stored-product insects in stored grain. J. Ga. Entomol. Soc. 17 (1), 106–112.

Tilton, E.W., Vardell, H.H., 1982. An evaluation of a pilot plant microwave vacuum drying unit for stored-product insect control. J. Ga. Entomol. Soc. 17 (1), 133–138.

van Dyck, W., 1965. La destruction des insectes et des acariens dans les grains et ala farine qu moyen d'un champ electrique a haute frequence. Rev. Agric. 18 (4), 455–462.

van den Bruel, W.E., Bollaerts, D., Pietermaat, F., van Dyck, W., 1960. Etude des facteurs determinant les possibilites d'utilisation du chauffage dielectrique a haute frequence pour la destruction des insectes et des acariens dissimules en profondeur dan les denrees alimentaires empaquetees. Parasitica 16 (2), 29–61.

Watters, F.L., 1962. Control of insects in foodstuffs by high-frequency electric fields. Proc. Entomol. Soc. Ontario 92, 26–32.

Webber, H.H., Wagner, R.P., Pearson, A.G., 1946. High-frequency electric fields as lethal agents for insects. J. Econ. Entomol. 39, 487–498.

Whitney, W.K., Nelson, S.O. and Walkden H.H., 1961. Effects of High-Frequency Electric Fields on Certain Species of Stored-Grain Insects. Market. Res. Rep. 455 (USDA, MQRD, AMS).

SEED TREATMENT APPLICATIONS 5

5.1 BACKGROUND INFORMATION

Problems with poorly or slowly germinating seeds have plagued growers of numerous plant species in agricultural production of field crops, horticultural crops, ornamentals, nursery materials for landscaping, and forest trees. Even researchers attempting to grow weeds for studies on weed control have problems in getting the seeds of many species to germinate well. Numerous seed treatments have been studied in efforts to solve some of these problems. In some types of seed, dormancy is imposed by a seed coat that is impermeable to water (Barton et al., 1965b). Other kinds of dormancy are of a biochemical nature and are not well understood in many instances. We began studying the effects of radio-frequency (RF) dielectric heating treatments to improve seed germination nearly 60 years ago, and the idea had held the attention of scientists for about 30 years before then (Nelson and Walker, 1961).

For several years, beginning in 1931, Riccioni (1942) conducted a detailed study of electrical seed treatment employing various electrical discharge and high-frequency methods. These advanced to the point where treated seed of 16 varieties of wheat were field-tested in 23 provinces of Italy in 1938, and evidence of improved growth and yields justified the construction of an industrial installation for the purpose of treating seed. Seeds were treated as they dropped through the electric discharges between a series of three pairs of spherical electrodes. The frequency was 1000 Hz, and a potential of about 40 kV was employed to create the discharge between the spherical electrodes. The installation had a capacity of 120 q (about 440 bu or 12,000 kg) per 24 h. World War II interrupted this work, and, while some interest in renewing it developed in 1948, no more reports have been noted. A report from the USSR indicated that an electrical treatment similar to some of Riccioni's work was being used in the experimental treatment of cottonseed (Kalantarov, 1961). Seed treated in a high-voltage impulse discharge was reported to have better germination, earlier emergence, and improved growth, yield and other desirable characteristics.

In 1940, Ark and Parrry published an extensive review of experimental work using high-frequency electric fields in agriculture. Reports they cited included seed-treatment studies that revealed accelerated growth of sweet corn during the early germination period (McKinley, 1930), increased germination of onion, carrot, and wheat seed treated at 53 MHz (Siniuk and Vilanik, 1935), accelerated germination and changes in starch, sugar, and albumin in RF-treated seeds (Frolov, 1935), earlier fruiting in plants developing from 38- to 43-MHz-treated seed of cucumber and tomato, and stimulation of germination of oats, wheat, and milo (grain sorghum) (Pospelov et al., 1935).

Dielectric Properties of Agricultural Materials and Their Applications. DOI: http://dx.doi.org/10.1016/B978-0-12-802305-1.00005-1

57

Later studies also showed that germination of several vegetable seeds was increased by treatments at 43–44 MHz (Jonas, 1952, 1953). Results depended upon field intensity, power and energy input, and temperatures of the seed. Accompanying changes in the sugars in the seed were noted (Jonas, 1952). Other studies have shown indications of increased germination, stimulated germination and early growth following exposure of seed to RF electric fields of various frequencies (Nelson and Walker, 1961; Iritani and Woodbury, 1954; Matson et al., 1956).

Some studies in the frequency range 500 Hz to 220 kHz failed to show any differences in germination of maize seed (Cholet, 1954). Germination of controls was nearly 100%, however, and percent germination was the only criterion employed. The influence of seed moisture content on the levels of exposure that seeds can tolerate without damage has been noted by a number of investigators (Iritani and Woodbury, 1954; Nelson and Walker, 1961; Nelson and Wolf, 1964; Soderholm, 1957).

Permissible temperatures produced by RF treatment without damage to germination increase as moisture content of the seed decreases. The use of RF electric fields was studied for devitalizing weed seed, and attempts were made to compare the damage in weed seeds and crop seeds, but because of the methods employed results were not conclusive (Lambert et al., 1950).

Studies have also been conducted to learn whether loose smut in wheat and barley might be controlled by dielectric heating treatments, but results were not very promising because treatments that controlled the fungi damaged seed germination (Kleis et al., 1951; Nelson and Walker, 1961). The effectiveness of RF electric fields for increasing the germination of alfalfa and red clover seed in seed lots containing high percentages of hard seed has been demonstrated in several studies (Eglitis and Johnson, 1957; Iritani and Woodbury, 1954; Nelson et al., 1964; Nelson and Wolf, 1964; Nelson and Walker, 1961). Unlike mechanically scarified seed, the RF-treated seed maintained its quality well in storage (Eglitis and Johnson, 1957; Nelson and Wolf, 1964; Nelson and Stetson, 1985; Nelson, 1976; Nelson et al., 1984).

5.2 ALFALFA SEED STUDIES

Alfalfa, or lucerne, *Medicago sativa* L., is an important high-protein, perennial forage crop that is grown in temperate regions throughout the world. In the production of alfalfa, a natural condition of seed-coat impermeability presents a problem for farmers and seedsmen. This condition, referred to as the hard-seed problem, occurs frequently in legumes. This problem is not one of seed viability because hard seeds will eventually germinate and grow. Instead, "hard seed" means that the seed coat is impermeable to moisture. Hard seeds cannot be depended upon to germinate quickly when planted to produce an acceptable stand of the crop. Therefore, if percentages of hard seed are very high, more seed is required to produce an immediate stand. Hard seeds that germinate later may result in seedlings that are too small to compete effectively with other vegetative growth. In some crops, the hard-seed condition persists until subsequent years when the seed may germinate and grow after the land has been planted with a different crop, resulting in an undesirable volunteer-crop problem.

For many years seedsmen have used a process known as scarification to increase the permeability of seed coats in seed lots with high percentages of hard seed. This is an abrasive process that has a damaging effect on seed, and seed lots that have been mechanically scarified cannot be stored for the next season without high risk of serious deterioration in seed quality. Seedsmen, therefore, prefer to reduce hard-seed percentages by blending seed lots that contain large amounts of hard seed with other seed lots that have smaller quantities of hard seed. This practice is only a partial solution because the hard-seed condition is so prevalent. Each year seedsmen scarify thousands of bushels of alfalfa seed to reduce hard-seed contents to desired levels. Normally, they like to lower hard-seed percentages to the 10–15% range. If this reduction is accomplished by mechanical scarification, it is important that all scarified lots be used for planting because they cannot safely be held over for the next planting season.

Hard seeds can serve a useful purpose if the initial stand fails because of drought or frost following germination. The hard seeds may then establish the seeding when more favorable weather conditions prevail. Hard-seed percentages as high as 40–60%, however, are definitely undesirable. Both seedsmen and growers discriminate against such seed lots, so some process is needed that will safely render hard seeds permeable to moisture.

Because of the damage produced by mechanical scarification, other means of making hard seeds permeable have been tried, including acid scarification, hot-water treatment, freezing and thawing, pressure treatment, impaction or percussion, and dry-heating treatments (Porter, 1949; Quinlivin, 1971; Barton, 1965b). Several of these methods have been effective on certain kinds of seed, but the methods have been too cumbersome to be useful in commercial seed processing operations. Infrared heat treatment has also successfully increased the permeability of hard seeds of alfalfa in laboratory work (Rincker, 1954, 1957; Works and Erickson, 1963; Works, 1964), but these methods have not been put into commercial practice either.

RF dielectric heating treatments have been studied for reducing hard-seed percentages. Iritani and Woodbury (1954) found that RF treatment at 10 MHz effectively increased the germination of alfalfa and red clover seed lots by lowering hard-seed percentages. Eglitis and Johnson (1957), using a frequency of 27 MHz, increased germination of alfalfa seed lots from levels of about 60% to about 90% with exposures that lasted 25–30 s. The optimum temperature to which seed was heated by RF exposures to reduce hard-seed percentages was 56°C. The treatment produced no alteration in the rate of growth or appearance of the plants, and beneficial effects of RF treatment were still evident after treated seed had been stored for 1 year. Because of the promising results obtained by earlier investigators, an extensive series of experiments was conducted for the further study and evaluation of RF treatment of alfalfa seed for reduction of hard-seed percentages. The findings of this research are briefly summarized.

5.2.1 BASIC FACTORS

Alfalfa seeds subjected to RF electrical treatment become permeable because of some action brought about by the dielectric heating. Frequency and field intensity, therefore, are factors which can be altered to affect the rate of energy absorption by the seed. The dielectric loss factor, which influences the heating rate, is dependent upon frequency and temperature, and also is strongly influenced by the moisture content of the seed. Variability in the response of alfalfa seed lots to RF treatment might also be expected from biologic sources. Differences in response can be expected among seed of different alfalfa cultivars, or varieties. Variation in response of seed lots within the same variety might be anticipated because of differences in the environmental conditions under which the seed was grown and stored after harvest. The influence of seed moisture content on the dielectric properties has already been noted. Seed moisture content also has other significant effects on the way in which seed lots respond to RF energy exposures. Because the changes in moisture permeability characteristics of hard seeds, when exposed to RF electric fields, depend upon many factors, a systematic study of the various factors was necessary.

5.2.2 EXPERIMENTAL FINDINGS

When seed is exposed to RF fields of sufficiently high frequency and intensity its temperature will rise due to dielectric heating. If subjected to a sequence of exposures of increasing duration, germination will increase to some maximum as exposure increases. Then, with continued increase in exposure, germination will decline (Nelson and Walker, 1961). The seed temperature increases correspondingly as the length of exposure increases. These results are illustrated in Table 5.1.

The hard-seed percentage decreases correspondingly with the increase in normal seedling germination to the point of optimum exposure and continues to decrease, but beyond the optimum point, the high temperatures damage seed viability as indicated by the increased percentages of dead seeds.

5.2.2.1 Frequency

Comparisons of RF treatments on alfalfa seed at frequencies of 5, 10, and 39 MHz showed that the hard-seed reduction and germination that resulted from use of all three frequencies were about the same for corresponding temperatures produced in the seed lots by the treatments (Nelson and Wolf, 1964). These tests included comparisons of the three frequencies on alfalfa seed samples over a wide range of moisture content (2.8–9.8%). Direct comparisons of 39- and 2450-MHz exposures on germination of alfalfa seed of three different varieties were also studied (Stetson and Nelson, 1972). The two frequencies were equally effective for reducing hard-seed percentages and increasing germination when the resulting seed temperatures were comparable. Exposures of 39 MHz and 2450 MHz were compared for their effects on legume seed-coat permeability

Table 5.1 Germination of Nevada-Grown "Ranger" Alfalfa Seed Samples After Exposures to 39-MHz Electrical Treatment at 5.8% Moisture Content and 24°C Initial Temperature with an Electric Field Intensity of 2.4 kV/cm

Exposure Time, s	Final Temperature, °C	Germination Test Results[a], %			
		Normal Seedlings	Hard Seed	Abnormal Seedlings	Dead Seed
0	24	49 e	48 a	2	1 c
4.0	60	59 d	37 b	2	2 c
4.8	66	74 c	21 c	3	2 c
5.6	72	89 ab	6 d	3	2 c
6.4	77	91 a	4 d	3	2 c
7.2	83	82 b	3 d	5	10 b
8.0	92	72 c	2 d	3	23 a

[a]*Means within a column followed by the same letter or no letter are not significantly different at the 5% probability level as separated by Duncan's multiple range test.*

(Ballard et al., 1976; Nelson et al., 1976b). Alfalfa seed lots responded similarly to treatment at the two frequencies at exposures below the level at which germination is reduced, but some evidence of differences in seed viability at very damaging exposures was indicated for the two frequencies. At such damaging exposures, differences were of no practical significance, but they might be explained by differences in the dielectric properties of constituent parts of the seed at the two frequencies. The choice of operating frequency for the treatment of alfalfa seed to improve germination seems to be of little consequence as far as the seed response is concerned. That choice is more likely to be determined by process requirements or economic factors.

5.2.2.2 Field intensity

Treatments of alfalfa seed at 39 MHz with field intensities ranging from 0.7 to 2.1 kV/cm all produced about the same increases in germination (Nelson and Wolf, 1964). Much longer exposure times were required at the lower field intensities than at high field intensities to produce comparable seed temperatures. When final temperatures were similar, germination responses to different field intensities were the same. Similar results were obtained with 10-MHz exposures at different field intensities. In later experiments at 39 and 2450 MHz, use of different field intensities did not significantly alter the germination response of treated alfalfa seed as long as the temperatures produced in the seed were comparable.

5.2.2.3 Moisture content

Moisture in seeds is always an important factor. In hard seeds, like those that occur in alfalfa, the hilum operates as a hygroscopically activated valve, permitting moisture to escape from the seed but closing to prevent the seed from taking up moisture from a humid environment (Hyde, 1954). When seeds are treated with RF energy, moisture content is important for two reasons. Because the dielectric loss factor of seeds increases with moisture content (Nelson, 1965a, 1973), their moisture content determines the rate at which they will absorb energy from the RF electric fields. Then, too, the temperatures that seeds can tolerate without loss of viability are dependent upon their moisture content. Experiments with three alfalfa seed lots, representing two varieties, in which moisture contents were adjusted to levels that ranged from 1.4% to 15.6%, showed that the degree of hard-seed reduction and increase in germination were highly correlated with moisture content (Nelson and Wolf, 1964). RF treatment is more successful in lowering hard-seed percentages when moisture content is low. The percentage of normal seedlings that germinated from treated samples increased regularly as moisture content, at the time of treatment, was lowered in successive increments. Also, the final seed temperatures produced by optimum treatment levels increased as moisture content decreased, ranging from 49°C for the highest moisture content to 110°C for extremely dry seed. The moisture-content range used in the study was much wider than that expected to be encountered in the practical processing of alfalfa seed. The range of moisture contents observed in 27 alfalfa seed lots collected at commercial seed plants in the alfalfa-seed-producing areas of the West and Pacific Northwest in the United States in 1964 was from 5.6% to 8.3% (Nelson et al., 1968). Temperatures produced in seed samples by optimum RF exposures of these 27 seed lots ranged from 57°C to 80°C. The moisture content, therefore, is an important factor that determines the proper exposure for a seed lot when it is to be treated with RF energy to increase germination.

5.2.2.4 Varietal and other variations

The studies included a relatively wide range of hard-seeded alfalfa seed lots from the crop years 1957 to 1970, the 13-year period during which the studies were conducted. Nine varieties were represented, and every seed lot responded to proper RF treatment with increased germination and lowered hard-seed content. The seed sources included the States of Idaho, Nevada, South Dakota, Utah, and Washington, and the Canadian Province of Saskatchewan. Although significant reductions in hard-seed percentages were always obtained, considerable variation in germination response to RF treatment was noted. Some differences in the response of different varieties have been observed (Nelson et al., 1968), but there was also variation in response among seed lots of a given variety. Occasionally, seed lots retain a hard-seed percentage of more than 20% after RF treatment, but, generally, hard-seed retention after proper exposure falls in the range 5−15%.

5.2.2.5 Temperature effects

The most important temperature influence appears to be that of the final temperature to which the seed is raised by RF treatment. For seed lots of normal moisture content, about 6−7%, a treatment that results in a temperature of 75°C is close to the optimum exposure for increasing germination by lowering hard-seed content. The effects of seed-lot temperature just before exposure to RF fields appear to be rather unimportant, as far as the germination response to treatment is concerned. The optimum final temperatures for RF-treated alfalfa seed samples conditioned to 4°C, 26°C, and 49°C prior to RF exposure were the same (Nelson and Wolf, 1964), and these results were confirmed later with other seed lots conditioned at −18°C, −7°C, 4°C, and 24°C before RF treatment (Stetson and Nelson, 1972). When seed temperatures are low, more energy is required, of course, to raise the seed temperature to the level required for optimum germination response than when initial temperatures are higher. Holding alfalfa seed at temperatures up to 71°C for 1 h after RF treatment revealed no further reductions in hard-seed percentages, and seed was not damaged.

5.2.2.6 Other effects

Because the RF treatment renders the seed coat of a large portion of the hard seeds permeable to water, the water sorption of seeds should be expected to increase also. Experiments, in which the water sorption (percentage of increase in weight caused by sorption of water) of seed samples was measured and standard germination test data were obtained, showed that the water sorption was highly correlated with the reduction in the hard-seed percentage (Nelson and Wolf, 1964). Water sorption increased regularly as hard-seed percentage decreased. The same relationship between water sorption and hard-seed content was noted in later experiments. In those experiments the respiration (oxygen uptake) of dry seed samples and the electrical conductivity of the leachate solution, in which other subsamples of the seed were soaked, were also measured (Nelson et al., 1964). High correlations between hard-seed content and both oxygen uptake and leachate solution conductivity were noted. Both oxygen uptake and leachate solution conductivity increased as the hard-seed content was lowered by successive RF treatments of increasing exposure. In these same studies, the emergence of alfalfa seedlings in greenhouse sand benches and in the field correlated well with laboratory germination tests on seed lots that had been given RF exposures. Examination of seedlings revealed no adverse effects on seedling vigor and no increase in abnormal seedlings as a result of RF exposure of proper levels. Damage from overexposure was evident in both greenhouse and field emergence tests and in laboratory germination tests. Repeated germination tests after periods of storage up to 4 years after treatment revealed no deterioration in seed quality due to RF treatment, and the benefits of treatment for hard-seed reduction were evident after 4 years. Thus, RF treatment for lowering hard-seed contents in alfalfa seed lots promises advantages compared to mechanical scarification, which damages the quality of seed when it is held in storage.

Some effort has been devoted to learning how hard seeds are modified by RF treatments that increase their permeability to moisture (Ballard et al., 1976; Nelson et al., 1976b). Microscopic examination of stained seeds did not reveal any visible differences between seed samples that had been treated to provide very high germination (2% hard seed) and untreated samples from the same lot with 58% hard seed. The staining of seed tissues with imbibed ferrous salt solutions was used to identify the paths of entry of moisture into seeds that were rendered permeable by RF electrical treatment (Ballard et al., 1976). These studies showed that such seeds became permeable in the strophiolar region as do seeds that become permeable naturally over time.

5.2.3 ASPECTS OF PRACTICAL APPLICATION

Research has shown that hard-seed percentages in alfalfa seed lots can be safely reduced by exposure of seed to RF electric energy over a wide range of frequencies. The maintenance of high quality in RF-treated seed lots for several years after treatment is an important advantage over mechanical scarification processes currently in use. Because alfalfa seed is a commodity of very high value compared with most other agricultural products, it seems that the costs of equipment and energy for an RF seed-processing system could easily be justified on an economic basis. So far, two principal factors have prevented the practical application of the method. One is the reluctance of seed processors to accept a new process that has not been completely developed on a commercial scale. The other is the limited market, from an electronic equipment supplier's viewpoint, for such RF seed-processing systems. Because of the limited market, a manufacturer has difficulty in providing the equipment at a price that will be attractive to an interested seed company. Another factor that may cause some concern for both the interested seed company and the electronic equipment manufacturer is the possibility that a less expensive treatment of some type might be developed that may be as successful as the RF treatment. Two such treatments, which may or may not be less expensive, are an infrared radiation treatment and a hot-air treatment. With laboratory equipment, some types of these treatments have been compared with RF treatments for their effectiveness in reducing hard-seed percentages (Nelson et al., 1964, 1968; Stetson and Nelson, 1972) and found to be about equally effective. Perhaps some further tests are necessary, and some careful economic estimates may be required, based on particular seed-processing plant requirements, before the question about alternative treatments can be answered. In the current market, the preference of growers and seedsmen for seed lots that do not have high percentages of hard seeds does not seem to be reflected in the price structure. Lots with low percentages of hard seed are preferred, but seedsmen also buy the other lots and scarify them. Occasionally, seed lots are ruined by scarification, and, whenever possible, only that seed is scarified which can be sold with reasonable certainty before the next season. A safe, effective process should, therefore, be of value to the seed processors. Some potential exists for the development of a practical-scale RF treatment for alfalfa seed, but all of the factors just mentioned require careful evaluation in view of present and predicted future market situations.

5.3 SWEETCLOVER SEED

Historically, the value of sweetclover for soil reclamation and for restoring soil fertility was recognized in the United States before the turn of the century, but its widespread use for soil improvement developed between about 1910 and 1940 (Smith and Gorz, 1965). Sweetclover also has long been valued as an apiarian crop in the production of honey and as an important pasture crop. However, its principal economic importance was for use as a companion crop with oats and other small grains. Seeded with small grain in the spring, sweetclover continues its growth and development after the grain is harvested, developing large root systems in the fall and providing additional organic matter when turned under as a green manure crop late in the fall. As a legume, sweetclover lives in a symbiotic relationship with nitrogen-fixing bacteria, and because it also has the ability to grow under adverse conditions, it has remarkable soil-improving value through accumulation of large amounts of nitrogen and organic matter which are readily available to succeeding crops (Smith and Gorz, 1965). Beginning in the 1950s, the use of sweetclover declined as economic commercial fertilizers became generally available, so it is not currently used as extensively as it was earlier. However, as fertilizer costs continue to rise and energy stocks from which they are manufactured become more dear, sweetclover and other soil-improving crops may regain some of their earlier prominence.

Mature well-ripened seed of sweetclover has a high proportion of 'hard seeds', seeds with seed coats that are impermeable to water. Therefore, such seeds cannot germinate until the seed coats become permeable to water. When planted in soil, hard seeds eventually become permeable after weathering one or more seasons (Martin and Watt, 1944), but many remain hard for periods of up to several years. Hard seeds of sweetclover have been known to remain viable in soil for at least 23 years and can present a problem when volunteer second-year sweetclover plants show up in grain fields (Stoa, 1941). Because sweetclover seed lots generally have high percentages of hard seeds, nearly all sweetclover seed sold for planting is first 'scarified', that is, subjected to an abrasive process that scratches the seed coat enough to make the hard seeds permeable. Some hard seeds are rendered permeable by injury during the normal harvesting processes, but such injured seeds and also those rendered permeable by scarification suffer serious losses of viability if stored until the next planting season. Therefore, other methods of reducing hard-seed percentages in sweetclover seed have been studied.

Thirteen lots of sweetclover seed were exposed to 39-MHz dielectric heating treatments to study the effectiveness of the RF treatment for reducing hard-seed content (Nelson and Stetson, 1982). Seed lots included Common Yellow and "Madrid" sweetclover, *Melilotus officinalis* (L.) Lam., Common White and "Spanish", *Melilotus alba* Desr., and "Golden Annual", *Melilotus suaveolens* Ledeb. Samples of each seed lot were exposed for varied time periods of several seconds to 39-MHz electric fields of intensities between 1.2 and 2.4 kV/cm and subsequently tested for germination by standard techniques. Greenhouse sand emergence and water sorption tests were conducted for some seed lots.

Most seed lots did not respond favorably when treated at normal moisture contents (7−10%), but when moisture content was reduced to the 2−6% range by desiccation before treatment, significant reductions in hard-seed content and significant increases in normal-seedling germination generally were obtained. Substantial amounts of hard seed remained in the biennial sweetclover seed lots. The single seed lot of Golden Annual sweetclover seed gave the best response to the RF treatment. Tests of biennial sweetclover seed samples retained in storage at 4°C and 50% relative humidity for 17 years after treatment revealed that much of the hard seed in untreated control samples had become permeable, but that treated samples still had lower hard-seed percentages and produced greater numbers of normal seedlings than untreated samples. No significant damage by treatment was evident after this extended storage that was not evident in overexposed samples in the initial germination tests.

5.4 OTHER SMALL-SEEDED LEGUMES AND SOME CEREALS

Seed lots of seven small-seeded legume species and winter oats, winter rye, and perennial ryegrass were exposed to 39-MHz electric fields of known field intensities for various time periods and tested for germination response (Nelson et al., 1976a). The legumes included alfalfa, *M. sativa* L., red clover, *Trifolium pratense* L., alsike clover, *Trifolium hybridum* L., white Dutch clover, *Trifolium repens* L., ladino clover, *T. repens* L., sweetclover, *M. alba* Deer., and birdsfoot trefoil, *Lotus corniculatus* L. The cereals were winter oats, *Avena sativa* L., winter rye, *Secale cereale* L., and one grass, perennial ryegrass, *Lolium parenne* L.

The RF treatments effectively increased germination of alfalfa, red clover, and ladino clover through reduction of hard-seed percentages. Significant increases in germination for these three species, as a result of electrical treatment, were retained in temperature- and humidity-controlled storage for 14 years. Differences between treated and untreated samples held in a seed warehouse without temperature and humidity control tended to disappear after a few years because hard seeds became permeable naturally as the seeds aged. RF treatment was less effective on the alsike clover seed lot and failed to improve the germination of the birdsfoot trefoil. Lowering the moisture content of the hard-seeded legume lots improved the response to RF treatment. Extreme desiccation of seed lots lowered the quality of ladino and red clover seed lots in long-term storage, but did not affect the quality of alfalfa, birdsfoot trefoil, and sweetclover seed during the same period. When seed samples were exposed to RF treatment at proper levels, the treatment did not accentuate deterioration of quality in seed lots held for 6−14 years after treatment as determined by percentages of dead seed. There was some natural deterioration in seed quality during the storage period. In 6 years of warehouse storage, the decline in seed quality was much greater for alsike clover, white Dutch clover, and birdsfoot trefoil than it was for ladino clover, red clover, and alfalfa. RF treatment at proper levels did not affect the longevity of a rye seed lot that was held and tested 6 years after treatment. RF treatments overcame postharvest dormancy in winter oats.

5.5 VEGETABLE SEED

Hard seed, dormancy, slow rates of germination, and delayed and irregular emergence are problems frequently encountered in the production of certain vegetables. Numerous studies of these problems have been conducted, and several types of seed conditioning and treatment have been tried with varying degrees of success in efforts to alleviate some of these problems (Barton, 1965a,b; Edmond and Drapala, 1960; Kyle and Randall, 1963; Lebedeff, 1943; Nelson, 1965b; Nutile and Nutile, 1947). One method investigated involves exposure of seed to RF electric fields producing dielectric heating of the seed. This type of treatment is usually accomplished by placing or passing the seed between electrodes connected to an electronic power oscillator. Frequencies of the alternating fields have generally ranged between 1 and 100 MHz. With electric field intensities, or voltage gradients, of about 1 kV/cm or more, seeds absorb energy from the field rapidly at these frequencies. Since RF treatments of some crop seeds had improved germination, a series of cooperative tests was conducted, mainly with vegetable seeds to evaluate possibilities for improving their germination characteristics by exposing seed to RF electric fields (Nelson et al., 1970). Seed lots of several vegetables, including garden beans, *Phaseolus vulgaris* L., cabbage, *Brassica oleracea* var. *capitata* L., cantaloupe, *Cucumis melo* L., cucumber, *Cucumis sativus* L., lettuce, *Lactuca sativa* var. *capitata* L., okra, *Hibiscus esculentus* L., onion, *Allium cepa* L., garden peas, *Pisum sativum* L., pepper, *Capsicum frutescens* L., spinach, *Spinacea oleracea* L., and tomato, *Lycopersicon esculentum* Mill., and two seed lots of Kentucky bluegrass, *Poa pratensis* L., were treated and evaluated in the study. Samples of seed of known cultivars and moisture content were exposed to RF electric fields of about 40 MHz and field intensities of 1−2 kV/cm for times ranging from a few seconds to a half minute or so that produced seed temperatures between room temperature and 80°C or 90°C. Results of treatment were subsequently evaluated by standard germination tests and some soil emergence tests. In many of the tests, sprouted seeds were counted daily during periods of high germinating activity. Results of these tests were described in detail (Nelson et al., 1970), but are only summarized here.

Brief exposures to the RF electric fields substantially increased the germination of okra, garden peas, and garden beans by reducing the percentage of hard seeds. RF electrical treatment accelerated germination in seed lots of tomato, Kentucky bluegrass, and spinach, being particularly effective and consistent with spinach. Emergence of spinach from the soil in greenhouse tests was noticeably accelerated, but treatments did not alter the final germination or emergence percentages unless samples were damaged by overtreatment.

5.6 TREE SEED

5.6.1 PINE SEED

Seeds of some pine species will germinate rapidly and normally when they are first extracted from their cones if they are anatomically and physiologically mature and have not been subjected to rapid drying. Seed lots of some pine species, when stored for short periods, commonly germinate satisfactorily, but uniformity and rate of germination are improved by cold stratification prior to planting. However, if pine seed is extracted at high temperatures or stored for a long time, germination becomes highly variable (Heit, 1967). The type and degree of dormancy responsible for this behavior vary among species, among geographic sources of the same species, and even among seed lots from the same source (Fowler and Dwight, 1964; Krugman and Jenkinson, 1974; McLemore and Barnett, 1966).

Seed of Jeffrey pine, *Pinus jeffreyi* Grev. and Balf., sugar pine, *Pinus lambertiana* Dougl., ponderosa pine, *Pinus ponderosa* Laws. var. *ponderosa*, eastern white pine, *Pinus strobus* L., loblolly pine, *Pinus taeda* L., generally require stratification to overcome embryo dormancy following extraction and several years of cold storage. In *P. lambertiana* and *P. strobus*, dormancy is often deep after storage, and a prolonged period of cold stratification is usually required before germination will occur. Digger pine, *Pinus sabiniana* Dougl., which has a relatively thick, impermeable seed coat, often requires physical rupture of the seed coat, as well as stratification to overcome both physical and physiologic dormancy (Mirov, 1936).

Stratification commonly consists of soaking the seeds in water for 1 or 2 days, draining off the water, and holding them in a moist environment at 3−5°C for a specified time period. Germination of *P. sabiniana* seed can be speeded by cracking the thick seed coat prior to the low temperature stratification (Krugman and Jenkinson, 1974). Stratification also increases percentages of normal seedlings in some species by promoting the further development of small embryos present at the time of collection. Without stratification, many stored pine-seed lots will produce a higher-than-normal percentage of abnormal seedlings (Krugman, 1966).

Stratification, although essential for achievement of the best germination performance from stored pine seed, is both time consuming and troublesome to carryout prior to seeding. Certain types of electrical treatments have shown promise for improving the germination of agricultural seeds, particularly alfalfa and some of the other small-seeded legumes (Nelson, 1965b). Therefore, studies were conducted to learn whether such treatments might show promise for improving the germination performance of pine seed. Three types of electrical treatments were considered—RF dielectric heating, infrared radiation, and the gas-plasma radiation produced in a partially evacuated tube in which a glow discharge is established by applying an electric voltage of about 2000 V across the rarified air in the discharge tube. All three of these electrical treatments had shown about equal effectiveness for increasing alfalfa seed germination by rendering the seed coat permeable to water (Nelson et al., 1964). Radio-frequency electrical treatments have also increased the germination of seeds of species in which impermeable seed coats are not responsible for low germination (Nelson and Walker, 1961; Nelson et al., 1970).

Germination responses of pine seed lots to RF dielectric heating, infrared, and gas-plasma-radiation treatments, and to stratification, were somewhat variable, but so was the germination of untreated seed lots (Nelson et al., 1980). All three kinds of treatment produced significant increases in germination of eastern white pine seed as compared to the untreated and nonstratified control, but the infrared treatment gave the best results. Response of eastern white pine seed to stratification was better than the responses to any of the other treatments. Radio-frequency treatments increased germination of sugar pine seed slightly, but many of the germinated seedlings were abnormal. Radio-frequency treatments produced moderate increases in germination percentages of some Digger pine and some loblolly pine seed lots, and accelerated substantially the germination of loblolly pine seed. Conventional stratification increased the germination of sugar pine and Digger pine seed more than did the RF treatments. Radio-frequency and gas-plasma treatments did not improve germination of single seed lots of ponderosa and Jeffrey pine, and neither did stratification.

5.6.2 OTHER TREE AND WOODY PLANT SEEDS

Radio-frequency treatment of Douglas-fir, *Pseudotsuga menziesii* var. *menziesii* (Mirb.) Franco, seeds with microwave energy at 2.45 GHz was reported to increase germination from levels of about 50% in untreated seed to more than 90% (Jolly and Tate, 1971). Germination was also accelerated by exposure to the microwave treatments. Substantial increases in germination of white spruce, *Picea glauca* (Moench) Voss, and red spruce, *Picea rubens* Sargent, following treatment at 2.45 GHz were also reported (Kashyap and Lewis, 1974).

In work with seeds of smaller woody plants, RF treatment at frequencies of 10 and 39 MHz significantly increased germination of honey mesquite, *Prosopis juliflora* (Swartz) DC, and huisache, *Acacia farnesiana* (L.) Willd, but responses of different huisache seed lots were not consistent (Nelson et al., 1978). Increased germination as a result of seed treatment with intense infrared radiation for short time periods was obtained with western white pine, *Pinus monticola* Dougl., and ponderosa pine (Works and Boyd, 1972). Although responses with ponderosa pine were not consistent, the infrared treatments consistently enhanced the germination of western white pine seed lots. Additional studies would be required to properly evaluate the potential of any of these electrical treatments for improving the germination performance of tree seeds and woody plant seeds before any of them could be considered for practical application.

5.7 SUMMARY

The germination responses of seeds of more than 80 plant species exposed to RF dielectric heating treatment have been summarized elsewhere (Nelson and Stetson, 1985). Most were treated at a frequency of about 40 MHz, but 10- and 2450-MHz microwave treatments were included in several studies. Seed lot and treatment descriptions and seed lot germination responses were summarized concisely in tabular form for reference (Nelson and Stetson, 1985). Information was included on the type of germination tests conducted. Cooperators were identified, and references were cited when more detailed information was published about the studies.

No improvement in germination was achieved with many kinds of seed, but many others responded favorably, either with increases in germination or accelerated germination. Particular success was achieved in the treatment of alfalfa seed and that of some other small-seeded legumes that exhibited impermeable seed coat problems.

Responses among different kinds of seed varied widely. Some of the small-seeded legumes, such as alfalfa, *Medicago sativa* L., red clover, *Trifolium pratense* L., and arrowleaf clover, *Trifolium vesiculosum* Savi, responded consistently with increased normal-seedling germination and reduced hard-seed content. Others, such as sweetclover did not respond nearly so well. Hard-seed percentages were markedly and consistently reduced in alfalfa seed lots with substantial hard-seed contents, and the RF treatments rendered the impermeable seed coats permeable in the region of the strophiole where the seeds become permeable to water naturally in time (Nelson et al., 1976b).

Several of the vegetable and ornamental species did not show any improvement, but some, such as okra, garden peas, and garden beans responded favorably. Germination of spinach was consistently accelerated (Nelson et al., 1970). As a group, grasses did not respond very well. However, the germination of Kentucky bluegrass seed was accelerated by the RF treatments, and interesting germination increases were obtained with side-oats grama, annual ryegrass, and switchgrass. In general, seeds of woody plants and tree species did not respond very well, although germination was increased or accelerated in some of the pine species. Results with field crops such as corn, cotton, and wheat were mixed, with acceleration and increased growth being noted in some lots but not in others. These particular seeds normally germinate well, but increases in rate of germination and seedling development would be welcome improvements. Practical aspects of RF electrical seed treatment have been discussed elsewhere (Nelson, 1976). Practical application of new methods must be justified economically, and some of the factors are difficult to evaluate. However, the increased germination achievable in some kinds of seed and the proven long-range safety of the treatment for alfalfa (Nelson et al., 1984) and sweetclover (Nelson et al., 1982) indicate that the method might be considered for further study and economic analysis where practical benefits might be realized.

REFERENCES

Ark, P.A., Parrry, W., 1940. Application of electrostatic fields in agriculture. Q. Rev. Biol. 15 (2), 172–191.

Ballard, L.A.T., Nelson, S.O., Buchwald, T., Stetson, L.E., 1976. Efects of radiofrequency electric fields on permeability to water of some legume seeds, with special reference to strophiolar conduction. Seed Sci. Technol. 4, 257–274.

Barton, L.V., 1965a. Seed dormancy: general survey of dormancy types in seeds, and dormancy imposed by external agents. In: Ruhland, W., Lang, A. (Eds.), Encyclopedia of Plant Physiology, vol. 15. Springer Verlag, Berlin, Hiedelberg, New York, NY, pp. 699–720.

Barton, L.V., 1965b. Dormancy in seeds imposed by the seed coat. In: Ruhland, W., Lang, A. (Eds.), Encyclopedia of Plant Physiology, vol. 15. Springer Verlag, Berlin, Hiedelberg, New York, NY, pp. 727–745.

Cholet, P.H., 1954. Investigation of Certain Frequency Dependent Electrical Properties of Biological Materials. Ph. D. Thesis. Syracuse University, Syracuse, New York.

Edmond, J.B., Drapala, W.J., 1960. Studies of the Germination of Okra Seed, Technical Bulleltin 47, Mississippi State University.

Eglitis, M., Johnson, F., 1957. Control of hard seed of alfalfa with high-frequency energy. Phytopathology 47 (1), 9.

Fowler, D.P., Dwight, T.W., 1964. Provenance differences in the stratification requirements of white pine. Can. J. Bot. 42, 669–673.

Frolov, M.V., 1935. Effect of high-frequency field on sterilization of seed and flour. Elektrificatsiia Selskogo Khoziaistya 1, 36–37 (in Russian).

Heit, C.E., 1967. Propagation from seed, Part 10. Storage method for conifer seeds. Am. Nurseryman 126 (8), 14–15, 38–54.

Hyde, E.O.C., 1954. The function of the hilum in some Papilionacae in relation to the ripening of the seed and the permeability of the testa. Ann. Bot. 18, 241–256.

Iritani, W.M., Woodbury, G.W., 1954. Use of Radio-Frequency Heat in Seed Treatment, Research Bulletin No.25, Agricultural Experiment Station, University of Idaho.

Jolly, J.A., Tate, R.L., 1971. Douglas-fir tree seed germination enhancement using microwave energy. J. Microw. Power 6 (2), 125–130.

Jonas, H., 1952. Some effects of radio frequency irradiations on small oil-bearing seeds. Physiol. Plant 5, 41–51.

Jonas, H., 1953. R-F irradiation of seed. Electronics 26 (4), 161–163.

Kalantarov, M.I., 1961. Electrification of agriculture. Proc. Acad. Sci. Azerbaijan SSR 17 (1), 25–29.

Kashyap, S.C., Lewis, J.E., 1974. Microwave processing of tree seeds. J. Microw. Power 9 (2), 99–107.

Kleis, R.W., Lucas, E.H., Brown, H.M., de Zeeuw, D.J., 1951. Dry treatment of wheat for loose smut (a preliminary report). Q. Bull. Mich. Agric. Exp. Station Mich. State Coll. 34 (2), 162–165.

Krugman, S.L., 1966. Artificial Ripening of Sugar Pine Seeds, USDA Forest Service Research Paper PSW-32, p. 7.

Krugman, S.L., Jenkinson, J.L., 1974. *Pinus* L. Pine, Seeds of Woody Plants of the United States. USDA Forest Service, Washington, DC.

Kyle, J.H., Randall, T.E., 1963. A new concept of the hard seed character in *Phaseolus vulgaris* L. and its use in breeding and inheritance studies. Proc. Am. Soc. Hortic. Sci. 83, 461–475.

Lambert, D.W., Worzella, W.W., Kinch, R.C., Cheadle, J.N., 1950. Devitalization of cereal and weed seeds by high frequency. Agron. J. 42 (6), 304–306.

Lebedeff, G.A., 1943. Heredity and Environoment in the Production of Hard Seeds in Common Beans (*Phaseolus vulgaris* L.), Research Bulletin 4, Puerto Rico University, Agricultural Experiment Station.

Martin, J.N., Watt, J.R., 1944. The strophiole and other seed structures associated with hardness in *Melilotus alba* L. and *M. officinalis* Willd. Iowa State Coll. J. Sci. 18 (4), 457–469.

Matson, W.E., Nilan, R.A., Betts, A.L. and Spcencer, J.V., 1956. The Application of Radio Frequency Energy in Treatment of Agricultural Products. Annual Report,Washington Farm Electrification Committee, Pullman, Washington.

McKinley, G.M., 1930. Some biological effects of high-frequency electrostatic fields. Proc. Pennsylvania Acad. Sci. 4, 43–46.

McLemore, B.F., and Barnett, J.P., 1966. Loblolly Seed Dormancy Influenced by Cone and Seed Handling Procedures and Parent Tree. USDA Forest Service Research Note SO-41, p. 4.

Mirov, N.T., 1936. A note on germination methods for coniferous species. J. Forestry 34, 719–723.

Nelson, S.O., 1965a. Dielectric properties of grain and seed in the 1 to 50-mc range. Trans. ASAE 8 (1), 38–48.

Nelson, S.O., 1965b. Electromagnetic Radiation Effects on Seeds, Conference Proceedings—Electromagnetic Radiation in Agriculture: 60–63.

Nelson, S.O., 1973. Electrical properties of agricultural products—a critical review. Trans. ASAE 16 (2), 384–400.

Nelson, S.O., 1976. Use of microwave and lower frequency RF energy for improving alfalfa seed germination. J. Microw. Power 11 (3), 271–277.

Nelson, S.O., Stetson, L.E., 1982. Germination response of sweetclover seed to 39-MHz electrical treatments. Trans. ASAE 25 (5), 1412–1417.

Nelson, S.O., Stetson, L.E., 1985. Germination responses of selected plant species to RF electrical seed treatment. Trans. ASAE 28 (6), 2051–2058.

Nelson, S.O., Walker, E.R., 1961. Effects of radio-frequency electrical seed treatment. Agric. Eng. 42 (12), 688–691.

Nelson, S.O., Wolf, W.W., 1964. Reducing hard seed in alfalfa by radio-frequency electrical seed treatment. Trans. ASAE 7 (2), 116–119, 122.

Nelson, S.O., Stetson, L.E., Stone, R.B., Webb, J.C., Pettibone, C.A., Works, D.W., et al., 1964. Comparison of infrared, radiofrequency, and gas-plasma treatments of alfalfa seed for hard-seed reduction. Trans. ASAE 7 (3), 276–280.

Nelson, S.O., Stetson, L.E., Works, D.W., 1968. Hard-seed reduction in alfalfa by infrared and radio-frequency electrical treatments. Trans. ASAE 11 (5), 728–730.

Nelson, S.O., Nutile, G.E., Stetson, L.E., 1970. Effects of radiofrequency electrical treatment on germination of vegetable seeds. J. Am. Soc. Hortic. Sci. 95 (3), 359–366.

Nelson, S.O., Heckert, R.M., Stetson, L.E., Wolf, W.W., 1976a. Radiofrequency electrical treatment effects on dormancy and longevity of seed. J. Seed Technol. 1 (1), 31–43.

Nelson, S.O., Ballard, L.A.T., Stetson, L.E., Buchwald, T., 1976b. Increasing legume seed germination by VHF and microwave dielectric heating. Trans. ASAE 19 (2), 369–371.

Nelson, S.O., Bovey, R.W., Stetson, L.E., 1978. Germination response of some woody plant seeds to electrical treatment. Weed Sci. 26 (3), 286–291.

Nelson, S.O., Krugman, S.L., Stetson, L.E., Belcher Jr., E.W., Works, D.W., Stone, R.B., et al., 1980. Germination response of pine seed to radiofrequency, infrarred, and gas-plasma-radiation treatments. Forest Sci. 26 (3), 377−388.

Nelson, S.O., Stetson, L.E., Works, D.W., 1982. Germination responses of sweetclover seed to infrared, radiofrequency, and gas-plasma electrical treatments. J. Seed Technol. 7 (1), 10−22.

Nelson, S.O., Stetson, L.E., Wolf, W.W., 1984. Long-term effects of RF dielectric heating on germination of alfalfa seed. Trans. ASAE 27 (1), 255−258.

Nutile, G.E., Nutile, L.G., 1947. Effect of relative humidity on hard seeds in garden beans. Proc. Assoc. Off. Seed Anal. 37, 106−114.

Porter, R.H., 1949. Recent developments in seed technology. Bot. Rev. 15 (4), 221−344.

Pospelov, A.P., Zhilenkov, I.V., Burnatzky, D.P., 1935. Effect of ultra-short electromagnetic waves on the processes of germinating seeds. Zapiski Voronezhskogo Selskokhoziaistvennogo Instituta 1 (16), 295−303 (in Russian).

Quinlivin, B.J., 1971. Seed coat impermeability in legumes. Aust. Inst. Agric. Sci. 37 (4), 293−295.

Riccioni, B., 1942. Il trattamento elettrico del seme di grano, "sistema Riccioni" dal laboratorio all' agricoltura pratica. Istituto Itliano D'Arti Grafiche, Bergamo, Milano, Roma.

Rincker, C.M., 1954. Effect of heat on impermeable seed of alfalfa, sweet clover, and red clover. Agron. J. 46 (6), 247−250.

Rincker, C.M., 1957. Heat-Treating Hard Alfalfa Seed, Bulletin 352, Wyoming Agricultural Experiment Station.

Siniuk, U.N., Vilanik, U.K.V., 1935. Action of ultra-short waves on increase and acceleration of germination in seeds. Elektrificatsiia Selskogo Khoziaistya 1, 38−40 (in Russian).

Smith, W.K., Gorz, H.J., 1965. Sweetclover improvement. In: Norman, A.G. (Ed.), Advances in Agronomy, vol. 17. Academic Press, New York, NY, pp. 163−231.

Soderholm, L.H., 1957. Effects of dielectric heating and cathode rays on germination and early growth of wheat. Agric. Eng. 38 (5), 302−307.

Stetson, L.E., Nelson, S.O., 1972. Effectiveness of hot-air, 39-MHz dielectric, and 2450-MHz microwave heating for hard-seed reduction in alfalfa. Trans. ASAE 15 (3), 530−535.

Stoa, T.E., 1941. Volunteer sweetclover. North Dakota Agric. Exp. Station Bmonthly Bull. 3 (6), 3−6.

Works, D.W., 1964. Infrared irradiation for water-impermeable seeds. Trans. ASAE 7 (3), 235−237.

Works, D.W., Boyd Jr., R.J., 1972. Using infrared irradiation to decrease germination time and to increase percent germination in various species of western conifer trees. Trans. ASAE 15, 760−762.

Works, D.W., Erickson, L.C., 1963. Infrared Irradiation—An Effective Treatment for Hard Seeds in Small Seeded Legumes, Research Bulletin No. 57, Agricultural Experiment Station, Universy of Idaho.

PRODUCT CONDITIONING APPLICATIONS

6.1 BACKGROUND INFORMATION

Experiments were conducted to determine the influence of radio-frequency (RF) dielectric heating and microwave heating on some quality attributes of a few agricultural products. One application was related to RF drying of chopped alfalfa forage. Another dealt with the influence of RF and microwave heating of raw soybeans on their nutritional value. A third study was conducted to evaluate the effects of RF dielectric heating on the maintenance of quality in stored pecan nuts.

6.2 DRYING OF CHOPPED ALFALFA

Alfalfa forage accounts for a large portion of the value of the annual hay or forage crops in the United States. The quality of the crop greatly affects its economic value. Higher quality hay is obtained when exposure to the weather is reduced. By reducing field-drying time, weather hazards for the hay crop are minimized. The possible use of RF dielectric heating for drying was intriguing because no conductive or convective means of heat transfer is required. Heat is developed in the material very rapidly, depending on the dielectric properties of the material. For these reasons, laboratory experiments on drying chopped alfalfa forage were conducted with dielectric heating exposures at 43 MHz (Stetson et al., 1969).

Samples of chopped alfalfa were exposed in a polystyrene chamber between parallel-plate electrodes with air circulation at different air-flow rates, RF electrode voltages, and times of exposure. Following RF treatment, drying of samples was continued in a forced-air oven. Drying curves were presented and constants for a moisture-ratio model were determined. Estimates of energy requirements and costs were calculated for drying. Carotene and protein content were determined for the chopped alfalfa samples at the end of the process. Findings showed that chopped alfalfa can be dried rapidly by dielectric heating, but for economic reasons, the method cannot be recommended for field-drying applications. Short exposures of chopped alfalfa to RF electric fields moderately increased its subsequent drying rates. Dielectric heating did not affect the crude protein content of dried chopped alfalfa. Carotene retention in dried chopped alfalfa was increased by RF treatment, and the carotene content was doubled by exposures to RF electric fields as short as 1 min. While not practical for field application, dielectric heating might be useful as a means for inactivating carotene-destroying enzymes in laboratory procedures or field testing.

Dielectric Properties of Agricultural Materials and Their Applications. DOI: http://dx.doi.org/10.1016/B978-0-12-802305-1.00006-3

6.3 IMPROVING NUTRITIONAL VALUE OF SOYBEANS

Trypsin is an essential enzyme for the efficient digestion of proteins in the digestive system of man and monogastric animals. Raw soybeans, *Glycine max* (L.) Merrill, contain a trypsin inhibitor that must be inactivated to obtain the maximum nutritional value of soy products in both animal and human diets. Inactivation of this antitryptic factor is currently accomplished industrially by steam "toasting" treatments after the soybeans have been dehulled, split, flaked, and defatted. Various heating methods have been studied for inactivation of the trypsin inhibitor, but moist heating, or the presence of excessive moisture during heating, has generally been deemed necessary for effective trypsin-inhibitor inactivation except in RF dielectric heating (Borchers et al., 1972; Pour-El et al., 1981).

Dielectric-heating treatments at 43 MHz for about 1 min raised the temperature of soybeans to about 130°C, lowered the trypsin-inhibitor activity, and resulted in nutritional values equal to those of steam-autoclaving treatments as assessed by rat-feeding trials (Borchers et al., 1972). These treatments also produced much less browning or discoloration of the meal ground from the soybeans when compared to steam autoclaving. This result was desirable from an aesthetic view-point and thus has economic considerations. Effective inactivation of the trypsin inhibitor has also been reported as a result of microwave heating of soybeans for 3 min (Hamid et al., 1975). The experimental work summarized here was conducted to obtain further information on the effects of dielectric heating on the biologic properties of soybeans. Treatments at both 42 MHz and the microwave heating frequency of 2450 MHz were included in the study (Pour-El et al., 1981; Nelson et al., 1981).

Dielectric heating treatments, at 42 and 2450 MHz, of intact soybeans containing only their innate moisture can produce products of potentially high nutritional value. The biologic properties of soybeans treated by dielectric heating are dependent on the minimum energy absorbed (MEA), which can be calculated from the temperature rise and moisture loss of the product during treatment. Trypsin inhibitor, urease, and lipoxygenase activities are all reduced to low levels by treatments characterized by MEA values in the 60–80 cal/g range, whereas peroxidase activity remains relatively high in soybeans exposed at these same levels, all of which are desirable charac-teristics. Protein solubility and dispersibility are reasonable indicators of trypsin inhibitor activity for dry soybeans exposed to dielectric heating, but urease activity is not a good indicator. MEA values, and perhaps moisture loss alone, may be adequate indicators of trypsin inhibitor activity in dielectric heat-treated soybeans. Some of the products that might be produced by this treatment may have unique properties and may have functional properties of similar novel character. This method of treatment deserves further study to explore these points in greater depth.

6.4 QUALITY MAINTENANCE IN PECANS

The quality of freshly harvested pecans deteriorates rapidly unless they are held in refrigerated stor-age. Deterioration in quality is usually attributable to oxidation of the unsaturated fatty acids in pecan kernels with subsequent production of off-flavors, and to the development of a dark red color in the testa as a result of oxidative transformations of colorless compounds into their respective col-ored pigments (Senter et al., 1978). Quality is maintained better in unshelled pecans than it is after shelling. If holding periods exceed a few months, in-shell pecans must be stored at temperatures of 0°C or lower to preserve their quality (Woodroof, 1967).

Heating treatments of pecans have been evaluated as a means of preserving quality without the expense of refrigerated storage (Wells, 1951). Steam conditioning of in-shell pecans for cracking was shown to stabilize the sensory quality of kernels during storage in addition to improving processing efficiency (Forbus and Senter, 1976). Studies were conducted to evaluate the effects of steam heating and dielectric heating on the sensory and color quality of pecan kernels during a 16-week accelerated storage test (Senter et al., 1984a,b; Nelson et al., 1985). Results of these studies are summarized here.

Moisture content of the untreated control pecan kernel halves decreased gradually over the 16-week storage period from 3.3% to about 2.5% (Senter et al., 1984b; Nelson et al., 1985). The steam treatment elevated the moisture content to about 4.5%, and it then decreased over the storage period to about the same level as the untreated controls. The dielectric heating treatments lowered the moisture content to 2.7%, 1.8%, and 1.1% for exposures of 1, 2, and 2.5 min, respectively, and the kernels regained moisture during the storage period. At the end of the 16-week period, the pecans treated for 1 and 2 min had a moisture content similar to that of the untreated control, but those treated for 2.5 min had a moisture content of 1.8%, which was significantly lower than that of the control. Peroxide values, which indicate development of rancidity, increased in the control pecans from a value of 0.15 meq O_2/kg oil at the beginning of the storage period to about 1.1 meq O_2/kg 16 weeks later (Senter et al., 1984b). Peroxide values for all treatments tended to increase somewhat over the storage period, but they remained lower than that of the control. Hedonic ratings by a sensory evaluation panel revealed no significant differences in scores for a reference sample from $-30°C$ storage, the untreated control, and samples from the steam and 1-min dielectric heating treatments at the beginning of the storage period. The 2- and 2.5-min dielectric heating treatments imparted a toasted flavor; so the hedonic scores on these samples were somewhat lower in initial tests and also at 4 weeks into the storage period. At 8 weeks, significant differences between the four treatments had disappeared, the untreated control had dropped to a lower score than the treated samples, and the reference sample was rated superior to the treated samples. This trend continued throughout the 16-week storage period with hedonic scores trending downward, except on the 2.5-min dielectric heating treatment. At 16 weeks the untreated control was rated 3.4 (moderately disliked), whereas the treated samples ranged from 4.5 for the steam-heated sample to 5.3 (where 5 is neither liked nor disliked) for the 2.5-min dielectric heating treatment (Nelson et al., 1985).

Color characteristics were evaluated in terms of the Hunter color values. A detailed account of color changes during the storage period was provided (Senter et al., 1984a). The overall impression was an initial darkening of the pecan kernels by the steam treatment, some darkening of all kernels during storage, and noticeably darker kernels for the steam-treated samples throughout the storage period. Less darkening is an advantage because lighter kernels are desired for aesthetic reasons.

Both the steam and the dielectric heating treatments had a stabilizing influence on pecan quality during the 16-week accelerated storage period at 21°C and 65% relative humidity because they retained a desirable flavor better than that of the untreated control. The reduction in rancidity as a result of the heating treatments was confirmed by peroxide values. Steam heating treatments produced a darkening of the kernels, but dielectric heating to the same temperature range (about 90°C) and even higher temperatures (156°C) did not darken the kernels. The heating treatments showed promise for maintaining quality of pecans in storage, and the dielectric heating treatments offer the advantage of maintaining the desirable lighter color of the pecan kernels.

REFERENCES

Borchers, R., Manage, L.D., Nelson, S.O., Stetson, L.E., 1972. Rapid improvement in nutritional quality of soybeans by dielectric heating. J. Food Sci. 37, 333–334.

Forbus Jr., W.R., Senter, S.D., 1976. Conditioning pecans with steam to improve shelling efficiency and storage qualtiy. J. Food Sci. 41 (4), 794–798.

Hamid, M.A.K., Mastowy, N.J., Bhartia, P., 1975. Microwave bean roaster. J. Microw. Power 10 (1), 109–114.

Nelson, S.O., Pour-El, A., Stetson, L.E., Peck, E.E., 1981. Effects of 42- and 2450-MHz dielectric heating on nutrition-related properties of soybeans. J. Microw. Power 16 (3–4), 313–318.

Nelson, S.O., Senter, S.D., Forbus Jr., W.R., 1985. Dielectric and steam heating treatments for quality maintenance in stored pecans. J. Microw. Power 20 (2), 71–74.

Pour-El, A., Nelson, S.O., Peck, E.E., Tjiho, B., 1981. Biological properties of VHF- and microwave-heated soybeans. J. Food Sci. 46 (3), 880–885, 895.

Senter, S.D., Forbus Jr., W.R., Smit, C.J.B., 1978. Leucoanthocyanidin oxidation in pecan kernels: relation to discoloration and kernel quality. J. Food Sci. 43 (1), 128–134.

Senter, S.D., Forbus Jr., W.R., Nelson, S.O., Horvat, R.J., 1984a. Effects of dielectric and steam heating treatments on the pre-storage and storage color characteristics of pecan kernels. J. Food Sci. 49 (6), 1532–1534.

Senter, S.D., Forbus Jr., W.R., Nelson, S.O., Wilson Jr., R.L., Horvat, R.J., 1984b. Effects of dielectric and steam heating treatments on the storage stability of pecan kernels. J. Food Sci. 49 (3), 893–895.

Stetson, L.E., Ogden, R.L., Nelson, S.O., 1969. Effects of radiofrequency electric fields on drying and carotene retention of chopped alfalfa. Trans. ASAE 12 (3), 407–410.

Wells, A.W., 1951. The Storage of Edible Nuts, US Department of Agriculture, Transportation and Storage Official Report 240.

Woodroof, J.G., 1967. Tree-Nuts—Production, Processing, Products. AVI Publishing Co., Inc., Westport, CT.

GRAIN AND SEED MOISTURE SENSING APPLICATIONS

7.1 BACKGROUND INFORMATION

The moisture content of cereal grain and oil seed determines their suitability for harvest and storage and is usually measured whenever grain and seed are traded. Because standard and reference methods for determining moisture content in grain and seed involve tedious laboratory procedures and long oven-drying periods, rapid methods for moisture measurement have been essential in the grain and seed trade. Electrical measurements have been developed that provide rapid grain and seed moisture testing. This chapter covers key historical information over the past century related to the development of electrical grain and seed moisture meters.

Moisture content is sensed through correlations with the electrical characteristics, or dielectric properties, of these materials. This chapter covers the use of electrical resistance or conductance of grain samples, radio-frequency (RF) capacitance measurements, and microwave measurements of dielectric properties for sensing the moisture content of grain and seed. It deals with the principles involved rather than descriptions or comparisons of particular instruments that have been developed for measurement of grain and seed moisture content.

The dielectric properties, or permittivities, of cereal grains and oilseeds vary with the frequency of the applied electric field, the moisture content of these materials, their temperature, and bulk density. Grain and seed permittivities have, therefore, been useful for the rapid sensing of moisture content. Instruments operating at frequencies of 1–20 MHz have been used for this purpose for many years. More recently, higher frequencies have been investigated, and the use of grain and seed permittivities at microwave frequencies show promise for the simultaneous sensing of moisture content and bulk density in static and flowing granular materials. Because of the advantages offered by measurement at the higher frequencies, commercial development of new moisture meters for grain and seed will improve the reliability and utility of such instruments in the grain and seed industries.

7.2 EARLY HISTORY

Early in the twentieth century, it was discovered that the natural logarithm of electrical resistance of wheat decreased linearly with increasing moisture content (Briggs, 1908a,b). Grain moisture meters were later developed based on this principle. Resistance or conductance of grain samples between metallic electrodes or between crushing-roller electrodes was sensed and correlated with grain

Dielectric Properties of Agricultural Materials and Their Applications. DOI: http://dx.doi.org/10.1016/B978-0-12-802305-1.00007-5

moisture content. In the late 1920s, the use of capacitance measurements for moisture determination in grain was studied, and moisture meters were developed that utilized relationships between instrument meter readings and reference methods (Berliner and Ruter, 1929; Burton and Pitt, 1929). Many different electrical grain-moisture meters have been developed that use the capacitance-sensing principle. Considerable effort has been reported in comparing the different types of moisture meters and assessing their success in providing results comparable to standard moisture determination methods.

A portable battery-operated moisture meter measuring dc conductance of grain was designed for field use (Brockelsby, 1951). The general principles employed by electrical moisture testers for grain and seed have been discussed, and differences were noted in the performance of conductance- and capacitance-type moisture meters (Zeleny, 1954, 1960). Sources of errors in the use of electrical moisture testers were discussed by several researchers (Hunt et al., 1963; Hunt, 1963; Hunt and Neustadt, 1966). The influence of various factors on the accuracy of capacitance-type moisture meters was taken into account in the design and testing of a capacitance-type meter operating at a frequency of 2 MHz (Matthews, 1963). Detailed studies were conducted on capacitance sensing of moisture content with frequencies up to 10 MHz for rice dryer control (Ban and Suzuki, 1977). Sources of error and performance of electrical moisture meters of the conductance and capacitance types were studied (Hart and Golumbic, 1963, 1966).

Extensive studies were conducted comparing moisture content values provided by various electrical or electronic moisture meters with those of approved oven-testing procedures or other standards (Cook et al., 1934a,b; Hlynka et al., 1949). Such comparisons were reported for a conductance-type meter and two capacitance-type meters (USDA, 1963; Hart and Golumbic, 1966). In Europe, comparisons of several conductance- and capacitance-type meters were reported (Stevens and Hughes, 1966), and studies on the performance of dielectric-type capacitance-sensing moisture meters compared to standard methods were also reported (Jacobsen, 1971; Moller, 1971). A conductance-type moisture meter that passed grain samples between roller electrodes was the official meter for the USDA from 1936 until 1963, when a capacitance-type meter was adopted. No attempt is made here to compare the accuracies of these various moisture meters. Interested readers are referred to the cited references for such detail.

Not until mid-century were measurements begun to provide values for the permittivities, or dielectric properties, of grain and seed upon which the rapid sensing of moisture content depends (Nelson et al., 1953; Knipper, 1959). Studies of the dependence of permittivities of grain and seed on influencing factors, including frequency, moisture content, density, and temperature (ASABE, 2013; Nelson, 1965, 1981b, 1982), have enabled the continued improvement of grain and seed moisture meters to meet the needs of the agricultural industry.

7.3 DIELECTRIC PROPERTIES

The complex permittivity relative to free space is represented here as $\varepsilon = \varepsilon' - j\varepsilon''$, where ε' is the dielectric constant and ε'' is the dielectric loss factor. Early measurements of the dielectric properties of many kinds of grain and seed in the frequency range 1–50 MHz revealed high correlations between the grain and seed moisture content and their permittivities or dielectric properties (ASABE, 2013; Nelson, 1965). Examples are shown in Figure 7.1 for hard red winter wheat (*Triticum aestivum* L.) and in Figure 7.2 for soybeans (*Glycine max* L.).

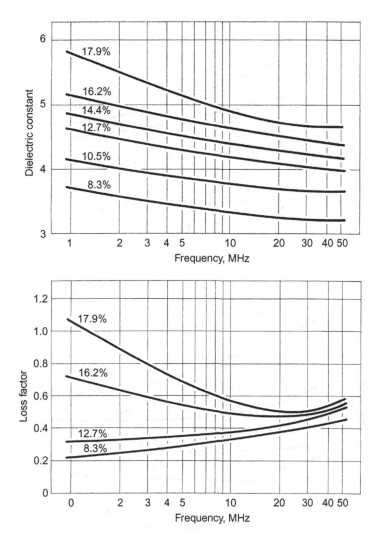

FIGURE 7.1

Permittivities of "Nebred" hard red winter wheat at 24°C and indicated moisture contents. Test weight: 768 kg/m³ (59.7 lb/bu) at 13% moisture content (Nelson, 1965).

All moisture contents are expressed here in percent by weight on the wet basis.

For both wheat (Figure 7.1) and soybeans (Figure 7.2), dielectric properties are clearly correlated with moisture content and can, therefore, be used for moisture sensing. Permittivity measurements on hard red winter wheat over wide ranges of frequency and moisture content are summarized with contour plots of the dielectric constant and loss factor as functions of moisture content and frequency in Figure 7.3.

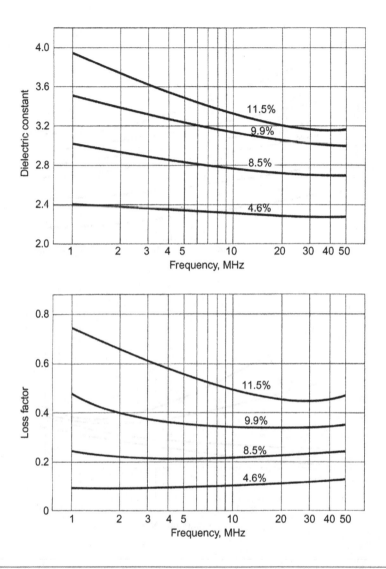

FIGURE 7.2

Permittivities of "Hawkeye" soybeans at 24°C and indicated moisture contents. Test weight: 738 kg/m^3 (57.3 lb/bu) at 7.5% moisture content (Nelson, 1965).

Close estimates for the dielectric constant and loss factor of wheat can be obtained at any moisture content between 3% and 24% at any frequency from 250 Hz to 10 GHz by reference to the values for the contour lines of these two graphs. The behavior of the dielectric constant is regular with respect to both moisture content and frequency, but the variation of the dielectric loss factor is much less regular due to the influence of dielectric relaxation and conduction processes.

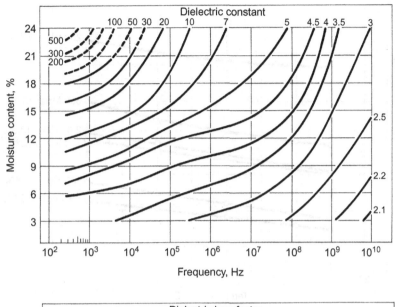

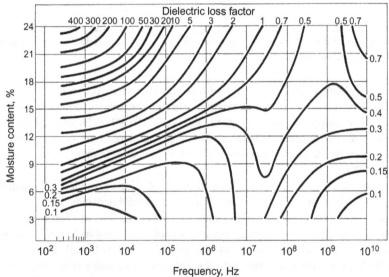

FIGURE 7.3

Dielectric constant (ε') and loss factor (ε'') for hard red winter wheat at 24°C at frequencies from 25 Hz to 10 GHz and moisture contents from 3% to 24% at natural densities. Mean values for seven cultivars (Nelson, 1981b).

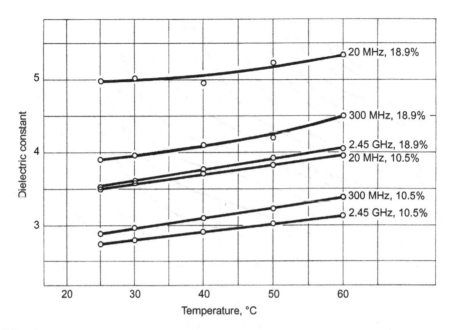

FIGURE 7.4

Temperature dependence of the dielectric constant (ε') of shelled yellow-dent field corn at indicated frequencies and moisture contents (Nelson, 1978).

Temperature is another factor that influences the dielectric properties of grain and seed. Figure 7.4 shows the variation in the dielectric constant for shelled corn, *Zea mays* L., of two different moisture contents at frequencies of 20, 300, and 2450 MHz (Nelson, 1979).

The increase in dielectric constant with increasing temperature is reasonably linear, although the deviation from linearity tends to increase at higher moisture contents and lower frequencies (Lawrence et al., 1990).

In addition, as shown in Figure 7.5, the dielectric constant of shelled corn increases linearly with bulk density at all moisture levels for normally encountered densities.

This linearity with density was also shown at frequencies of 300 MHz and 2.45 GHz (Nelson, 1979). Over wider ranges of bulk density, the dielectric constant is not linear, but the square and cube roots of the dielectric constant are essentially linear with bulk density (Nelson, 1983, 1984b). These findings are also consistent with the well-known complex refractive index and Landau and Lifshitz, Looyenga dielectric mixture equations, respectively (Nelson, 1992, 2005).

Utilizing linear relationships between frequency, moisture, temperature, and density and the dielectric constants or functions of the dielectric properties, mathematical models for the dielectric constant of several cereal grains and soybeans were developed from which close estimates of dielectric constants can be calculated for frequencies between 20 MHz and 2.45 GHz over wide ranges of moisture content at 24°C (ASABE, 2013; Nelson, 1984a, 1987), or as functions of moisture, density, and temperature at frequencies of 20, 300, and 2450 MHz (Nelson, 1979).

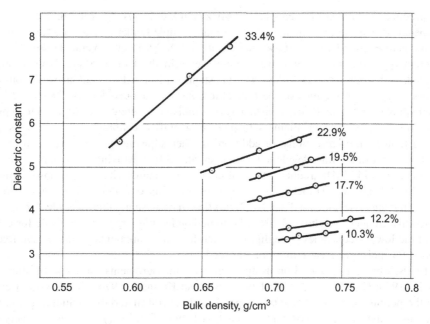

FIGURE 7.5

Density dependence of the dielectric constant (ε') of shelled yellow-dent field corn at 24°C, 20 MHz, and indicated moisture contents (Nelson, 1979).

7.4 MOISTURE CONTENT SENSING

Because of the close relationships between dielectric properties and moisture content illustrated in Figures 7.1−7.3, many commercial instruments have been developed for measuring grain and seed moisture content. Over the past 60 years, most of these grain moisture meters utilized frequencies between 1 and 20 MHz and sensed changes in capacitance of parallel-plate or coaxial sample holders when grain samples were introduced between the electrodes (Nelson, 1977; Lawrence and Nelson, 1998). However, dc conductance meters were in common use earlier and are still used on grain dryers and for specialty applications (Nelson and Lawrence, 1989). Moisture meters based on RF measurements on capacitive sample holders require corrections for the influence of temperature and bulk density, or test weight (Nelson, 1977, 1981b, 1982). These corrections have been applied through calibration charts or built into the instruments for automatic application. Accuracy of the moisture measurements required in the grain and seed trade has been achieved by continual improvement and calibration testing, but there is still some dissatisfaction with reliability in the higher moisture ranges, that is, above 20−25% for cereal grains. Recalibrations are also required occasionally because of differences in growing locations and variations in seasonal growing conditions (Hurburgh et al., 1985).

An intensive study, with large numbers of samples for various grains and seed, of the variation in moisture readings due to influencing factors has been reported, and techniques were proposed for their corrections with measurements at 149 MHz (Funk et al., 2007). Measurements were taken with an impedance analyzer and shielded parallel-plate sample holder as part of a transmission line, following methods reported earlier (Lawrence et al., 1999, 2001) on several thousand samples of many different types of grain grown over several years in the United States. The resulting data were analyzed, and variations in the dielectric constants with moisture content were compared for various types of grain. Corrections of the dielectric constant for bulk density (Nelson, 1992) were applied, and adjustments of dielectric constant-versus-moisture content line slopes, intercept offsets, and translations were derived in attempts to provide a unified calibration for all types of grain. Incorporating temperature corrections, in addition to other adjustments and corrections, a "unified grain moisture algorithm" was developed (Funk et al., 2007). Moving from the 1- to 20-MHz frequency range to 149 MHz reduces the influence of conductivity at the lower frequencies (Nelson, 1979; Nelson and Trabelsi, 2006; Nelson and Stetson, 1976), but some conductivity influence still remains (Nelson and Trabelsi, 2006) that may present calibration challenges.

At microwave frequencies above 3 GHz, the ionic conduction largely responsible for calibration variations at the lower frequencies is negligible. Therefore, measurements at microwave frequencies are of interest for moisture sensing in grain and seed.

In the 1970s, work was reported on using microwave measurements for grain moisture determination (Brain, 1970; Okabe et al., 1973; Kraszewski and Kulinski, 1976; Kraszewski et al., 1977). Studies on the permittivities, or dielectric properties, of several materials, including grain, revealed that a function of the dielectric constant ε' and loss factor ε'', $(\varepsilon' - 1)/\varepsilon''$, is a relatively density-independent function for use in predicting moisture content of particulate materials (Jacobsen et al., 1980; Meyer and Schilz, 1980, 1981; Kent and Meyer, 1982). The ratio of attenuation, which depends mainly on dielectric loss factor, and phase shift, which is mainly dependent on dielectric constant, was also investigated as a density-independent function for microwave sensing of moisture content (Jacobsen et al., 1980; Kent and Meyer, 1982; Kress-Rogers and Kent, 1987). Further studies with microwave measurements confirmed the usefulness of this ratio for sensing moisture content independent of bulk density fluctuations in grain, and indicated a possibility for a single calibration for cereal grains (Kraszewski, 1988; Nelson and Kraszewski, 1990). Additional studies on soft and hard red winter wheat confirmed the density-independent nature of simultaneous attenuation and phase measurements for moisture sensing, the provision of bulk density from the same measurements, and the usefulness of a single calibration for both hard and soft classes of wheat (Kraszewski and Nelson, 1992).

Many studies followed these initial efforts, further developing principles for microwave moisture sensing in grain and seed independent of bulk density (Nelson et al., 1998; Trabelsi et al., 1998a, c, 1997; Trabelsi and Nelson, 2004). Density-independent functions of the dielectric properties for predicting moisture content from measured dielectric properties of wheat and corn were compared in several studies (Trabelsi et al., 1998b; Kraszewski et al., 1998a; Trabelsi and Nelson, 1998). Means of compensating measurements for the influence of temperature variation were also studied (Kraszewski et al., 1997, 1998b,c; Trabelsi et al., 2001a). These studies included the development of unified or potentially universal calibrations for corn, wheat, barley, oats, grain sorghum, rapeseed, and soybeans (Trabelsi et al., 1999, 2001c; Trabelsi and Nelson, 2007).

Principles of the permittivity-based calibration functions (Trabelsi et al., 1997, 1998b) were described previously for use in measuring both moisture content and bulk density of grain and seed (Trabelsi et al., 2001b; Nelson and Trabelsi, 2004; Kraszewski, 2001), but a brief description is

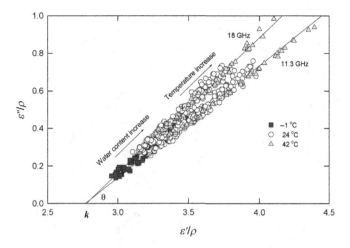

FIGURE 7.6

Complex-plane plot of the dielectric constant and loss factor, divided by bulk density (ρ) of hard red winter wheat of various moisture contents and bulk densities at indicated temperatures for two frequencies, 11.3 and 18.0 GHz (Trabelsi et al., 1998).

provided here. Measurements on a large number of samples of hard red winter wheat of different moisture contents, bulk densities, and temperatures are summarized in complex-plane permittivity plots for measurements at 11.3 and 18.0 GHz in Figure 7.6.

Note that, for permittivities determined by these measurements at a given frequency, all of the points fall along a straight line, and that differences in either moisture content or temperature amount to translations along that same line. The lines for each frequency intersect the ε'/ρ axis at a common point k, which represents the value for 0% moisture content or the value at very low temperatures. Any change in microwave frequency amounts to a rotation of the straight line about that intersection point. Thus, for a given frequency, the equation of the line is expressed as $\varepsilon''/\rho = a_f[(\varepsilon'/\rho) - k]$, where a_f is the slope at a given frequency. Slope varied linearly with frequency over the 11- to 18-GHz range (Trabelsi et al., 1998b). Solving the equation of the straight line for ρ, we have:

$$\rho = \frac{a_f\varepsilon' - \varepsilon''}{a_f k} \tag{7.1}$$

For a given frequency, a_f is a constant, and for a given material, k is a constant. Thus, an estimate of the bulk density is provided in terms of the permittivity alone without explicit knowledge of the temperature or moisture content (Trabelsi et al., 1998a, 2001b). Considering that $\tan \delta = \varepsilon''/\varepsilon'$, where δ is the loss angle of the dielectric, expresses the distribution ratio of dissipated and stored energy in a dielectric, and that $\tan \delta$ varies with bulk density, it was divided by bulk density. Using this expression for ρ, we can write:

$$\frac{\tan \delta}{\rho} = a_f k \left(\frac{\varepsilon''}{\varepsilon'(a_f\varepsilon' - \varepsilon'')} \right) \tag{7.2}$$

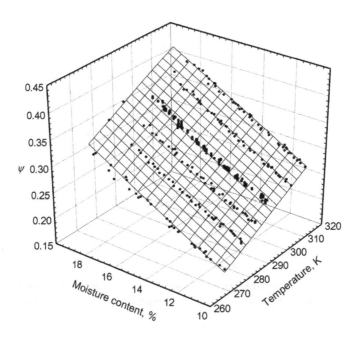

FIGURE 7.7

Moisture and temperature dependence of density-independent moisture calibration function ψ at 14.2 GHz for hard red winter wheat (Kraszewski et al., 1998b).

For a given frequency and particular kind of material, $a_f k$ is a constant, and a new density-independent moisture calibration function can be defined as (Trabelsi et al., 1998b):

$$\psi = \sqrt{\frac{\varepsilon''}{\varepsilon'(a_f\varepsilon' - \varepsilon'')}} \qquad (7.3)$$

This calibration function has been studied for a large set of measurements on hard red winter wheat over practical ranges of moisture content, bulk density, and temperature (Trabelsi et al., 2001a). By plotting ψ against moisture content and temperature, the points define a plane in three-dimensional space (Figure 7.7), and the following equation was obtained: $\psi = bM + aT + c$, for which values of the constants a, b, and c were determined by regression analysis.

The equation for moisture content:

$$M = (\psi - aT - c)/b \qquad (7.4)$$

is then given in terms of the density-independent calibration function ψ, which, at any given frequency, depends only on the grain permittivity. The dielectric constant and loss factor can be determined by any suitable microwave measurement.

Further research with this density-independent moisture calibration function has shown that very similar values of regression constants were obtained for kinds of grain as different as wheat and corn (Trabelsi et al., 2001a). In other comparisons, the same constants performed very well for

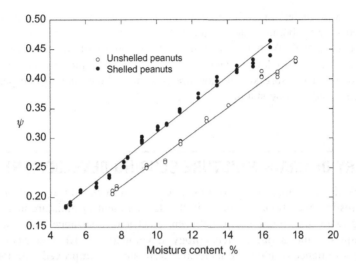

FIGURE 7.8

Density-independent moisture calibration function ψ for unshelled pod peanuts, ψ_p, and for shelled peanut kernels, ψ_k, at 8 GHz and 24°C (Trabelsi et al., 2010).

wheat, oats, and soybeans, and for corn, wheat, and soybeans, which have very different characteristics with respect to kernel shape, size, and composition (Trabelsi and Nelson, 2004; Trabelsi et al., 2001c). These findings support the possibility of a universal calibration.

The technique described for sensing grain moisture content independent of bulk density has also been expected to be useful in providing reliable moisture information on flowing grain. Recent studies have shown that the technique does provide the same result when lots of shelled yellow-dent field corn, wheat and soybeans were measured under static or flowing conditions (Nelson et al., 2014). Therefore, the technique is available for monitoring moisture content in flowing grain.

Microwave measurements on unshelled peanuts (pod peanuts) and shelled peanuts (peanut kernels) have demonstrated that the moisture content of the peanut kernels, which is required in practice, can be determined from rapid measurements on the unshelled peanuts (pod peanuts) (Trabelsi and Nelson, 2006). The density-independent moisture calibration function (Eq. (7.3)) was determined separately for unshelled and shelled peanuts, as shown in Figure 7.8, where separate lines of nearly the same slope are shown for the unshelled peanuts (pod peanuts) of moisture content M_p and the shelled kernels of moisture content M_k.

Linear regressions for the unshelled pod peanuts and for the kernels provide two independent equations of the form $\psi = aM + b$, where values of M and the regression constants a and b for the unshelled pod peanuts are different from those for the shelled peanut kernels. For any given value of ψ, different moisture contents are provided for the unshelled (pod peanuts) and shelled peanuts (kernels). The relationship between these two moisture contents can be obtained by equating ψ_k and ψ_p and solving for kernel moisture M_k in terms of the unshelled pod peanut moisture content M_p:

$$M_k = 0.917 M_p - 1.2083 \qquad (7.5)$$

Microwave moisture meters operating at 5.8 GHz and based on these principles were built and successfully tested in the laboratory and in the field at several peanut-buying points in 2008 and 2009 (Trabelsi et al., 2010). With the need for critical control of moisture content in peanuts, because of aflatoxin threats to human and animal health, much interest has been expressed in moisture meters that can measure kernel moisture content without shelling the peanuts for use at peanut buying points and grading stations.

7.5 SUMMARY OF GRAIN MOISTURE SENSING DEVELOPMENT

The literature cited on grain and seed moisture measurement through electrical properties describes an interesting evolutionary process in the development of moisture meters over the past century. Starting with the relationship between electrical resistance and moisture content, improved techniques came about as technology developed for RF electrical measurements. Accuracies and convenience of moisture meter use were steadily improved with the incorporation of new ideas and advances in electronics. Much of the progress resulted from research and development in private industry, which is naturally not well documented in public literature. The availability of dielectric properties data on grain and seed—and information on the dependence of these properties on moisture content and other variables—aided the process of improving moisture meters. In general, the variability, accuracies, and reliability of various moisture meters and measurement techniques are not discussed here, but the reader is referred to the cited references for that information. Accuracies of 0.3−0.5% moisture content have been expected in comparing moisture testers to standard reference methods for many years. New techniques have given comparable standard errors in such comparisons (Funk et al., 2007; Kraszewski et al., 1998a,d; Trabelsi et al., 2001a,c).

Dielectric properties of many types of grain and seed are now available for frequencies spanning the range 250 Hz to 15 GHz (ASABE, 2013). Moisture content is highly correlated with the dielectric properties of grain and seed at any frequency, so moisture can be sensed with instruments operating at any frequency (Nelson, 1973). Frequencies for use in grain moisture meters were selected mainly for convenience of design and economical construction. Most of them used frequencies between 1 and 20 MHz, where design has been generally simpler than at higher frequencies. Studies on the dielectric properties of grain and their variation with moisture content at different frequencies established that the dielectric constant was the best single property to use as an indicator of grain moisture content, but that inclusion of the loss factor can be helpful in limiting the disturbance of other variables (Nelson, 1982, 1977). Development of the "unified grain moisture algorithm" at 149 MHz (Funk et al., 2007) utilized only the dielectric constant, noting that the loss factor 'appeared to be minimized' at that frequency. However, the loss factor continues to decrease with increasing frequency well into the microwave range (Nelson and Trabelsi, 2006).

At microwave frequencies, both the real and imaginary parts of the relative permittivity, that is, the dielectric constant and the loss factor, are conveniently utilized in the density-independent moisture calibration function already described. With microwave measurements, it appears that the adjustments and corrections for different types of grain are unnecessary in the calibration process, and temperature can be included in that process as already described (Eq. (7.4)). In addition, at microwave frequencies, errors due to uneven moisture distribution in the kernels will be less than those encountered at frequencies below the microwave range (Sokhansanj and Nelson, 1988). The density-independent nature of moisture sensing in grain and seed provided by microwave frequencies should also make the monitoring of moisture in flowing grain much more successful than it has been with capacitive sensing at lower frequencies. The capability of sensing peanut kernel moisture content without shelling the peanuts, which has been proven with microwave measurements, may also find applications with other nuts and similar products.

The permittivities or dielectric properties of cereal grains and oilseeds vary with the frequency of the applied electric fields, the moisture content of these materials, their temperature, and bulk density. Thus, grain and seed permittivities are useful for the rapid sensing of moisture content, and instruments operating at frequencies of 1−20 MHz have been used for this important application for many years. Recent research has indicated that higher frequencies, which are less susceptible to variations because of ionic conductivity, offer promise for improving the performance of grain moisture meters. Measurements at 149 MHz, with adjustments and corrections for bulk density and for unifying responses to different types of grain, may alleviate some calibration problems.

The use of grain and seed permittivities measured at microwave frequencies shows promise for simultaneous sensing of moisture content and bulk density in both static and flowing materials, providing moisture content independent of bulk density. Microwave measurements can also sense the kernel moisture content in unshelled peanuts. Because of advantages offered by measurement at the higher frequencies, commercial development of new moisture meters for grain and seed can be expected to improve the reliability and utility of such instruments in the grain and seed industries.

7.6 SINGLE KERNEL OR SEED MOISTURE SENSING

Commonly used grain moisture meters measure the moisture in bulk samples. It was found earlier that the moisture content of individual kernels in well conditioned samples of corn, *Z. mays* L., can vary as much as ±0.8% from the average bulk sample moisture content (Kandala, 1987). Larger variations can be expected in unconditioned samples. Variation in kernel moisture content depending on the location of kernels on the corn ear at harvest has been reported (Nelson and Lawrence, 1991). Kernel moisture variation with location on the ear in hybrid, yellow-dent field corn was measured by individual kernel oven moisture content determination as ears dried during the harvest season. Differences in kernel moisture content (m.c.) at three locations on the ear, that is, near the butt end, at the mid-section, and near the tip, were highly significant, with m.c. highest at the butt and decreasing toward the tip. Butt-to-tip kernel m.c. differences were small at high moisture levels (35−40% m.c., wet basis), but reached a maximum, which averaged about 6% m.c., at 24% m.c. for the ear. As the ears dried to safe-storage moisture levels, the butt-to-tip m.c. differences diminished to about 2% m.c. No significant differences were noted between hybrids with respect to variation of kernel m.c. with location on the ear. The kernel m.c. at the middle of the ear gave the best estimate of the bulk-sample m.c. for the ear.

Blending of high-moisture corn with corn of safe-storage moisture levels is practiced to save on costs of drying, but this can tend to promote the growth of microorganisms that spoil the corn. Some problems with spoilage of corn in transit to foreign ports may be attributed to this blending of corn lots. Therefore, there has been interest for a long time in detection of blended lots. An instrument that sensed the dc conductance of single corn kernels, when passing between crushing rollers, was described and used to study equilibration rates of corn mixed from lots of two different moisture contents (Watson et al., 1979). The fluctuating dc signal from a Tag-Heppenstall grain moisture meter, normally calibrated to provide an average moisture reading for a grain sample passing between crushing rollers (from dc conductance measurements), was recorded and analyzed for information on moisture distributions in blended corn lots (Martin et al., 1986). High correlations were found between the signal standard deviation and the difference in moisture content of corn in blended lots. Further studies indicated that standard deviations of digitally processed signals showed dependence on the quantity of wet corn in the blend, the difference in moisture contents of the two component corn lots, and the amount of broken kernels and fine material in the sample (Martin et al., 1987).

7.6.1 SINGLE-KERNEL GRAIN MOISTURE MEASUREMENTS ON CORN

Techniques for sensing the moisture content of individual corn kernels have been investigated in connection with the problem of detecting lots of corn blended from low and high-moisture corn.

Measurement of moisture content in single-kernel corn by capacitive sensing has been described in a series of papers (Kandala et al., 1987, 1988a,b, 1989). These measurements of moisture content of single kernels between plates of small parallel-plate capacitors at frequencies between 1 and 5 MHz showed a steady progression of improved accuracy as different measured variables were considered. Initial measurements of capacitance at 1 MHz, along with kernel weight, thickness, and projected area, demonstrated the feasibility of moisture content measurement by this method (Kandala et al., 1987). Moisture contents of individual kernels were predicted within ±1% moisture content on about 80% of the kernels over the moisture range from 12% to 20%, wet basis. With the addition of capacitance measurement at a second frequency, 4.5 MHz, this percentage was increased to the 80–90% level over a moisture content range of 1–24% (Kandala et al., 1988a). With the improved measurement of kernel projected area, using a digital imaging system, and the measurement of the dissipation factor as well as the capacitance of the parallel-plate capacitor at two frequencies, 1 and 4.5 MHz, the percentage of kernels properly classified within ±1% moisture content was improved to 89% over the moisture content range 9.5–26% without the need for measurement of kernel thickness (Kandala et al., 1988b). Finally, with measurement of the complex impedance (capacitance, phase angle, and dissipation factor) and proper combination of the variables, the percentage of properly classified kernels was increased to 91% without the need for measurements of kernel weight or projected area (Kandala et al., 1989).

7.6.1.1 Sensor principles for the parallel-plate high-frequency impedance measurement technique

A parallel-plate capacitor serves as a useful sensor for RF measurements and can be used in the high-frequency range for determining the moisture content of the kernel. The parallel-plate arrangement (Figure 7.9) provides a uniform electric field between the plane, parallel-plate electrodes of the empty capacitor.

Therefore, the measurement is not sensitive to small variations in kernel location as long as the kernel remains between the plates and away from the periphery of the plates. Theoretical

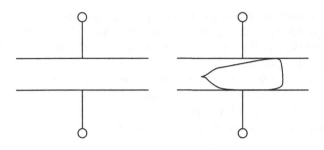

FIGURE 7.9

Parallel-plate capacitive sensor, empty, and with corn kernel between the plates (Nelson et al., 1992c).

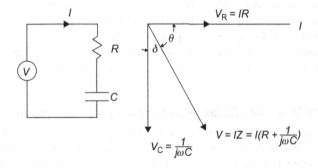

FIGURE 7.10

Series-equivalent RC circuit for impedance measurement and associated phasor diagram (Nelson et al., 1992c).

considerations and experimental measurements showed that, for sufficient sensitivity, the corn kernel should be in contact with both electrodes of the capacitor (Kandala et al., 1987).

Representing the kernel and parallel-plate assembly by a series-equivalent circuit with resistance R and capacitance C (Figure 7.10), its impedance is $Z = R + 1/(j\omega C)$, where $j = \sqrt{-1}$, and $\omega = 2\pi f$ is the angular frequency, where f is the frequency of the imposed alternating field.

The current and voltage phase relationships are shown in Figure 7.10. The phase angle, θ, and its complement, the loss angle δ, are functions of the energy loss accounted for by the corn kernel between the electrodes. Thus, the impedance of the capacitor loaded with a kernel depends upon the dielectric properties of the kernel. For a lossy dielectric, it is convenient to use the complex permittivity relative to free space (complex dielectric constant), $\varepsilon = \varepsilon' - j\varepsilon''$, where the real part, ε', is the dielectric constant, and the imaginary part, ε'', is the dielectric loss factor. The loss tangent of the dielectric is then $\tan \delta = \varepsilon''/\varepsilon'$, where this loss angle will differ from the loss angle of the capacitor with the kernel between the plates, because the kernel does not completely fill the space between the plates. Fortunately, the exact function relating the dielectric properties of the kernel and those of the capacitor with the kernel need not be known for practical use of this sensor. By measuring the impedance of the capacitor with a kernel at two frequencies, it has been possible to determine the moisture content of corn kernels within 1.0% moisture content over the range 9.5–26% (Kandala et al., 1989). Working with a parallel-plate assembly and an impedance analyzer, Kandala et al. (1989) measured capacitance C, phase angle θ, and dissipation factor $D = \tan \delta = \omega CR$ with a corn kernel between the plates at frequencies of 1 and 4.5 MHz. An equation relating these measurements to the moisture content of the kernel was developed as follows:

$$M = A_0 + A_1(C_1 - C_2) + A_2(C_1 - C_2)^2 + A_3 \left[\frac{(\theta_1 - \theta_2)}{(C_1 - C_2) + (D_1 - D_2)} - (\theta_1 - \theta_2)(C_1 - C_2) \right] \quad (7.6)$$

where M represents the wet basis moisture content (%), the constants A_0, A_1, A_2, and A_3 are determined by least squares computation from calibration data, and the subscripts 1 and 2 on the variables refer to measurements taken at the two frequencies, 1 and 4.5 MHz, respectively. In subsequent work with popcorn (Kandala et al., 1992; Kandala and Nelson, 1990), Eq. (7.1) also provided moisture contents of individual kernels from RF impedance measurements at 1 and 4.5 MHz. Studies with the same technique on in-shell pecans and pecan kernel halves resulted in different forms of the equation for predicting moisture content, depending on whether measurements were taken on the whole nut or the half kernel after being shelled (Nelson et al., 1992a). Moisture determinations in single seeds and kernels made with this dual-frequency measurement are essentially independent of kernel size and shape (Nelson et al., 1992c).

7.6.1.2 Single corn kernel conductance and impedance measurements

In 1986, a commercial single-kernel moisture tester for rice, wheat, and barley was introduced in Japan (Shizuoka Seiki[1] Model CTR-800). A modified model of this tester, which utilizes dc conductance of kernels as they pass individually between crushing rollers, was calibrated and evaluated for use with corn (Nelson and Lawrence, 1989). The standard error of performance over the moisture content range from 12% to 30% was 0.99% moisture content, compared to 0.71% moisture content over the 9.5–26% moisture · range for the parallel-plate measurement system.

The performance of this dc conductance sensing instrument and of the parallel-plate complex impedance sensing system described above were directly compared on corn samples from the 1988 harvest. Hybrid yellow-dent field corn lots for this comparison were collected during the harvest in Georgia, Illinois, Iowa, and Nebraska. The two electrical methods for sensing single-kernel moisture content were compared with a standard forced-air oven method for determining the moisture content of single kernels of corn (Nelson and Lawrence, 1989). One method used the complex impedance of a parallel-plate capacitor with the kernel between the plates, measured at frequencies of 1 and 4.5 MHz. The other used the dc conductance of the kernel as it passed between crushing rollers. Both methods, tested with hybrid yellow-dent field corn samples ranging from 11% to 27% moisture content (wet basis), produced nearly comparable results. Standard errors of performance were 0.66% and 0.89% moisture content, for the parallel-plate impedance and crushing-roller

---

[1]Mention of trade names or commercial products in this publication is solely for the purpose of providing specific information and does not imply recommendation or endorsement by the US Department of Agriculture.

conductance methods, respectively, as determined by tests on 55 samples with 30 kernels per sample. Variations in measured moisture content of freshly dried samples, by the dc conductance and RF impedance methods, caused by nonequilibrium moisture distribution in the kernel, disappeared 3–4 h after drying.

A prototype instrument was designed and constructed to demonstrate the feasibility of the nondestructive measurement of moisture content in single corn kernels by RF impedance measurements at frequencies of 1.0 and 4.5 MHz (Kandala et al., 1993). The principles of the measurement were explained, and the mechanical single-kernel feeder and parallel-plate electrode assembly were described. The impedance-magnitude and phase-measuring electronic circuits designed and constructed for the instrument were also described. In initial calibration and performance tests with corn ranging from 11% to 25% moisture (wet basis), more than 80% of the kernels were measured to within 1% moisture content of reference values determined by oven-drying procedures. In performance tests on individual kernels of yellow-dent hybrid field corn, the instrument provided a standard error of performance of 0.89% moisture content, so the technique showed promise for practical use with further development.

The use of impedance measurements on small parallel-plate sensors holding a single kernel between the plates for single-kernel moisture measurement was also explored for popcorn (Kandala et al., 1992). An empirical equation, developed earlier for predicting the moisture content of single kernels of field corn from RF measurements, was used successfully for popcorn kernels. Measurements of capacitance, dissipation factor, and/or phase angle with an impedance analyzer at 1 and 4.5 MHz on a small parallel-plate capacitor, holding the kernel between the plates, predicted kernel moisture contents for 97% of the popcorn kernels tested in the moisture range 11–25% (wet basis), within 1% moisture content. The same type of measurement was also used successfully in determining moisture content of small samples (15–30 kernels) of popcorn between somewhat larger circular parallel-plate electrodes (Kandala et al., 1994).

7.6.2 SINGLE SOYBEAN SEED MOISTURE MEASUREMENTS

The moisture content in soybeans, *G. max* (L.), and other seed and grain commodities, is the most important factor determining suitability for storage. To prevent spoilage, the moisture content must be low enough so that fermentation processes and fungal development cannot proceed under the conditions expected in storage. Generally, the moisture content is assumed to be uniform among seeds and kernels in seed and grain lots, but often this assumption is not justified. When lots from different sources are blended, mixing is often not complete enough to achieve the desired equilibration of moisture content, and spoilage can result even though average moisture content is believed to be at safe levels. Differences of as much as 2% in individual kernel moisture content have been observed in samples of corn from commercial storage (Christensen and Kaufmann, 1969), and differences of 0.8% moisture content were observed among corn kernels equilibrated for long periods in sealed quart jars (Kandala et al., 1989). Kernel and seed moisture levels can differ by 6% moisture content or more at harvest (Nelson and Lawrence, 1991), and uneven heated-air drying can produce large differences in moisture between kernels with consequent hysteresis effects that prevent equilibration to the same moisture content.

Therefore, the variation in seed and kernel moisture content, and the importance of such variation, under various conditions, has been of interest. Thus, interest in single-kernel moisture meters for research, and potentially practical, use has developed. A crushing-roller-type, single-kernel instrument was developed for commercial use in Japan (Nelson and Lawrence, 1989). A nondestructive technique for single-kernel moisture sensing, with RF impedance measurements, was also successful for field corn (Kandala et al., 1989), peanuts (Kandala and Nelson, 1990; Nelson et al., 1990b), and popcorn (Kandala et al., 1992). With kernels of each of these commodities, standard errors of performance of between 0.4% and 0.7% moisture content (Nelson et al., 1992b) have been obtained, which are comparable to those achieved by the crushing-roller type instruments (Nelson et al., 1990c, 1991). Research was undertaken to determine how well the nondestructive dual-frequency impedance measurement technique could determine the moisture content of individual soybean seeds.

Impedance measurements were taken at frequencies of 1 and 5 MHz on a small parallel-plate (2-cm diameter) capacitor holding single soybean seeds between, and in contact with, the plates on samples from two cultivars each from soybean-producing areas in the Midwest and the Southeast. Moisture contents ranged from 7% to 20% (wet basis). Calibration equations were developed from these measurements and oven moisture determinations on individual soybeans. Two calibration equations were developed, one with a single variable involving only capacitance measurements at the two frequencies, and another with three variables including capacitance differences and complex admittance differences. The two equations performed comparably in these tests with standard errors of 0.5−0.6% moisture content. Thus, the technique could provide rapid nondestructive tests for moisture content of single soybeans and offered promise for practical use.

Similar tests were made to compare the performance of the parallel-plate capacitance measurements and the crushing-roller conductance measurements for determining the moisture content of single soybean seeds (Nelson and Lawrence, 1994). Both dc conductance measurements on individual soybeans passing between crushing-roller electrodes and RF capacitance measurements at 1 and 5 MHz on parallel-plate electrodes, holding the soybean between and in contact with the plates, can provide moisture content readings with standard errors of about 0.6−0.7% moisture content when compared to single-soybean oven moisture determinations over the moisture range from 7% to 20% (wet basis). The RF measurements, which were calibrated against a single-soybean oven moisture determination, had a mean bias of less than -0.1% moisture. The dc conductance instrument had an unexplained bias of about 1% moisture content, which could also be reduced to a low level by calibration against the same oven reference. The RF measurement is nondestructive, whereas the soybeans are partially crushed for the dc conductance measurement.

7.6.3 MEASURING MOISTURE CONTENT IN SINGLE KERNELS OF PEANUTS

Moisture content is an important property to be managed in the harvesting, storage, marketing and processing of peanuts, *Arachis hypogaea* L. Peanuts are allowed to dry on the vines after being dug at harvest. In the Southeast, they are permitted to dry to an average moisture content of 18−22%, wet basis, before being combined. After combining, the drying process must begin immediately to reduce the moisture content to less than 10.5% for grading and sale (PAC, 1988). Peanuts can be stored safely at this moisture level if adequate ventilation is provided. Electronic moisture meters, calibrated against standard oven moisture determinations, are used for the measurement of peanut moisture. The peanut pods are shelled and the cleaned peanut kernels are tested for moisture content in electronic grain moisture meters calibrated for peanut kernels. These instruments sense electrical properties of the peanuts that are highly correlated with moisture content. Most moisture meters sense the capacitance of a sample chamber containing a bulk sample of peanut kernels, and that capacitance is a function of the dielectric properties of all the peanuts in the sample chamber. Thus, these meters give a reading that is an average for all of the peanuts in the sample. A moisture meter that could provide the moisture content of the individual kernels in a peanut sample would be helpful in determining the range of kernel moisture. At harvest, a very wide range of kernel moisture contents exists, and that range is reduced upon drying. However, information on individual kernel moisture contents would be helpful in the drying, storage, and processing of peanuts. It could be a significant tool in quality control.

A moisture profile meter for peanut samples was described earlier (Hutchison and Holaday, 1978). The dc conductance of peanut kernels—as they passed individually between metal crushing rollers—was used to provide information on the range of peanut kernel moisture contents present in a sample. A nondestructive measurement of peanut kernel moisture content by measurements on a small parallel-plate capacitor holding the kernel between the plates was also reported (Kandala and Nelson, 1990). This technique, which used the complex impedance of the parallel-plate and peanut combination measured at two frequencies, 1 and 4.5 MHz, determined single-kernel moisture content within $\pm 1\%$ moisture content on 97% of the kernels measured over a moisture range between 5% and 15% (wet basis). Continued studies of this technique for measuring single-kernel moisture content in peanuts showed that a standard error of performance of 0.5% moisture content was achieved when compared to the standard oven moisture determination. The technique, in which kernel moisture content was calculated from differences in capacitance, dissipation factor, and/or phase angle at the two frequencies, provided reliable values independent of kernel size as verified by measurements on jumbo and medium-size peanut kernels. It also performed well on peanut samples of both 1988 and 1989 crop years. The method appeared to be adaptable for development of practical instruments for the measurement of single-kernel peanut moisture.

7.6.4 COMPARISON OF FOUR SINGLE-KERNEL MOISTURE SENSING TECHNIQUES FOR CORN

Four electrical techniques for sensing the moisture content of individual corn kernels were compared in a study in which results were compared with standard oven drying moisture determinations and Karl Fisher moisture tests (Nelson et al., 1990a). Direct comparisons of kernel moisture content determinations by single-kernel oven drying, nuclear magnetic resonance (NMR) measurements, microwave resonant cavity measurements, parallel-plate RF complex impedance measurements, and crushing-roller dc conductance measurements were made on samples from the same corn lots equilibrated at moisture contents between 10% and 40% (wet basis). All four of the electrical methods considered for sensing the moisture content of individual corn kernels were shown to permit moisture content measurements of sufficient accuracy for practical use. Although insufficient experience with any of these methods had been gained to fully evaluate them for practical use, some observations were appropriate.

Fundamentally, all four of the electrical methods are sensing electromagnetic properties of the grain kernel. The dc conductance is sensitive to the ionic conductivity of the grain, which is influenced mainly by its water content. However, other factors related to chemical composition, such as salt content, have an influence on ionic conductivity, and, if variable among grain lots from different origins or different storage environments, could contribute to errors in moisture detection.

The RF impedance and microwave cavity measurements are sensitive mainly to the mass and shape of the kernel and its dielectric properties. The influence of mass and shape have been largely eliminated by appropriate measurement techniques (Kandala et al., 1989; Kraszewski et al., 1991). The contribution to the dielectric properties at RF and microwave frequencies is mainly a result of dipole rotation of polar molecules, and water is the only appreciable polar constituent of grain. However, at frequencies below about 10 GHz in the microwave range, ionic conductivity is still a contributing factor (Hasted, 1973), and the lower the frequency the larger is the influence of ionic conductivity. Thus, microwave frequencies offer an advantage if ionic conductivity variation is a problem.

NMR signals result from magnetic resonance of the hydrogen nuclei, regardless of chemical structure. Thus, other constituents of grain, such as oils, can contribute to NMR responses. However, because of differences in signal decay times, the influence of oils can be minimized by appropriate delays with pulsed NMR instruments. The NMR response is dependent on the mass of the kernel, because larger kernels have more hydrogen nuclei, so the weight of the kernel is a factor that must be taken into account in single-kernel moisture measurement.

Distribution of moisture in the kernel is another factor to consider. Variations in kernel moisture distribution—because of recent drying or moistening of the kernels—should have no effect on NMR signals. Disruption of kernel moisture equilibrium does have some influence on the dielectric properties of grain measured within the first 3 h after surface wetting or drying of grain (Sokhansanj and Nelson, 1988); however, the variation with time after moistening or drying wheat kernels was very slight at microwave frequencies (higher than about 1 GHz). Pronounced variations at frequencies of 1 and 18 MHz, because of wetting or drying, had also disappeared after 3–4 h. Moisture contents determined by RF impedance and dc conductance measurements on corn kernels that were dried in a hot-air oven from about 30% to 19% moisture content increased during the first few hours after drying, but stabilized after 3–4 h (Nelson et al., 1990c). The moisture measurements by the dc conductance method increased 2.5% moisture during this period compared to 1.4% for the RF impedance method, indicating that the conductance measurement was more sensitive to the nonuniform kernel moisture distribution produced by drying than was the RF impedance measurement (Nelson and Lawrence, 1989). For applications where this factor is important, such as grain-drier monitoring or tempering of grain for milling, microwave frequencies offer an advantage.

A practical instrument must be able to measure the moisture content of kernels rapidly to provide the required information on moisture distribution of a sample within an appropriate time period. The dc conductance, RF impedance, and microwave cavity measurements require only fractions of a second, so the mechanical singulation and synchronization of the measurements are the limiting time factors for single-kernel moisture measurements. These measurements are conceptually feasible at a rate of several kernels per second. For resonant microwave cavity measurements, the corn kernel must be measured in two orientations, because of its irregular shape. Therefore, resonant cavity measurements would require more time than the dc conductance or RF impedance measurements, but they need not be substantially slower. NMR measurements generally require several seconds per scan and a number of scans for each measurement; so it is the slowest of the four methods. Correction for temperature variation may need to be applied for all of these electrical measurements.

Costs are usually the dominant factor in developing and producing instruments for agricultural applications because the products are generally of relatively low unit value. Instruments using the four techniques discussed here would probably fall into the following classes in order of increasing costs: dc conductance, RF impedance, microwave cavity, and NMR techniques. With the exception of the dc conductance measurements, which were done with a slightly modified commercial single-kernel moisture tester, relatively expensive laboratory instruments were used in these studies. Less expensive measurement equipment and circuits would have to be designed and developed for practical single-kernel moisture testing applications.

The RF impedance, microwave cavity, and NMR techniques are all potentially nondestructive tests for individual kernels. The dc-conductance instrument crushes the kernel to varying degrees, depending on the moisture level of the kernels. Nondestructive tests are preferable in most applications, but the loss of small samples could be insignificant in many practical cases. The dc conductance tester could become commercially available with minor design changes for suitable handling of single corn kernels. All the other methods require developmental work before practical instruments could be made available.

7.6.5 SINGLE NUT AND KERNEL PECAN MOISTURE SENSING

According to Woodroof (1967), "Controlling moisture is the most important factor in harvesting, storing or processing pecans, even for as short a time as one week." The moisture content of pecans, *Carya illinoensis* (Wangenh.) K. Koch, is high when harvested, particularly when shakers are used to remove them from the trees, as is common practice in commercial pecan production. Kernel moisture contents of pecans shaken from the trees can be as high as 24% (wet basis), with corresponding entire nut moisture contents of 30% or more (Woodroof, 1967).

Nuts that fall from the trees and lie on dry soil or leaves, or are held for 2−4 weeks in dry storage, equilibrate to about 4.5% moisture content for the kernels and about 8.5−9% for the entire nuts (Woodroof, 1967). Thus, pecans arriving at the processing plants can have a wide range of moisture contents. To preserve good quality, pecans need to be dried to kernel moisture contents of 4.5% as soon as possible after harvest (Woodroof, 1967; Heaton et al., 1977). This drying is accomplished by forcing warm air at 38°C (100°F), or a lower temperature, and a relative humidity of 60%, or lower, through the nuts (Heaton et al., 1977).

When substantial numbers of dry nuts are mixed with those freshly shaken from the trees, more drying capacity and facilities are used than would be necessary if the nuts could be sorted into different moisture content categories before drying. Sorting before drying would also save the energy lost in raising the temperature of pecans that do not need artificial drying. After drying, most pecans are placed in refrigerated storage until they can be scheduled for shelling or until they are needed for in-shell marketing. Storage at 0−1°C (32−34°F) at 65% relative humidity retains the kernel moisture at 4.5% (Heaton et al., 1977). For storage longer than a few weeks, holding at subfreezing temperatures is required to preserve quality.

Prior to cracking in-shell pecans in the shelling process, shellers attempt to increase the kernel moisture content to about 8% to prevent shattering of the kernels (Heaton et al., 1977). Upon completion of the shelling and separation processes, the moisture added for shelling must be promptly removed and kernel moisture content reduced to 3.5−4% to maintain the quality (Woodroof, 1967; Heaton et al., 1977). Drying time can be shortened by the use of heated air, but air temperatures should not exceed 49°C (120°F) (Woodroof, 1967). In addition to drying the kernels to the desired moisture level, it is also important not to overdry them because of resultant quality degradation and loss in weight of saleable product.

Because of the importance of moisture content in the harvesting, storage, and processing of pecans, a rapid moisture sensing technique that could be adapted to online operations would be helpful in lowering energy consumption and preserving high quality. The high correlation between dielectric properties of pecans and their moisture content (Nelson, 1981a) offers such an opportunity for sensing moisture content through the interaction of RF and microwave electric fields with the pecans. Success in the measurement of single corn kernel moisture contents by these methods (Kandala et al., 1989; Nelson et al., 1990c; Kraszewski et al., 1990) encouraged the exploration of these techniques for individual pecan nuts and kernel halves. The basis for applying such techniques and the results of initial measurements for sensing pecan moisture content by high-frequency and microwave techniques were reported (Nelson et al., 1992a).

Two approaches were explored, (i) the influence of the pecan on the RF impedance of a parallel-plate capacitor with the nut between the plates, and (ii) the influence of the pecan on the resonant frequency and transmission characteristics of a microwave resonant cavity. In both cases, the effects on the electrical parameters of the sensors are influenced mainly by the amount of water in the nut. This results from the relative values of the electrical permittivities of water and the dry matter of the pecan. Total nut moisture content and kernel moisture content are both of interest. For most storage and processing decisions, however, the pecan kernel moisture is the main criterion. Equilibrium moisture contents of the kernel and the nut shell are quite different (Woodroof, 1967; Chhinnan and Oliver, 1980). Under the same temperature and relative humidity conditions, the shell equilibrates at a higher moisture content than the kernel. Because the electrical sensing methods are influenced by the total moisture content, a measurement of the kernel moisture for in-shell pecans must depend upon a correlation between total nut moisture content and kernel moisture content. Among different pecan cultivars, there is variation in the kernel-to-shell weight ratio. On average, the kernel accounts for 40−50% of the total nut weight and only 22−25% of the total nut moisture (Woodroof, 1967). The distribution of moisture between kernel and shell in pecans was determined over the total nut moisture range from 7.8% to 12.2%, and the proportion of the total moisture contained by the kernel varied from 22% to 25% (Heaton and Woodroof, 1965).

The moisture content of mature in-shell pecans was measured by RF impedance measurements of a 2-cm diameter parallel-plate capacitor with a pecan between the plates and by rectangular-waveguide microwave resonant cavity measurements with a pecan in the resonant cavity (Nelson et al., 1992a). Impedance measurements were also taken on pecan kernel halves. Nuts of six pecan cultivars, four from the United States and two from Israel, with kernel moisture contents ranging from 3% to 11%, were included in the study. Although equilibrium moisture contents (m.c.) of pecan kernels and shells differ greatly, and proportions of total nut weight represented by the kernel and shell vary significantly among cultivars, a high correlation was found between total nut m.c. and kernel m.c. Standard errors of calibration (SEC) of 0.6% m.c. and 1.5% m.c. were obtained for determination of the entire nut moisture content with electrical measurements on the whole nuts for the RF impedance and microwave cavity measurements, respectively. RF impedance measurements on kernel halves and on the in-shell pecans predicted kernel moisture contents equally well with SEC values of 0.6% m.c. for both measurements. Sensing of pecan moisture content by RF complex impedance measurements or by microwave resonant cavity measurements offers potential methods for on-line monitoring of moisture content for use in sorting pecans or in controlling drying equipment. Both entire-nut and kernel moisture contents of in-shell pecans can be sensed rapidly and nondestructively by measuring the complex impedance of a parallel-plate capacitor holding the nut between the two electrodes at frequencies of 1 and 5 MHz (Nelson et al., 1992a). Kernel moisture content can also be sensed with individual kernel halves between the electrodes (Nelson and Lawrence, 1995). Further research and development was recommended to better evaluate the techniques for practical application.

A dual-frequency parallel-plate RF impedance sensor for moisture content in individual pecan nuts was implemented in an experimental, automated computer-controlled pecan feeder, singulation, and conveying system for on-line testing in Israel (Schmilovitch et al., 1996). Equipment was developed under a joint research project between engineers in Israel and in the United States. The on-line system provided stationary contact for the nut between the electrodes of the parallel-plate sensor for the imped-ance measurements and moisture determination. An inexpensive electronic RF impedance-measuring unit was developed and tested under on-line conditions. The electronic unit was calibrated to provide kernel moisture contents in individual pecans, and subsequently measured pecans in on-line operation with standard errors of 1.1% moisture content compared with those determined by the vacuum-oven ref-erence method (Schmilovitch et al., 1996). The prototype automatic conveyor and moisture sensing sys-tem was capable of measuring moisture content of individual pecans at rates up to three nuts per second.

7.6.6 SINGLE-KERNEL MICROWAVE-RESONATOR MOISTURE SENSING

Another technique studied for the purpose of measuring single-kernel or seed moisture content is the use of microwave resonant cavities (Nelson et al., 1992c; Kraszewski et al., 1989, 1990, 1991; Kraszewski and Nelson, 1993a,b, 1994). A dielectric region completely surrounded by a conducting surface constitutes a resonant cavity. If a hollow rectangular waveguide is sealed with a short-circuiting conductive wall perpendicular to the direction of propagation, the incident and reflected waves are superimposed and create a standing wave. The tangential electric field and normal mag-netic field are zero at this wall and at distances of integral half wavelengths from it. In such a nodal plane, a second short-circuiting conductive wall can be located without disturbing the standing wave. Waves reflected by the first short-circuiting wall will be reflected back by the second short-circuiting wall to coincide with the incident wave, and a resonant cavity results in which waves are reflected back and forth, and a standing wave pattern persists if energy is supplied to offset losses in the cavity. If the cavity is excited at the proper frequency, through a coupling hole for example, the fields can build up within the cavity in certain restricted configurations or modes of operation. If a rectangular waveguide section is fitted with shorting plates at each end of the waveguide (Figure 7.11), a resonant cavity is formed with air as the dielectric.

The difference between the resonant cavity and a low-frequency LCR resonant circuit is that the circuit has only one resonant frequency, whereas the cavity has many resonant frequencies. The cavity will be resonant at frequencies having an integral number of half wavelengths in the wave-guide equal to the length, L. The guide wavelength λ_g (the wavelength in the waveguide) has the following relationship to the free-space wavelength λ_0 (Harrington, 1961):

$$\lambda_g = \frac{\lambda_0}{\sqrt{1 - (1/\lambda_c)^2}} \tag{7.7}$$

where λ_c is the cut-off wavelength of the waveguide ($\lambda_c = 2a$ for rectangular waveguide, a is the waveguide dimension shown in Figure 7.11).

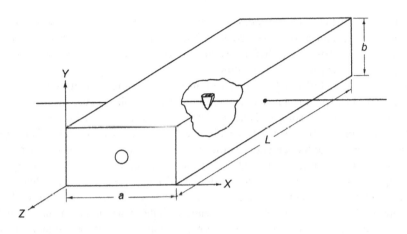

FIGURE 7.11

Rectangular waveguide resonant cavity with coupling hole in shorting end-plate and corn kernel suspended at the cavity center (Kraszewski and Nelson, 1992).

Thus, resonance will occur at frequencies for which $p\lambda_g/2 = L$, where p is an integer. The distribution of resonant frequencies for a rectangular waveguide of given length is shown in Figure 7.12.

The resonant frequency, f_0 can be determined as $f_0 = c/\lambda_0$, where c is the velocity of electromagnetic energy propagation in free space, after λ_0 is calculated as:

$$\lambda_0 = \frac{1}{\sqrt{\left(\frac{1}{\lambda}\right)^2 + \left(\frac{p}{2L}\right)^2}} \tag{7.7}$$

The lowest resonant frequency for a rectangular cavity is that for $L = \lambda_g/2$. This mode is known as the TE_{101} mode, indicating that the electric field vector, E, is transverse to the z direction of propagation in the waveguide, with the first subscript indicating one half-cycle for the E-field pattern in the y direction across the a dimension of the waveguide, the second subscript indicating no E-field in the x direction across the b dimension, and the third subscript indicating the value of p, the number of half wavelengths along the z-axis within the length L of the cavity. Since the electric field must vanish at the conducting boundaries, the E-field will have a maximum value at the center of the cavity for the TE_{101} mode or for any TE_{10p} mode for which p is an odd integer. In most practical applications, however, by proper choice of coupling elements and operating frequency range, one can effectively limit the number of modes to the single one desired. This is always the case when a cavity is used for measurement purposes. In a measurement system, the resonance of the cavity will appear as a peak in transmission through the cavity. To measure the resonant frequency, the frequency of a signal coupled to the cavity is varied until resonance is observed. The second parameter of the resonance curve, as shown in Figure 7.13, is its shape.

The Q-factor of a cavity has essentially the same meaning as the Q of a resonant circuit at lower frequencies. The coupling devices connected to the cavity, energy losses in the cavity walls, or losses in the sample introduced into the cavity lower the apparent Q-factor. Thus, when a dielectric

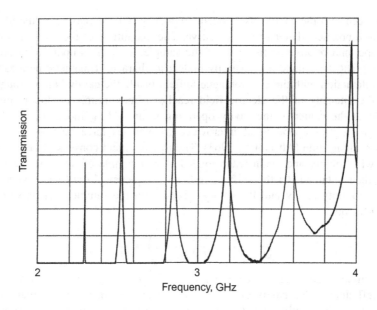

FIGURE 7.12

Distribution of resonant frequencies for a WR-284 rectangular waveguide cavity 304.8 mm long, showing the TE_{10p} modes for $P = 2, 3, 4, 5, 6,$ and 7 (Nelson et al., 1992c).

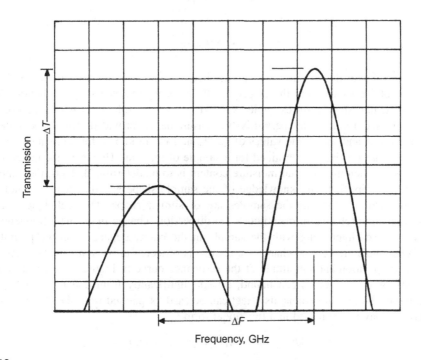

FIGURE 7.13

Resonance curves for empty cavity (right) and cavity loaded with sample (left) (Nelson et al., 1992c).

sample is introduced into the cavity, its resonant frequency will decrease, and the Q-factor will be lowered, causing a broader, flatter resonance curve. The magnitude of these changes depends upon dimensions, shape, and relative permittivity of the sample, as well as upon its location inside the cavity. Assuming that the sample has a strictly dielectric character (magnetic permeability equal to that of air), its interaction with the electromagnetic field inside the cavity is the strongest when it is located where the intensity of the electric field reaches a maximum. As already mentioned, that condition occurs at the center of the cavity operating in any TE_{10p} mode for which p is an odd integer. Resonant cavities operating at microwave frequencies have been used for permittivity measurement for a long time (Altschuler, 1963). The measurement consists of determining the cavity parameters when empty and again when loaded with a sample of material. Two measurement parameters are then obtained: (i) the shift of resonant frequency $\Delta f = f_0 - f_s$, where subscripts 0 and s refer to the empty cavity and the cavity loaded with the sample, respectively; and (ii) the change in the Q-factors,

$$\frac{1}{Q_s} - \frac{1}{Q_0} = \frac{1}{Q_0}\left(\frac{Q_0}{Q_s} - 1\right) = \frac{\Delta T}{Q_0} \tag{7.8}$$

where the transmission factor $\Delta T = (10^k - 1)$, where $k = (S_{210} - S_{21s})/20$, and S_{21} is the voltage transmission coefficient of the cavity expressed in decibels. It can be shown that these two measurement parameters are directly related to the sample permittivity, shape, and volume by the following equations (Altschuler, 1963; ASTM, 1981):

$$\Delta f = 2(\varepsilon' - 1)Kf_0\left(\frac{v_s}{v_0}\right) \tag{7.9}$$

$$\Delta T = 4\varepsilon''K^2Q_0\left(\frac{v_s}{v_0}\right) \tag{7.10}$$

where K is a shape factor dependent upon sample shape, orientation and permittivity, and v_s and v_0 are the volumes of the sample and the empty cavity, respectively. The shape factor can be determined for some regular shapes, for example, $K = 3/(\varepsilon + 2)$ for a sphere. In general, these relationships have been used for material permittivity measurements, provided that the sample is well defined in dimensions and shape, the values of f_0, Q_0, and v_0 are known for a given cavity, and the values of K and v_s are precisely determined for a sample of material. However, the material permittivity is not needed when the material moisture content is to be determined. Thus, the experimental procedure may be simplified, and knowledge of the sample shape and dimensions is not required prior to measurement, as long as the samples are of similar shape. The technique is based on the fact that dissipation (change in magnitude) and dispersion (change in resonant frequency) of the electromagnetic waves interacting with the sample in the cavity depend upon the permittivity of the sample. When moisture content changes, a change is reflected in wave parameters, Δf and ΔT, determined by the position and amplitude of the resonance curve in Figure 7.13. Because the relative permittivity of water, even when bound, differs significantly from that of most hygroscopic dielectric materials, like grain kernels, its effect can be easily separated from the effect of dry material. In general, it may be expressed in a functional form as:

$$\Delta F = \Phi_1(m_d, m_w) \quad \Delta T = \Phi_2(m_d, m_w)$$

where m_d and m_w are the mass of dry material and mass of water, respectively. In practice, these relationships are determined empirically. In most practical cases, the two equations can be solved simultaneously to express the mass of water and the mass of dry substance in terms of two measured parameters in the form:

$$m_d = \Psi_1(\Delta F, \Delta T) \quad m_w = \Psi_2(\Delta F, \Delta T)$$

Substituting the analytical expressions corresponding to Ψ_1 and Ψ_2 into the definition of moisture content, M in percent, wet basis, one obtains the following relationship:

$$M = \left(\frac{m_w}{m_d + m_w}\right) 100 = \left(\frac{\Psi_2(\Delta F, \Delta T)}{\Psi_1(\Delta F, \Delta T) + \Psi_2(\Delta F, \Delta T)}\right) 100 \qquad (7.11)$$

which contains only the wave parameters determined experimentally and is independent of the size of the sample. It has been shown (Kraszewski et al., 1989) that use of Eqs (7.9)−(7.11), permits a determination of moisture content for nearly spherical seeds (soybeans, *G. max* L.) independent of size variation among the seeds. For irregularly shaped seeds or kernels such as corn, this ratio is also useful, but two measurements of ΔF and ΔT, with kernel orientations separated by a 90° rotation with respect to the E-field vector, are necessary, and the averaged values of these variables can be used to provide a moisture determination independent of size and shape (Kraszewski et al., 1990, 1991). For regularly shaped kernels, such as peanuts, single measurements of ΔF and ΔT can be used to obtain both moisture content and mass of the kernels simultaneously (Kraszewski and Nelson, 1993a,b). Even moisture content and mass of wheat kernels were determined with uncertainties of less than 0.8% moisture content and 1 mg, respectively (Kraszewski and Nelson, 1994).

REFERENCES

Altschuler, H.M., 1963. Dielectric constant. In: Sucher, M., Fox, J. (Eds.), Handbook of Microwave Measurements. Polytechnic Press of the Polytechnic Inst., Brooklyn, New York, NY (Chapter IX).

ASABE, D293.4, 2013. Dielectric Properties of Grain and Seed. American Society of Agricultural and Biological Engineers in ASABE Standards 2013, St. Joseph, MI.

ASTM, 1981. Complex Permittivity of Solid Electrical Insulating Materials at Microwave Frequencies and Temperatures to 1650°C in Standard Test Methods D-2520-81. American Society for Testing and Materials, Philadelphia, PA.

Ban, T., Suzuki, M., 1977. Studies on Electrical Detection of Grain Moisture Content in Artificial Drying. Institute of Agricultural Machinery, Omiya, Japan (in Japanese).

Berliner, E., Ruter, R., 1929. Uber Feuchtigkeitsbestimmungen in Weizen und Roggen mit dem D K-Apparat. Zeitshrift fur das Gesamte Muhlenwesen 6 (1), 1−4.

Brain, M., 1970. Measurement and control of moisture in wheat. Meas. Control 3, T181−T187.

Briggs, L.J., 1908a. An electrical resistance method for the rapid determination of the moisture content of grains. Science 28 (727), 810−813.

Briggs, L.J., 1908b. An Electrical Resistance Method for the Rapid Determination of the Moisture Content of Grain, Bureau of Plant Industry Circular No. 20, USDA, Washington, DC.

Brockelsby, C.F., 1951. An instrument for estimating the moisture content of grain and other materials by the measurement of conductance. Cereal Chem. 28 (2), 83−94.

Burton, E.F., Pitt, A., 1929. A new method for the rapid estimation of moisture in wheat. Can. J. Res. 1, 155−162.

Chhinnan, M.S., Oliver, T.J., 1980. Equilibrium Moisture Content Equations for Sorption and Desorption Isotherms of Pecan Kernels and Shells, ASAE Paper No. 80-3532, American Society of Agricultural Engineers, St. Joseph, MI.

Christensen, C.M., Kaufmann, H.H., 1969. Grain Storage: The Role of Fungi in Quality Loss. University of Minnesota Press, Minneapolis, MN.

Cook, W.H., Hopkins, J.W., Gedes, W.F., 1934a. Rapid determination of moisture in grain. II. Calibration and comparison of electrical moisture meters with vauum oven for hard red spring wheat. Can. J. Res. 11 (4), 409−447.

Cook, W.H., Hopkins, J.W., Geddes, W.F., 1934b. Rapid determination of moisture in grain. III. Calibration and comparison of electrical moisture meters with vacuum oven for amber durum wheat, barley and oats. Can. J. Res. 11 (5), 547−563.

Funk, D.B., Gillay, Z., Meszaros, P., 2007. Unified moisture algorithm for improved RF dielectric grain moisture measurement. Meas. Sci. Technol. 18, 1004−1015.

Harrington, R.E., 1961. Time-Harmonic Electromagnetic Fields. McGraw-Hill Book Co., New York, NY.

Hart, J.R., Golumbic, C., 1963. Methods of moisture determination in seeds. Proc. Int. Seed Test. Assoc. 28 (4), 911−933.

Hart, J.R., Golumbic, C., 1966. The use of electronic moisture meters for determining the moisture content of seeds. Proc. Int. Seed Test. Assoc. 31 (2), 201−212.

Hasted, J.B., 1973. Aqueous Dielectrics. Chapman and Hall, London.

Heaton, E.K., Woodroof, J.G., 1965. Importance of proper drying of pecans. Proc. S. E. Pecan Growers Assoc. 52, 32−34.

Heaton, E.K., Shewfelt, A.L., Bodenhop, A.E., Beuchat, L.R., 1977. Pecans: Handling, Storage, Processing and Utilization, Research Bulletin 197, The University of Georgia, Georgia Station, Athens, GA.

Hlynka, I., Martens, V., Anderson, J.A., 1949. A comparative study of ten electrical meters for determining moisture content of wheat. Can. J. Res. 27, 382−397.

Hunt, W.H., 1963. Problems associated with moisture determination in grain and related crops, May 20−23. In: Wexler, A. (Ed.), Humidity and Moisture Measurement and Control in Science and Industry, International Symposium on Humidity and Moisture. Washington, D.C. Reinhold Publishing, New York, NY, pp. 123−125.

Hunt, W.H., Neustadt, M.H., 1966. Factors affecting the precision of moisture measurement in grain and related crops. J. Assoc. Off. Anal. Chem. 49 (4), 757−763.

Hunt, W.H., Martens, V., Hlynka, I., 1963. Determination of moisture in Canadian grain by electric moisture meter. In: Wexler, A. (Ed.), Humidity and Moisture Measurement and Control in Science and Industry, International Symposium on Humidity and Moisture. Washington, D.C. Reinhold Publishing, New York, NY, pp. 125−133. May 20−23.

Hutchison, R.S., Holaday, C.E., 1978. Development of a Moisture Profile Meter for Peanuts, ASAE Paper No.78-3063, American Society of Agricultural Engineers, St. Joseph, MI.

Hurburgh Jr., C.R., Hazen, T.E., Bern, C.J., 1985. Corn moisture measurement accuracy. Trans. ASAE 28 (2), 634−640.

Jacobsen, E., 1971. Feuchtigkeitsbestimmung von getreide, die Muhle 108 (36).

Jacobsen,, R., Meyer,, W., Schrage,, B., 1980. Density independent moisture meter at X-band. Proceedings of the 10th European Microwave Conference. Warsaw, Poland, pp. 216−220.

Kandala, C.V.K., 1987. Capacitive Sensing Technique for Estimating Moisture Content of Single Kernels of Corn. (M.S. thesis), The University of Georgia, Athens, GA.

Kandala, C.V.K., Nelson, S.O., 1990. Measurement of moisture content in single kernels of peanuts: a nondestructive electrical method. Trans. ASAE 33 (2), 567−571.

Kandala, C.V.K., Leffler, R.G., Nelson, S.O., Lawrence, K.C., 1987. Capacitive sensors for measuring single-kernel moisture content in corn. Trans. ASAE 30 (3), 793−797.

Kandala, C.V.K., Nelson, S.O., Leffler, R.G., Lawrence, K.C., 1988a. Moisture measurement in single kernels of corn by capacitive sensing at 1.0 and 4.5 MHz. Int. Agrophys. 4 (1−2), 3−12.

Kandala, C.V.K., Nelson, S.O., Lawrence, K.C., 1988b. Moisture determination in single kernels of corn. Trans. ASAE 31 (6), 1890−1895.

Kandala, C.V.K., Nelson, S.O., Lawrence, K.C., 1989. Non-destructive electrical measurement of moisture content in single kernels of corn. J. Agric. Eng. Res. 44, 125−132.

Kandala, C.V.K., Nelson, S.O., Lawrence, K.C., 1992. Nondestructive moisture determination in single kernels of popcorn by radio-frequency impedance measurement. Trans. ASAE 35 (5), 1559−1562.

Kandala, C.V.K., Nelson, S.O., Leffler, R.G., Lawrence, K.C., Davis, R.C., 1993. Instrument for single-kernel nondestructive moisture measurement. Trans. ASAE 36 (3), 849−854.

Kandala, C.V.K., Nelson, S.O., Lawrence, K.C., 1994. Nondestructive moisture determination in small samples of popcorn by RF impedance measurement. Trans. ASAE 37 (1), 191−194.

Kent, M., Meyer, W., 1982. A density-independent microwave moisture meter for heterogeneous foodstuffs. J. Food Eng. 1 (1), 31−42.

Knipper, N.V., 1959. Use of high-frequency currents for grain drying. J. Agric. Eng. Res. 4 (4), 349−360.

Kraszewski, A., 2001. Microwave aquametry: an effective tool for nondestructive sensing. Subsurface Sens. Technol. Appl. 2 (4), 347−362.

Kraszewski, A., Nelson, S.O., 1994. Microwave resonator for sensing moisture content and mass of single wheat kernels. Can. Agric. Eng. 36 (4), 231−238.

Kraszewski, A., Kulinski, S., Stosio, Z., 1977. A preliminary study on microwave monitoring of moisture content in wheat. J. Microw. Power 12 (3), 241−251.

Kraszewski, A., Nelson, S.O., You, T.-S., 1990. Use of a microwave cavity for sensing dielectric properties of arbitratily shaped biological objects. IEEE Trans. Microw. Theory Tech. 38 (7), 858−863.

Kraszewski, A., Trabelsi, S., Nelson, S.O., 1997. Moisture content determination in grain by measuring microwave parameters. Meas. Sci. Technol. 8, 857−863.

Kraszewski, A.W., 1988. Microwave monitoring of moisture content in grain—further considerations. J. Microw. Power 23 (4), 236−246.

Kraszewski, A.W., Kulinski, S., 1976. An improved microwave method of moisture content measurement and control. IEEE Trans. Ind. Electron. Control Instrum. 23 (4), 364−370.

Kraszewski, A.W., Nelson, S.O., 1992. Wheat moisture content and bulk density determination by microwave parameters measurement. Can. Agric. Eng. 34 (4), 327−335.

Kraszewski, A.W., Nelson, S.O., 1992. Resonant microwave cavities for sensing properties of agricultural products. Trans. ASAE 35 (2), 1315−1321.

Kraszewski, A.W., Nelson, S.O., 1993a. Moisture content determination in single peanut kernels with a micro-wave resonator. Peanut Sci. 20, 27−31.

Kraszewski, A.W., Nelson, S.O., 1993b. Nondestructive microwave measurement of moisture content and mass of single peanut kernels. Trans. ASAE 36 (1), 127−134.

Kraszewski, A.W., You, T.-S., Nelson, S.O., 1989. Microwave resonator technique for moisture content determination in single soybean seeds. IEEE Trans. Instrum. Meas. 38, 79−84.

Kraszewski, A.W., Nelson, S.O., You, T.-S., 1991. Moisture content determination in single corn kernels by microwave resonator technique. J. Agric. Eng. Res. 48, 77−87.

Kraszewski, A.W., Trabelsi, S., Nelson, S.O., 1998a. Comparison of density-independent expressions for moisture content determination in wheat at microwave frequencies. J. Agric. Eng. Res. 71, 227−237.

Kraszewski, A.W., Trabelsi, S., Nelson, S.O., 1998b. Temperature-compensated and density-independent moisture content determination in shelled maize by microwave measurements. J. Agric. Eng. Res. 72, 27−35.

Kraszewski, A.W., Trabelsi, S., Nelson, S.O., 1998c. Addendum—moisture content determination in grain by measuring microwave parameters. Meas. Sci. Technol. 9, 543–544.

Kraszewski, A.W., Trabelsi, S., Nelson, S.O., 1998d. Simple grain moisture content determination from microwave measurements. Trans. ASAE 41 (1), 129–134.

Kress-Rogers, E., Kent, M., 1987. Microwave measurement of powder moisture and density. J. Food Eng. 6, 345–376.

Lawrence, K.C., Nelson, S.O., 1998. Sensing moisture content of cereal grains with radio-frequency measurements—a review. In: Papers and Abstracts from the Third International Symposium on Humidty and Moisture, Teddington, London, England. vol. 2: pp. 195–203.

Lawrence, K.C., Nelson, S.O., Kraszewski, A.W., 1990. Temperature dependence of the dielectric properties of wheat. Trans. ASAE 33 (2), 535–540.

Lawrence, K.C., Funk, D.B., Windham, W.R., 1999. Parallel-plate moisture sensor for yellow-dent field corn. Trans. ASAE 42 (5), 1353–1357.

Lawrence, K.C., Funk, D.B., Windham, W.R., 2001. Dielectric moisture sensor for cereal grains and soybeans. Trans. ASAE 44 (6), 1691–1696.

Martin, C.R., Czuchajowska, Z., Pomeranz, Y., 1986. Aquagram standard deviations of moisture in mixtures of wet and dry corn. Cereal Chem. 63 (5), 442–445.

Martin, C.R., Czuchajowska, Z., Pomeranz, Y., 1987. Evaluation of digitally filtered aquagram signals of wet and dry corn mixtures. Cereal Chem. 64 (5), 356–358.

Matthews, J., 1963. The design of an electrical capacitance-type moisture meter for agricultural use. J. Agric. Eng. Res. 8 (1), 17–30.

Meyer, W., Schilz, W., 1980. A microwave method for density independent determination of the moisture content of solids. J. Phys. D Appl. Phys. 13, 1823–1830.

Meyer, W., Schilz, W., 1981. Feasibility study of density-independent moisture measurement with microwaves. IEEE Trans. Microw. Theory Tech. 29 (7), 732–739.

Moller, A., 1971. Measurement and control of moisture in cereals. Milling 153 (8), 24–28.

Nelson, S.O., 1965. Dielectric properties of grain and seed in the 1 to 50-mc range. Trans. ASAE 8 (1), 38–48.

Nelson, S.O., 1973. Electrical properties of agricultural products—a critical review. Trans. ASAE 16 (2), 384–400.

Nelson, S.O., 1977. Use of electrical properties for grain moisture measurement. J. Microw. Power 12 (1), 67–72.

Nelson, S.O., 1978. Radiofrequency and Microwave Dielectric Properties of Shelled Field Corn. Agricultural Research Service, USDA, ARS-S-184.

Nelson, S.O., 1979. RF and microwave dielectric properties of shelled, yellow-dent field corn. Trans. ASAE 22 (6), 1451–1457.

Nelson, S.O., 1981a. Frequency and moisture dependence of the dielectric properties of chopped pecans. Trans. ASAE 24 (6), 1573–1576.

Nelson, S.O., 1981b. Review of factors influencing the dielectric properties of cereal grains. Cereal Chem. 58 (6), 487–492.

Nelson, S.O., 1982. Factors affecting the dielectric properties of grain. Trans. ASAE 25 (4), 1045–1049, 1056.

Nelson, S.O., 1983. Observations on the density dependence of the dielectric properties of particulate materials. J. Microw. Power 18 (2), 143–152.

Nelson, S.O., 1984a. Moisture, frequency, and density dependence of the dielectric constant of shelled, yellow-dent field corn. Trans. ASAE 27 (5), 1573–1578, 1585.

Nelson, S.O., 1984b. Density dependence of the dielectric properties of wheat and whole-wheat flour. J. Microw. Power 19 (1), 55–64.

Nelson, S.O., 1987. Models for the dielectric constants of cereal grains and soybeans. J. Microw. Power Electromagn. Energy 22 (1), 35–39.

Nelson, S.O., 1992. Correlating dielectric properties of solids and particulate samples through mixture relationships. Trans. ASAE 35 (2), 625–629.

Nelson, S.O., 2005. Density–permittivity relationships for powdered and granular materials. IEEE Trans. Instrum. Meas. 54 (5), 2033–2040.

Nelson, S.O., Kraszewski, A.W., 1990. Grain moisture content determination by microwave measurements. Trans. ASAE 33 (4), 1303–1307.

Nelson, S.O., Lawrence, K.C., 1989. Evaluation of a crushing-roller conductance instrument for single-kernel corn moisture measurement. Trans. ASAE 32 (2), 737–743.

Nelson, S.O., Lawrence, K.C., 1991. Kernel moisture variation on the ear in yellow-dent field corn. Trans. ASAE 34 (2), 513–516.

Nelson, S.O., Lawrence, K.C., 1994. RF impedance and dc conductance determination of moisture in individual soybeans. Trans. ASAE 37 (1), 179–182.

Nelson, S.O., Lawrence, K.C., 1995. Nondestructive moisture determination in individual pecans by RF impedance measurements. Trans. ASAE 38 (4), 1147–1151.

Nelson, S.O., Stetson, L.E., 1976. Frequency and moisture dependence of the dielectric properties of hard red winter wheat. J. Agric. Eng. Res. 21, 181–192.

Nelson, S.O., Trabelsi, S., 2004. Principles for microwave moisture and density measurement in grain and seed. J. Microw. Power Electromagn. Energy 39 (2), 107–117.

Nelson, S.O., Trabelsi, S., 2006. Dielectric spectroscopy of wheat from 10 MHz to 1.8 GHz. Meas. Sci. Technol. 17, 2294–2298.

Nelson, S.O., Soderholm, L.H., Yung, F.D., 1953. Determining the dielectric properties of grain. Agric. Eng. 34 (9), 608–610.

Nelson, S.O., Lawrence, K.C., Kandala, C.V.K., Himmelsbach, D.S., Windham, W.R., Kraszewski, A.W., 1990a. Comparison of dc conductance, RF impedance, microwave, and NMR methods for single-kernel moisture measurement in corn. Trans. ASAE 33 (3), 893–898.

Nelson, S.O., Kandala, C.V.K., Lawrence, K.C., 1990b. Single-kernel moisture determination in peanuts by complex RF impedance measurement. Trans. ASAE 33 (4), 1308–1312.

Nelson, S.O., Lawrence, K.C., Kandala, C.V.K., 1990c. Comparison of RF impedance and dc conductance sensing for single-kernel moisture measurement in corn. Trans. ASAE 33 (2), 637–641.

Nelson, S.O., Lawrence, K.C., Kandala, C.V.K., 1991. Performance comparison of RF impedance and dc conductance measurements for single-kernel moisture determination in corn. Trans. ASAE 34 (2), 507–512.

Nelson, S.O., Lawrence, K.C., Kraszewski, A.W., 1992a. Sensing moisture content of pecans by RF impedance and microwave resonator measurements. Trans. ASAE 35 (2), 617–623.

Nelson, S.O., Lawrence, K.C., Kandala, C.V.K., 1992b. Moisture determination in single grain kernels and nuts by RF impedance measurements. IEEE Trans. Instrum. Meas. 41 (6), 1027–1031.

Nelson, S.O., Kraszewski, A.W., Kandala, C.V.K., Lawrence, K.C., 1992c. High-frequency and microwave single-kernel mositure sensors. Trans. ASAE 35 (4), 1309–1314.

Nelson, S.O., Kraszewski, A.W., Trabelsi, S., 1998. Advances in sensing grain moisture content by microwave measurements. Trans. ASAE 41 (2), 483–487.

Nelson, S.O., Trabelsi, S., Lewis, M.A., 2014. Density-Independent Microwave Moisture Sensing in Flowing Grain and Seed, ASABE Paper No. 141895707, American Society of Agricultural and Biologicical Engineers, St. Joseph, MI.

Okabe, T., Huang, M.T., Okamura, S., 1973. A new method for the measurement of grain moisture content by the use of microwaves. J. Agric. Eng. Res. 19, 59–66.

PAC, 1988. In Federal Register, Rules and Regulations. Peanut Advisory Committee, Atlanta, GA 53 (107), 20290–20306.

Schmilovitch, Z., Nelson, S.O., Kandala, C.V.K., Lawrence, K.C., 1996. Implementation of dual-frequency RF impedance technique for on-line moisture sensing in single in-shell pecans. Appl. Eng. Agric. 12 (4), 475–479.

Sokhansanj, S., Nelson, S.O., 1988. Transient dielectric properties of wheat associated with nonequilibrium kernel moisture conditions. Trans. ASAE 31 (4), 1251–1254.

Stevens, G.N., Hughes, M., 1966. Moisture meter performance in field and laboratory. J. Agric. Eng. Res. 11 (3), 210–217.

Trabelsi, S., Nelson, S.O., 1998. Density-independent functions for on-line microwave moisture meters: a general discussion. Meas. Sci. Technol. 9, 570–578.

Trabelsi, S., Nelson, S.O., 2004. Calibration methods for nondestructive microwave sensing of moisture content and bulk density of granular materials. Trans. ASAE 47 (6), 1999–2008.

Trabelsi, S., Nelson, S.O., 2006. Microwave sensing technique for nondestructive determination of bulk density and moisture content in unshelled and shelled peanuts. Trans. ASAE 49 (5), 1563–1568.

Trabelsi, S., Nelson, S.O., 2007. Unified microwave moisture sensing technique for grain and seed. Meas. Sci. Technol. 18, 997–1003.

Trabelsi, S., Kraszewski, A., Nelson, S.O., 1997. Simultaneous determination of density and water content of particulate materials by microwave sensors. Electron. Lett. 33 (10), 874–876.

Trabelsi, S., Kraszewski, A., Nelson, S.O., 1998a. Nondestructive microwave characterization for determining the bulk density and moisture content of shelled corn. Meas. Sci. Technol. 9, 1548–1556.

Trabelsi, S., Kraszewski, A., Nelson, S.O., 1998b. New density-independent calibration function for microwave sensing of moisture content in particulate materials. IEEE Trans. Instrum. Meas. 47 (3), 613–622.

Trabelsi, S., Kraszewski, A., Nelson, S.O., 1998c. A microwave method for on-line determination of bulk density and moisture content of particulate materials. IEEE Trans. Instrum. Meas. 47 (1), 127–132.

Trabelsi, S., Kraszewski, A.W., Nelson, S.O., 1999. Unified calibration method for nondestructive dielectric sensing of moisture content in granular materials. Electron. Lett. 35 (16), 1346–1347.

Trabelsi, S., Kraszewski, A., Nelson, S.O., 2001a. New calibration technique for microwave moisture sensors. IEEE Trans. Instrum. Meas. 50, 877–881.

Trabelsi, S., Kraszewski, A.W., Nelson, S.O., 2001b. Microwave dielectric sensing of bulk density of granular materials. Meas. Sci. Technol. 12, 2192–2197.

Trabelsi, S., Nelson, S.O., Kraszewski, A.W., 2001c. Universal calibration for microwave moisture sensors for granular materials. In: Proceedings of the 18th IEEE Instrumentation and Measurement Technology Conference, Budapest, Hungary, published by IEEE, Piscataway, New Jersey. vol.3: pp. 1808–1813.

Trabelsi, S., Lewis, M.A., Nelson, S.O., 2010. Microwave Moisture Meter for Rapid and Nondestructive Grading of Peanuts, ASABE Paper No. 1009183, American Society of Agricultural and Biological Engineers, St. Joseph, MI.

USDA, 1963. Comparison of Various Moisture Meters with the Oven Method in Determining Moisture Content of Grain, AMS-511, p. 11.

Watson, C.A., Greenway, W.T., Davis, G., McGinty, R.J., 1979. Rapid proximate method for determining moisture content in single kernels of corn. Cereal Chem. 56 (3), 137–140.

Woodroof, J.G., 1967. Tree-Nuts—Production, Processing, Products. AVI Publishing Co., Inc., Westport, CT.

Zeleny, L., 1954. Methods for grain moisture measurement. Agric. Eng. 35 (4), 252–256.

Zeleny, L., 1960. Moisture measurement in the grain industry. Cereal Sci. Today 5 (5), 130–136.

ASSESSMENT OF SOIL TREATMENT FOR PEST CONTROL

8

Over the past 50 years, the use of microwave energy has been proposed frequently as an alternative method for controlling pests in the soil, such as weed seeds, insects, nematodes, and soil-borne plant pathogens. Even earlier, radio frequency (RF) dielectric heating was considered for sterilizing soils for use in greenhouses. Since the question of the practicability of using microwave energy for soil treatment to control pests has been raised so often, it seems worthwhile to examine what is known about such applications and to offer an assessment of its potential.

8.1 SOIL MICROORGANISMS AND NEMATODES

Early studies with 27-MHz, high-frequency dielectric heating of different greenhouse soils showed that damping-off fungi infecting crimson clover could be controlled by 5-min exposures that raised soil temperatures to 86–101°C without destroying the nitrogen-fixing *Rhizobium* bacteria necessary for normal plant growth (Eglitis et al., 1956; Eglitis and Johnson, 1970). Experiments with 2450-MHz microwave heating of soils (Baker and Fuller, 1969) revealed large differences in the heating rates of soils and problems with uniformity of treatment. It was concluded that commercial microwave treatment of soils for control of soil-borne pathogenic fungi was impractical. In other work (Barker et al., 1972), for similar reasons, it was concluded that 2450-MHz treatments of nematode-infested soil samples had poor prospects for becoming a practical means of nematode control. The root-knot nematode was controlled in small samples of potting soil exposed to 2450-MHz energy in a microwave oven when temperatures lethal to the nematodes were achieved (O'Bannon and Good, 1971). In field tests with a 30-kW, 2450-MHz microwave source applying energy at 800 J/cm^2, Heald et al. (1974) reported that nematodes were controlled in a fine sandy loam soil infested with the reniform nematode at depths of 5 cm, but that the nematodes survived at depths of 10 and 15 cm.

Vela et al. (1976) reported, after conducting both laboratory and field studies, that soil microorganisms survived 2450-MHz exposures at much higher dosage levels than those required for control of weed seeds in soil. Later studies with other soil samples containing additional microorganisms showed that exposures of small samples in a microwave oven could be useful for sterilization, but that results depended on treatment time, amount of soil treated, and soil water content (Ferriss, 1984). Large-scale applications were judged unlikely. Other tests (van Wambeke et al., 1983; Benz et al., 1984) showed that, for soil samples treated in a microwave oven, seeds, fungi, and nematodes could all be controlled by exposures of a few minutes, but that the exposures required

Dielectric Properties of Agricultural Materials and Their Applications. DOI: http://dx.doi.org/10.1016/B978-0-12-802305-1.00008-7

depended on soil type, exposure period, depth in soil, and soil moisture content. Treatments of different soils at 2450 MHz were effective in controlling several plant pathogens, but penetration depths were reduced as soil moisture content increased (Van Assche and Uyttebroek, 1983). Soil-borne pathogens in mushroom casing and rock wool substrates were controlled by passing the materials through a 20-kW dielectric heater operating at 27.12 MHz (Diprose and Evans, 1988).

8.2 SOIL TREATMENT FOR WEED CONTROL

Interest in weed control by soil treatment with microwave energy in the early 1970s, and the hope of commercialization, stimulated many experiments in the laboratory and in the field on the lethality of such exposures for various kinds of seed. Davis et al. (1971, 1973) and Wayland et al. (1972) reported the germination of seeds of several crop and weed species after exposure in a microwave oven operating at 2450 MHz. Seeds were treated in a dry state and after several hours of water imbibition. They were found much more susceptible to damage from microwave heating in the imbibed states. Energy levels in joules per gram were given for the various treatments, but since they were based on energy absorbed by 50 mL of water in the same microwave oven for comparable exposure times, the data likely have little meaning except for relative energy absorption comparisons.

Wayland et al. (1973) irradiated wheat and radish seeds in paper envelopes 2.5 cm below the soil surface with 2450-MHz microwave energy from a 1.5-kW magnetron directed at the soil surface through an applicator at a reported level of 210 J/cm^2. They concluded that, for the power levels used, energy density and time of exposure were interchangeable, with respect to effectiveness, as long as total energy remained about the same. In field experiments with a 2450-MHz power applicator, seeds of several weed species were planted in the top 2 cm of irrigated and nonirrigated soils, and control was achieved for various species at energy densities of 180 and 360 J/cm^2 (Menges and Wayland, 1974). In other field experiments with a mobile microwave power unit consisting of four 1.5-kW magnetrons powered by a gasoline-engine-driven 60-Hz electric generator, preemergence and postemergence microwave treatments were administered to plots seeded with several weed species (Wayland et al., 1975). Energy densities of 183 J/cm^2 were required to provide 80–90% control in preemergence tests, while energy densities of 77–309 J/cm^2 were required in postemergence trials. Field experiments with the same microwave equipment were also conducted on plots with three replications for evaluation of weed seed, soil fungi, and nematode control (Cundiff et al., 1974). Treatments on dry sandy loam soil at energy densities of 633–1727 J/cm^2 failed to provide effective control of any of the pest organisms. No significant reductions in the presence of any of these pests resulted from even the highest energy-density microwave treatments.

In entirely independent laboratory studies, Hightower et al. (1974), using a horn antenna to irradiate seeds placed on top of soil samples, found that energy densities of the levels reported earlier (Wayland et al., 1973) produced no reductions in seed germination and that only slight reductions were produced by treatments at 300 J/cm^2. More detailed tests were then conducted with fescue grass seed, both dry and water soaked, on and under the soil surface, in samples of three different soils. The soil samples were irradiated with a dielectric-loaded waveguide applicator for improved impedance matching, higher power densities, and a more uniform radiation pattern. The results showed that energy densities of at least 1500 J/cm^2 were required for control of seed germination, that moist seeds in wet soil were more susceptible, but that the high energy requirement rendered the proposed application impractical.

In other laboratory tests on wild oats seeds exposed in glass test tubes inserted into a waveguide attached to a 1.5-kW, 2450-MHz source, moist seeds and seeds in soil were damaged more than dry seeds (Lai and Reed, 1980). Because of the high energy requirements, high costs of equipment, and low travel speed, field use was judged impractical. Olsen (1975) conducted an interesting theoretical analysis, based on physical principles and available data for the necessary properties of soils and seeds, and concluded that, for unimbibed seeds in a mineral soil, an energy density of at least 800 J/cm^2 would be required for germination control. An optimistic review of the reported work on microwave soil treatment for pest control was presented by Vela-Muzquiz (1983), indicating promise for ultrahigh frequency (UHF) radiation control of agricultural pests and at costs similar to those for chemical pest control. However, the viewpoint appears to be tempered by a strong concern for the environmental influence of continued use of chemical controls and an apparent naivete with respect to physical principles of electromagnetic energy absorption.

8.3 INITIAL ASSESSMENT

When the author was first contacted about the possibility of using microwave energy for devitalizing weed seeds in soil about 45 years ago, the innate response was that it would be impractical. Some simple calculations confirmed that impression. They were based on the assumption that the lethal mechanism was thermal, that selective dielectric heating was unlikely to be a major factor, and that it would probably be necessary to raise seed temperature to about 100°C for a short period to completely inhibit germination. These assumptions are still most likely reasonable. Calculating the energy required to raise the top 2 in of soil from 25°C to 100°C, assuming a soil bulk density of 1.6 g/cm^3 and a specific heat of 0.3 cal/(g°C) (1256 J/[kg K] in SI units), gives 766 J/cm^2. Converting this to a more familiar electric energy unit and a more practical scale, gives 8611 kW h/acre (21,277 kW h/ha). Because conversion of 60-Hz electric energy to microwave energy absorbed in the soil is only about 50% efficient in a field application, the electric energy requirement is about 17 MW h/acre. At a cost of $0.05/kW h, this would amount to $850/acre just for the electric energy alone. Costs for the high-power microwave equipment and gasoline- or diesel-driven electric generating equipment and other equipment required would have to be added to these costs to estimate a total cost per acre for microwave soil treatment. Examining the assumptions upon which the above estimates are based, there seems to be little opportunity for major reductions in either required energy or costs. In fact, the microwave energy cannot be confined to the upper 2 in, so to achieve the heating necessary, more energy would be required. Neither are the weed seeds confined to the 2-in surface layer, so a larger amount of soil would need to be raised to the 100°C level for effective control. The temperature level of 100°C is much higher than the 60°C level mentioned for soil temperatures in field trials (Wayland et al., 1973); however, Menges and Wayland (1974) did achieve such high temperatures near the soil surface in some trials. Treatment of soil samples containing oat, *Avena sativa* L., seed and seeds of indigenous weeds for 2 min or longer in a microwave oven operating at 2450 MHz produced soil temperatures of 90°C and reduced weed emergence to low levels (Barker and Craker, 1991).

Early work on determining thermal death points of weed seeds representing seven species and five families showed that seeds subjected to heat treatments for 15 min in sealed brass tubes immersed in an oil bath required temperatures varying from 85°C to 105°C for complete lethality (Hopkins, 1936). Exposures to high temperatures in soil for seeds of several weed species required 7 days at 70°C for control in dry soil, and some seeds survived several days at 60°C and 70°C even in moist soils (Egley, 1990). At much shorter exposure times to elevated temperatures, which are certainly required for any practical field treatment, these seeds could survive much higher temperatures.

Susceptibility of plant seeds to damage by dielectric heating exposures is heavily dependent on the moisture content of the seed at the time of exposure. Dry seed can tolerate higher temperature exposures than seed with higher moisture contents. Samples of hard red winter wheat exposed for 4–37 s to 39-MHz dielectric heating treatments had 50% reduction in seed germination at seed temperatures ranging from 65°C to 109°C with seed moisture content ranging from 18.3% to 6.7% (wet basis) (Nelson and Walker, 1961). Seed of several vegetables survived dielectric heating exposures that raised seed temperatures to well over 80°C and 90°C (Nelson et al., 1970). The importance of seed moisture content as it relates to responses of alfalfa seed, other small-seeded legumes, and seeds of woody plant species to dielectric heating exposures that raised temperatures well over the 80°C and 90°C levels is well documented (Nelson and Wolf, 1964; Stetson and Nelson, 1972a; Nelson et al., 1976, 1977, 1978, 1982). Dielectric heating treatments at 40 and 2450 MHz were found to be equally effective for increasing germination of alfalfa by lowering hard-seed contents (Stetson and Nelson, 1972); so the dielectric heating effects at the two different frequencies can be considered nearly equivalent. In all of this research, no responses were observed that could not be attributed to thermal causes. Consequently, the speculated "special effects" of UHF electromagnetic energy on seeds are not likely to provide the selectivity necessary for major improvement in the efficiency of microwave treatment for weed seed control in the soil.

8.4 BASIC PRINCIPLES

In considering the interaction of electromagnetic energy with matter, the electric permittivities or dielectric properties of the materials involved are of the utmost importance, as explained in Section 3.1. The dielectric properties or permittivities of the soil and the biologic materials in the soil, along with characteristics of the electromagnetic waves and physical properties of the soil and embedded organisms, determine the dissipation of electric energy in the materials by conversion of electric energy to heat energy in the materials. This is the phenomenon commonly referred to as dielectric heating, or microwave heating if microwave frequencies are used, and it is governed by Eqs (3.1) and (3.2). They are, respectively,

$$P = E^2 \sigma = 55.63 \times 10^{-12} f E^2 \varepsilon'' \tag{8.1}$$

$$\frac{dT}{dt} = P/(c\rho) \tag{8.2}$$

where P is the power dissipated per unit volume in the material, E is the electric field intensity in the material in volts/m (V/m), σ is the ac electrical conductivity associated with the dielectric loss in the material, f is the frequency of the alternating electric field in Hz, and ε'' is the dielectric loss factor of the material. Power dissipated over a period of time provides energy to raise the temperature of the material, and this time, t, rate of temperature, T, increase (°C/s) is given by Eq. (8.2), where c is the specific heat of the material in kJ/(kg°C), and ρ is its density (kg/m^3). If water is evaporated in the heating process, the energy required for the vaporization and release of the water must also be taken into account, and the temperature rise would be reduced accordingly.

The absorption of microwave energy propagating through a material depends upon the variables of Eq. (8.1). Thus, the dielectric loss factor of the material is important. The frequency of the wave is also a factor, and the power absorption also depends on the square of the electric field intensity. For a plane wave, traveling in the material, the electric field intensity E will vary as it travels through the material, as explained in Section 3.1. As the wave travels through a material that has a significant dielectric loss, its energy will be attenuated.

8.5 FURTHER ASSESSMENT

8.5.1 ATTENUATION

The dielectric properties of the soil are very important in evaluating the penetration of energy that can be achieved. The attenuation in decibels is provided by Eq. (3.10) when $(\varepsilon'')^2 \ll (\varepsilon')^2$. Thus, because ε'' for normal soils is relatively small with respect to ε', Eq. (3.10) can be used to provide good estimates of attenuation in soils if we know the values for the dielectric properties at the frequency of interest. Values for these properties have been extracted from some of the data in the literature on a few types of soil and are presented in Table 8.1.

Much of the data were presented graphically, some as permittivity values as functions of gravimetric moisture content (mass of water in percent of soil dry weight) and some in volumetric moisture (mass of water per unit volume of soil). Therefore, the data were converted to the dry-weight moisture content on the gravimetric basis, making an assumption for the likely density of the dry soil. Other data were presented in terms of the dielectric constant and attenuation in dB/cm and in terms of the dielectric constant and electrical conductivity. Values of ε'' were then calculated by the use of Eq. (3.10) or as $\varepsilon'' = \sigma/(\omega\varepsilon_0)$ for the appropriate frequencies.

Although the soils listed in Table 8.1 are all different, attenuation increases with moisture content of the soils and, in general, with frequency, as expected. However, soils are extremely variable in their makeup and properties, and data in Table 8.1 represent a very limited sampling. For plant growth, the available water, the range between field capacity and wilting point, varies greatly. For soils, these moisture contents can range from less than 4% and 2%, respectively, for a sandy soil to greater than 45% and 30%, respectively, for a clay soil (Perkins, 1987). For 10−20% moisture soils in Table 8.1, attenuation varies from 1.3 to 4.3 dB/cm. Nevertheless, it is instructive to consider a reasonably conservative attenuation of 2 dB/cm and examine the penetration characteristics for microwave energy incident upon the soil surface. A plane wave incident upon the soil surface will

Table 8.1 Permittivity, $\varepsilon' - j\varepsilon''$, and Attenuation, α, (dB/cm) Values for Soils at Indicated Microwave Frequencies and Dry-Basis Moisture Contents

Reference	Soil Texture	Frequency, GHz	Dry-Basis Soil Moisture Content 5%			10%			20%			25%		
			ε'	ε''	α	ε'	ε''	α	ε'	ε''	α	ε'	ε''	α
Hoekstra and Delaney (1974)	Clay	5.0	4.3	0.7	1.2	6.3	1.6	2.3	14.8	4.4	4.3	—	—	—
Hipp (1974)	Clay loam	2.5	5.4	0.8	0.8	8.8	1.7	1.3	16.0	3.7	2.1	—	—	—
Jesch (1978)	Sandy loam	2.0	4.0	1.0	0.9	6.6	2.5	1.8	16.4	7.1	3.2	22	9.4	3.6
		4.0	4.0	0.8	1.4	6.2	1.8	2.6	13.2	3.9	3.9	17	5.0	4.4
	Silt loam	2.0	3.4	0.6	0.6	5.6	2.4	1.8	—	—	—	—	—	—
		4.0	3.3	0.6	1.2	5.3	1.5	2.4	—	—	—	—	—	—
Ulaby et al. (1982)	Sandy loam	5.0	5.0	0.5	1.0	9.0	1.5	2.3	18.0	3.5	3.8	24	5	4.6

have some of the power reflected, and the rest, P_t, will be transmitted into the soil. The relationship is given by the following expression:

$$P_t = P_0(1 - |\Gamma|^2) \tag{8.3}$$

where P_0 is the incident power and Γ is the reflection coefficient. For an air–soil interface, the reflection coefficient can be expressed in terms of the complex relative permittivity of the soil as (Stratton, 1941):

$$\Gamma = \frac{1 - \sqrt{\varepsilon}}{1 + \sqrt{\varepsilon}} \tag{8.4}$$

The power density diminishes as an exponential function of the attenuation and distance traveled (Eq. (3.8)) as the wave propagates through the soil:

$$P = P_t\, e^{-2\alpha z} \tag{8.5}$$

with α expressed in nepers/m. For attenuation in decibels, dB/cm = 0.08686 × (nepers/m).

Considering again an attenuation of 2 dB/cm in soil, the attenuation at a 5-cm depth (about 2 in) will be 10 dB, which corresponds to a power density of just 10% of that at the surface. At a 4-in (10-cm) depth (20 dB attenuation), there would be only 1% of the power left, and at 15 cm (6 in) it would be reduced by 30 dB to only 0.1%. Because the soil heating is directly proportional to the power density, this means that, for heating as effective at a 2-in (5-cm) depth as at the surface, 10 times as much power would be required, at 4 in (10 cm), 100 times as much power would be needed, and at 6 in (15 cm), the ratio would be 1000:1. Thus, the attenuation problem alone renders the use of microwave soil treatment impractical. Even in an extremely dry soil, if attenuation were as low as 0.6 dB/cm, two times the power needed for surface soil treatment would be required for effective treatment at 2 in (5 cm) in depth, and four times as much power would be required to deliver the necessary energy at 4 in (10 cm).

8.5.2 SELECTIVE HEATING

If selective microwave heating of seeds relative to the soil were possible, the case for microwave control of weed seed germination would be improved. Therefore, consider the variables in Eqs (8.1) and (8.2) with respect to the relative heating effects of the microwave energy on the seeds and the soil. The frequency for the two materials will be the same. However, the dielectric loss factor for the seeds and the soil might be different, and in that instance, the electric field intensities in the seeds and the soil might also differ. Dielectric properties of a few crop and weed seed samples measured by the short-circuited waveguide technique (Nelson, 1972a, 1973) at frequencies of 1, 2.45, and 10 GHz are listed in Table 8.2.

All the samples were conditioned to equilibrium moisture content at 24°C and 40% relative humidity, and the moisture contents at the time of measurement were determined by drying samples for 24 h in a forced-air oven at 103°C. Bulk densities of the seed (seed and air–space mixtures) were those measured in the dielectric sample holders at the time of the permittivity measurements. Comparing the loss factors of the seed samples at 2.45 GHz with those of soils shown in Table 8.1, there appears to be little likelihood of selectively heating the seeds. In fact, on the basis of loss factor values only, the soils would be expected to heat more rapidly than the seeds. However, we must

Table 8.2 Microwave Permittivities, $\varepsilon' - j\varepsilon''$, of Weed and Crop Seed Samples at Indicated Wet-Basis Moisture Contents and Bulk Densities

Kind of Seed	Moisture Content, %	Density, g/cm³	Dielectric Constant, ε', and Loss Factor, ε'', at Indicated Frequencies, GHz					
			1		2.45		10	
			ε'	ε''	ε'	ε''	ε'	ε''
Wheat, *Triticum aestivum* L.	8.8	0.77	2.6	0.24	2.5	0.20	2.5	0.17
Grain sorghum, *Sorghum bicolor* L. Moench	8.7	0.74	2.6	0.26	2.6	0.25	2.5	0.27
White mustard, *Brassica hirta* Moench	6.7	0.70	2.2	0.11	2.2	0.10	2.2	0.09
Pigweed, *Amaranthus retroflexlus* L.	10.2	0.87	3.1	0.38	3.0	0.30	2.8	0.23
Curly dock, *Rumex crispus* L.	10.0	0.78	2.8	0.28	2.8	0.27	2.6	0.20
Junglerice, *Echinochloa colonum* (L.) Link	8.4	0.29	1.5	0.05	1.5	0.05	1.4	0.06

also consider the influence of the E^2 term in possible differential heating of the seeds and the soil. Also, the ε'' values in Table 8.2 are for the air–seed mixtures, and perhaps it would be more realistic to consider the dielectric properties of the seeds themselves embedded in a uniform medium with the dielectric properties of the soils as given in Table 8.1.

Limited data are available on the dielectric properties of wheat kernels at 9.4 GHz (Nelson, 1976) and corn and wheat kernels and soybeans at 11.5 and 22 GHz (Nelson and You, 1989). From these data, and the known frequency dependence of the dielectric properties of bulk samples of wheat (Nelson, 1982), it appears that the relative permittivity of the wheat kernel at 2.45 GHz will be about $5 - j0.5$. This loss-factor value, 0.5, is less than those of the soils listed in Table 8.1 for similar frequencies. To examine the relative electric field intensities in the seed, E_s and that in the soil medium, E_m, we can consider that, for a plane wave interacting with a spherical seed in a uniform, infinite soil medium, the electric field in the seed is (Stratton, 1941):

$$E_s = E_m \left(\frac{3\varepsilon_m}{2\varepsilon_m + \varepsilon_s} \right) \tag{8.6}$$

where ε_s and ε_m represent the complex relative permittivities of the seed and the soil medium, respectively. If we take the permittivity of the wheat kernel for a "spherical" seed and the values of the clay loam soil at 2.5 GHz and 10% moisture from Table 8.1 for the medium, we obtain a value of 1.37 for $(E_s/E_m)^2$, which gives us the ratio of the electric field contribution to the power dissipation per unit volume in the seed to that in the soil. However, this tendency for selectively heating the seed is offset by the lower $\varepsilon_s/\varepsilon_m$ ratio of 0.29. If we take the 2.0 GHz, 10% moisture sandy loam permittivity data from Table 8.1 as another example, we get 1.22 and 0.20 for the $(E_s/E_m)^2$ and $\varepsilon_s/\varepsilon_m$ ratios, respectively. The product of these two figures gives the estimated power absorption ratio for the seed in relation to the soil, and it would be 0.40 and 0.22 for the two soils respectively. This power absorption ratio for the same two soils at 20% moisture content would be

0.24 and 0.12, respectively. Thus, selective absorption of the microwave power by the seeds seems most unlikely to occur. For selective dielectric heating of the seeds, one must also consider the other two variables that, along with power absorbed, affect the heating rate in the two different materials (Eq. (8.2)). The specific heat of wheat at 10% moisture content is about 0.39 cal/(g°C) (1.65 J/[kg K]) (ASAE, 1984). The specific heat of dry mineral soils is about 0.20 cal/(g°C) (0.84 J/[kg K]), while at 20% and 30% moisture, the specific heat rises to 0.33 and 0.38 cal/(g°C), respectively (Lyon and Buckman, 1947). Estimating the specific heat for 10% moisture soil as 0.28 cal/(g°C) (1.17 J/[kg K]), the seed-to-soil specific heat ratio would be 0.39/0.28 = 1.39.

The density ratio is the other factor for consideration. Densities of soils vary, but taking an intermediate value of 1.6 g/cm^3, which was the density of the clay loam soil in Table 8.1, and 1.4 g/cm^3 for the density of the hard red winter wheat kernel at 10% moisture content (Nelson et al., 1976; Nelson and You, 1989), the seed-to-soil density ratio is about 0.88. Both the seed-to-soil specific heat and density ratios have an inverse influence, since c and ρ appear in the denominator of the right-hand side of Eq. (8.2), and the product of these ratios is 1.39 × 0.88 = 1.22. The reciprocal value is 0.82, which taken times the 0.40 and 0.22 power absorption ratios for 10% moisture soils, based on loss factor and electric field considerations, gives 0.33 and 0.18. This clearly indicates that selective dielectric heating of the seed, based on these dielectric properties, cannot be expected. Differences in the seed and soil moisture contents can change the values of all variables on which these estimates were based. In particular, if seeds take up moisture from the soil and begin to germinate, they may be more susceptible to heat damage. However, many seeds have moisture-impermeable seed coats, and they will not take up moisture until the seed coat permeability changes (Ballard et al., 1976; Nelson et al., 1976). Many weed seeds have dormancies of other types and do not germinate until certain conditions have been satisfied. Even without dormancy, it is not likely that most of the weed seeds would be in a susceptible stage at any given time for microwave heating to be effective.

8.5.3 SOIL INSECT AND NEMATODE TREATMENT

The same principles can be applied for assessment of the potential for microwave energy in controlling insects and nematodes in soil. The same problems prevail with attenuation of the energy as the waves travel into the soil. Less information is available on the dielectric properties of insects, and none has been noted on those properties of nematodes. Insects can be expected to have higher natural moisture contents than seeds, although little is known to the author on water content of soil-infesting insects in various stages of development. In work with adult rice weevils, *Sitophilus oryzae* L., which infest cereal grains, moisture content was determined to be about 49%, wet basis, and insect density was 1.29 g/cm^3 (Nelson, 1972b). Adult rice weevil permittivity was determined to be $32 - j13$ at 9.4 GHz (Nelson, 1976). Using Eqs (8.1) and (8.6), and arguments similar to those in the previous section, the insects treated in hard red winter wheat were predicted to absorb RF power at levels about 2–3.5 times greater than the wheat at frequencies in the 1- to 100-MHz range (Nelson and Charity, 1972). At the microwave frequency of 2.45 GHz, however, the power dissipation ratio dropped to about 1.1. These predictions were confirmed in laboratory exposures of rice-weevil-infested wheat to 39-MHz and 2.45-GHz dielectric heating, in which complete insect mortality was achieved by treatments raising grain temperatures briefly to 40°C for the 39-MHz treatments and to 80°C for the 2.45-GHz treatments (Nelson and Stetson, 1974). If the insect permittivity values, $32 - j13$, are used with the same soil permittivity data used in the previous section for the selective heating calculations with Eqs (8.1) and (8.2), insect-to-soil power absorption ratios of 1.99 and 1.86 are obtained for the 10%-moisture clay loam soil and the 10%-moisture sandy loam soil, respectively. In the 20% moisture soils, the respective power dissipation ratios are 0.99 and 1.04. These estimates indicate the possibility for some selective dielectric heating of insects in very dry soils. Specific heat of insects may be somewhat higher than that of the soils, because of their greater moisture content, and this would tend to reduce the heating rate of the insects with respect to the soil. The insect-to-soil density ratio, $1.29/1.6 = 0.81$ in this instance, would tend to provide a slight increase for the insect heating rate. These factors, all taken together, do not indicate any likelihood of a very significant differential heating advantage for insect control, except possibly in a very dry soil. The experimental results cited earlier for nematodes indicated control at a 5-cm depth in a fine sandy loam soil with a reported applied energy of 800 J/cm², but the nematodes survived at 10- and 15-cm soil depths, which might be expected because of microwave power attenuation in the soil.

8.6 DISCUSSION

There are still many unknown factors that can influence the absorption of microwave energy transmitted into the soil. As the soil absorbs the energy, its temperature rises. The dielectric properties of all materials are temperature dependent. The dielectric behavior of hygroscopic materials, with respect to change in temperature, depends on the frequency used and the nature of the water in the material. If water is bound, both the real and imaginary parts of the permittivity can be expected to increase with increasing temperature. If there is any significant amount of free water in the soil, its dielectric behavior will be influenced by the dielectric relaxation of free water, and that relaxation frequency, which is about 19 GHz at 25°C, shifts to higher frequencies as temperature increases. This would tend to increase the dielectric constant slightly, but the overall influence of increasing temperature would reduce the dielectric constant and would also tend to lower the loss factor value. At lower frequencies in the microwave range, the soil chemistry could exert an influence due to the effects of ionic conduction, which disappear at frequencies around 10 GHz (Hasted, 1973). At this time, there are no obvious temperature-induced changes in the dielectric properties that would significantly improve the opportunity for selective heating of the pest organisms with respect to the host soil materials.

The attenuation of energy with wave penetration into the soil appears to be a very limiting factor. Useful penetration would appear possible only in soils with a minimum amount of water, and even then, large amounts of power would be required to raise the soil temperature rapidly to levels effective in controlling pest organisms. The general order of susceptibility, in decreasing order, based on available experimental data and principles considered in this chapter, appears to be insects, weed seeds, nematodes, soil fungi, and soil bacteria. Because the beneficial effects of soil microbes are extremely important, their lower susceptibility to population reduction from microwave heating is fortunate. However, harmful plant pathogens would also benefit from this lower susceptibility. The large variation in soil characteristics, including texture, organic matter content, soil chemistry, and subsequent variation in density and specific heat, as well as the variations among the living organisms in the soil, can be expected to introduce large degrees of variation in the survival of these organisms when exposed in soil to microwave energy.

No dynamic conditions have been considered in the principles presented here. Obviously, heat transfer and diffusion of energy would come into play as soon as the temperature equilibrium is upset by microwave heating. Because of attenuation, large amounts of energy absorbed in the first few centimeters of penetration would result in heat conduction to deeper soil levels, but heat would also be lost to the atmosphere, and this would not appear to be an efficient process for heating the soil to the deeper levels required for control of pests. The costs of equipment for field application of microwave power have not been considered here because the basic assessment has not appeared to warrant the effort. However, at least two prototype field microwave power applicators were built in the 1970s and tested for field application (Wayland et al., 1973, 1978; Cundiff et al., 1974; Anonymous, 1973; Davis, 1975). An early prototype was equipped with four 1.5-kW, 2.45-GHz magnetrons, with movement down to 0.003 mile/h (0.005 km/h) provided by a cable winch, and with electric power supplied by a 20-kW gasoline-engine-driven generator (Cundiff et al., 1974). A later prototype had two 30-kW klystron microwave sources, and was powered by 155-kW diesel-operated generator (Anonymous, 1973). Apparently, the application was not deemed practical because the anticipated marketing of the method and the equipment did not materialize. To deliver enough energy for effective control of pests in a time compatible with field operations, very large power capabilities would be needed. For example, if 1500 J/cm^2 were the required energy for an application, to treat a band 1 m wide traveling at a speed of 1 km/h, the microwave power delivery to the soil would be 4167 kW. Even if the energy were applied in a narrow band of 10-cm width along the row in which crops were to be seeded, at least a 417-kW power output would be needed. A figure often used for estimating costs of microwave power source and applicator equipment for industrial use was $3/W ($3000/kW) of output rating. Economies of the sort that have brought down the costs of fractional-kilowatt magnetrons for domestic microwave oven use and low power magnetrons for industrial use have a long way to go for higher-power microwave sources.

Another aspect of microwave energy applications at the extremely high-power levels that would be required for soil pest control is that of safety for operating personnel. Since some energy is reflected by the soil surface, proper shielding design would be needed to insure that energy radiated to surroundings is maintained below safe levels for human exposure.

For application to the sterilization of greenhouse or potting soils, or similar uses, where quantities of material to be treated are lower, and efficient application of electromagnetic energy would be easier to achieve, potential practical use could be more likely. However, high-frequency or microwave dielectric heating would have to offer important advantages to justify costs of treatment. The saving of time would be the principal advantage apparent today, but other advantages may be important in particular applications. The often-speculated "nonthermal" effects of microwaves on living organisms have yet to be demonstrated convincingly for any useful pest-control purpose. The lethal mechanisms appear to be thermal in nature, and in many instances, differential or selective dielectric heating can account for observed results attributed to "nonthermal biological effects" (Johnson and Guy, 1972; Schwan, 1972; Stuchly, 1979, 1995; Polk and Postow, 1986; Michaelson and Lin, 1987). Even when conditions support the phenomenon of selective dielectric heating, it cannot be depended upon to be especially significant because of the rapid conduction of heat energy from the target organism into the host medium. Thus, the organisms cool to lower temperatures rapidly, and the benefit of the complementary time−temperature action is not retained.

8.7 CONCLUSIONS

Upon considering the basic principles of microwave energy absorption by dielectric materials and the experimental work that has been reported, there appears to be little probability for the practical application of microwave power for field use in controlling pests in the soil, such as insects, nematodes, weed seeds, and plant pathogens. The susceptibility to control by microwave heating of pests in soil, in decreasing order, appears to be insects, weed seeds, nematodes, fungi, and bacteria. Unless some nonthermal lethal effects are discovered that can be utilized, the energy and equipment costs for producing pest mortality or population reductions by thermal heating are far too great for serious consideration. In addition to the energy costs, the rapid attenuation of microwave energy to insignificant levels at shallow soil depths makes potential use of this form of energy impractical. Selective dielectric heating of pest organisms, because of differences in microwave permittivities of these organisms and those of the soil, appears to be highly unlikely, except possibly for insects in very dry soils. Thus, any serious future consideration of microwave electromagnetic energy for control of pests in soil must be subjected to careful and critical analysis and is most unlikely to be successful.

REFERENCES

Anonymous, 1973. Electronic weed "Zapper". Farm Industry News 6 (7), 27.

ASAE, D243.3, 1984. Thermal Properties of Grain and Grain Products. ASAE Standards American Society of Agricultural Engineers, St. Joseph, MI.

Baker, K.F., Fuller, H., 1969. Soil treatment by microwave energy to destroy plant pathogens. Phytopathology 59 (1), 193–197.

Ballard, L.A.T., Nelson, S.O., Buchwald, T., Stetson, L.E., 1976. Effects of radiofrequency electric fields on permeability to water of some legume seeds, with special reference to strophiolar conduction. Seed Sci. Technol. 4, 257–274.

Barker, K.R., Craker, L.E., 1991. Inhibition of weed seed germination by microwaves. Agron. J. 83, 302–305.

Barker, K.R., Gooding Jr., G.V., Eldre, A.S., Eplee, R.E., 1972. Killing and preserving nematodes in soil samples with chemicals and microwave energy. J. Nematol. 4 (2), 75–79.

Benz, W., Moosmann, A., Walter, H., Koch, W., 1984. Wirkung einer Bodenbenbehandlung mit Mikrowellen auf Unkrautsamen, Nematoden und bodenburtige Microorganismen. Mitteil. Biologish. Bunoesanst. Land Forstvirtsch 223, 128.

Cundiff, J.S., Johnson, A.W., Flowers, R.A., Glaze, N.C., 1974. Evaluation of Microwave Treatment of Field Soil for Control of Nematodes, Soil Fungi and Weeds, ASAE Paper No. 74-1562. American Society of Agricultural Engineers, St. Joseph, MI.

Davis, F.S., 1975. "Zapper" blasts weed seeds. N.Z. J. Agric. 131 (3), 53–54.

Davis, F.S., Wayland, J.R., Merkle, M.G., 1971. Ultrahigh-frequency electromagnetic fields for weed control: phytotoxicity and selectivity. Science 173, 535–537.

Davis, F.S., Wayland, J.R., Merkle, M.G., 1973. Phytotoxicity of a UHF field. Nature 241, 291–292.

Diprose, M.F., Evans, G.H., 1988. Soil partial sterilization by dielectric heating. In: Eng. Adv. Agric. Food Proc 1938–1988 Jubilee Conf., Inst. Agricultural Engineers, London, UK, pp. 363–364.

Egley, G.H., 1990. High-temperature effects on germination and survival of weed seeds in soil. Weed Sci. 38 (4–5), 429–435.

Eglitis, M., Johnson, F., 1970. Control of seedling damping off in greenhouse soils by radio frequency energy. Plant Dis. Rep. 54 (1), 268–271.

Eglitis, M., Johnson, F., Breakey Jr., E.P., 1956. Soil pasteurization with high frequency energy. Phytopathology 46, 635–636.

Ferriss, R.S., 1984. Effects of microwave oven treatment on microorganisms in soil. Phytopathology 74 (1), 121–126.

Hasted, J.B., 1973. Aqueous Dielectrics. Chapman and Hall, London.

Heald, C.M., Menges, R.M., Wayland, J.R., 1974. Efficacy of ultra-high frequency (UHF) electromagnetic energy and soil fumigation on the control of the reniform nematode and common purslane among southern peas. Plant Dis. Rep. 58 (11), 985–987.

Hightower, N.C., Burdette, E.C., Burns, C.P., 1974. Investigation of the Use of Microwave Energy for Weed Seed, Final Technical Report, Project E, 230-901. Georgia Institute of Technology, Atlanta, GA.

Hipp, J.E., 1974. Soil electromagnetic parameters as functions of frequency, soil density, and soil moisture. Proc. IEEE 62 (1), 98–103.

Hoekstra, P., Delaney, A., 1974. Dielectric properties of soils at UHF and microwave frequencies. J. Geophys. Res. 79 (11), 1699–1708.

Hopkins, C.Y., 1936. Thermal deathpoint of certain weed seeds. Can. J. Res. 14, 178–183.

Jesch, R.L. 1978. Dielectric Measurements of Five Different Soil Textural Types as Functions of Frequency and Moisture Content. NBSIR 78-896, Washington, DC, U.S. Department of Commerce, National Bureau of Standards.

Johnson, C.C., Guy, A.W., 1972. Nonionizing electromagnetic wave effects in biological materials and systems. Proc. IEEE 60 (6), 692–718.

Lai, R., Reed, W.B., 1980. The effect of microwave energy on germination and dormancy of wild oats seeds. Can. Agric. Eng. 22 (1), 85–88.

Lyon, T.L., Buckman, H.O., 1947. The Nature and Properties of Soils. The McMillan Co., New York, NY.

Menges, R.M., Wayland, J.R., 1974. UHF electromagnetic energy for weed control in vegetables. Weed Sci. 22, 584–590.

Michaelson, S.M., Lin, J.C., 1987. Health Implications of Radiofrequency Radiation. Plenum Press, New York, NY.

Nelson, S.O., 1972a. A system for measuring dielectric properties at frequencies from 8.2 to 12.4 GHz. Trans. ASAE 15 (6), 1094–1098.

Nelson, S.O., 1972b. Frequency Dependence of the Dielectric Properties of Wheat and the Rice Weevil. (Ph.D. dissertation), Iowa State University, Ames, IA.

Nelson, S.O., 1973. Microwave dielectric properties of grain and seed. Trans. ASAE 16 (5), 902–905.

Nelson, S.O., 1976. Microwave dielectric properties of insects and grain kernels. J. Microw. Power 11 (4), 299–303.

Nelson, S.O., 1982. Factors affecting the dielectric properties of grain. Trans. ASAE 25 (4), 1045–1049, 1056.

Nelson, S.O., Charity, L.F., 1972. Frequency dependence of energy absorption by insects and grain in electric fields. Trans. ASAE 15 (6), 1099–1102.

Nelson, S.O., Stetson, L.E., 1974. Comparative effectiveness of 39- and 2450-MHz electric fields for control of rice weevils in wheat. J. Econ. Entomol. 67 (5), 592–595.

Nelson, S.O., Walker, E.R., 1961. Effects of radio-frequency electrical seed treatment. Agric. Eng. 42 (12), 688–691.

Nelson, S.O., Wolf, W.W., 1964. Reducing hard seed in alfalfa by radio-frequency electrical seed treatment. Trans. ASAE 7 (2), 116–119, 122.

Nelson, S.O., You, T.-S., 1989. Microwave dielectric properties of corn and wheat kernels and soybeans. Trans. ASAE 32 (1), 242–249.

Nelson, S.O., Nutile, G.E., Stetson, L.E., 1970. Effects of radiofrequency electrical treatment on germination of vegetable seeds. J. Am. Soc. Hortic. Sci. 95 (3), 359–366.

Nelson, S.O., Ballard, L.A.T., Stetson, L.E., Buchwald, T., 1976. Increasing legume seed germination by VHF and microwave dielectric heating. Trans. ASAE 19 (2), 369–371.

Nelson, S.O., Kehr, W.R., Stetson, L.E., Stone, R.B., Webb, J.C., 1977. Alfalfa seed germination response to electrical treatments. Crop Sci. 17, 863–866.

Nelson, S.O., Bovey, R.W., Stetson, L.E., 1978. Germination response of some woody plant seeds to electrical treatment. Weed Sci. 26 (3), 286–291.

Nelson, S.O., Stetson, L.E., Works, D.W., Pettibone, C.A., 1982. Germination responses of sweetclover seed to infrared, radiofrequency and gas-plasma electrical treatments. J. Seed Technol. 7 (1), 10–22.

Olsen, R.G., 1975. A theoretical investigation of microwave irradiation of seeds in soil. J. Microw. Power 10 (3), 281–296.

O'Bannon, J.H., Good, M.M., 1971. Application of microwave energy to control nematodes in soil. J. Nematol. 3 (1), 91–94.

Perkins, H.F., 1987. Characterization Data for Selected Georgia Soils, Special. Publication 43, Georgia Agricultural Experiment Station. The University of Georgia, Athens, GA.

Polk, C., Postow, E., 1986. Effects of Electromagnetic Fields. CRC Press, Boca Raton, FL.

Schwan, H.P., 1972. Microwave radiation: biophysical considerations and standards criteria. IEEE Trans. Biomed. Eng. BME-19 (4), 304–312.

Stetson, L.E., Nelson, S.O., 1972. Effectiveness of hot-air, 39-MHz dielectric, and 2450-MHz microwave heating for hard-seed reduction in alfalfa. Trans. ASAE 15 (3), 530–535.

Stratton, J.A., 1941. Electromagnetic Theory. McGraw Hill Book Co., New York, NY.

Stuchly, M.A., 1979. Interaction of radiofrequency and microwave radiation with living systems—a review of mechanisms. Radiat. Environ. Biophys. 16, 1–14.

Stuchly, M.A., 1995. Health Effects of Exposure to Electromagnetic Fields. In: IEEE Aerospace Applications Conference Proceedings, vol. 1. Institute of Electrical and Electronics Engineers (IEEE), Piscataway, New Jersey, pp. 351–368.

Van Assche, C., Uyttebroek, P., 1983. Possibilities of microwaves in soil disinfestation. EPPO Bull. 13 (3), 491–497.

Vela, G.R., Wu, J.F., Smith, D., 1976. Effect of 2450 MHz microwave radiation on some soil microorganisms in situ. Soil Sci. 121 (1), 44–51.

Vela-Muzquiz, R., 1983. Control of field weeds by microwave radiation. Acta Hortic. 152, 201–208.

van Wambeke, E., Wijsmans, J., d'Hertefelt, P., 1983. Possibilities in microwave application for growing substrate disinfestation. Acta Hortic. 152, 209–217.

Wayland, J.R., Davis, F.S., Young, L.W., Merkle, M.G., 1972. Effects of UHF fields on plants and seeds of mesquite and beans. J. Microw. Power 7 (4), 385–388.

Wayland, J.R., Davis, F.S., Merkle, M.G., 1973. Toxicity of an UHF device to plant seeds in soil. Weed Sci. 21 (3), 161–162.

Wayland, J.R., Merkle, M.G., Davis, F.S., Menges, R.M., Robinson, R., 1975. Control of seeds with UHF electromagnetic fields. Weed Res. 15, 1–5.

Wayland, J.R., Davis, F.S., Merkle, M.G., 1978. Vegetation Control, US Patent No. 4,092,900.

Ulaby, F.T., Wu, L., Hallikainen, M., Dobson, M.C. 1982. Microwave Dielectric Behavior of Wet Soil, Part 1: Experimental Observations at 1.4 and 5.0 GHz. RSL Tech. Rep. 545-1. Lawrence: University of Kansas Center for Research.

QUALITY SENSING IN FRUITS AND VEGETABLES

9

9.1 BACKGROUND INFORMATION

New, nondestructive techniques for sensing the quality of fresh fruits and vegetables, or produce that has been stored before consumption, would be helpful to producers, handlers, and consumers. Subjective quality standards for many products tend to be highly arbitrary, different for individual consumers and for sellers and buyers, and often subject to supply and demand. Thus, objective measurements that can be made rapidly and nondestructively could provide improvements in sorting and handling operations, and could also be helpful to consumers in selecting produce of desired quality. Measurements with visible and infrared radiation can be useful in many instances for detecting surface characteristics associated with quality, and X-rays, sound and ultrasound can be useful in detecting some internal characteristics associated with product quality. However, electric fields can penetrate fruits and potentially provide information on the quality of the internal tissues that is of interest. For many fruits, the best objective quality indicator is the soluble solids content (SSC, mostly sugars), which currently requires samples from the internal tissues and is therefore a destructive test. Therefore, dielectric studies have been conducted to learn whether the penetration of electric fields can be utilized for nondestructive determination of product quality in certain fruits and vegetables.

9.2 STUDIES ON THE USE OF DIELECTRIC PROPERTIES

Information in the literature on dielectric properties of fruits and vegetables was summarized earlier (Nelson, 1973), but little information for fresh products was found, particularly at microwave frequencies. The dielectric properties of some fresh fruits and vegetables were measured at 2.45 GHz and 23°C by the short-circuited coaxial-line technique, and data were presented for the dielectric constant and loss factor of several peach cultivars, two sweet potato cultivars and single cultivars of potato, apple, cantaloupe, and carrot (Nelson, 1980). No differences in dielectric properties were distinguishable between mature-green and full-ripe peaches or between cured sweet potatoes and those that had been subjected to chilling injury after curing for induction of the hardcore condition. The dielectric constant was correlated with moisture content and also appeared to be influenced by tissue density. No correlation was observed between dielectric properties and soluble solids as measured by a refractometer. It was concluded that use of dielectric properties at a single microwave frequency offered little likelihood of success for detecting such fruit and vegetable qualities.

Dielectric Properties of Agricultural Materials and Their Applications. DOI: http://dx.doi.org/10.1016/B978-0-12-802305-1.00009-9

For better assessment of the usefulness of electromagnetic fields for measurement of fruit and vegetable quality, study of the frequency dependence of the dielectric properties of such materials over a wide range of frequencies was recommended.

Additional data were obtained by the same measurement technique on dielectric constant and loss-factor values for potato, sweet potato, peach, watermelon, cantaloupe, and cucumber tissue at frequencies of 2.45, 11.7, and 22.0 GHz (Nelson, 1983). Data on the moisture content, density, and SSC of the fruit and vegetable tissues, were also reported. Correlations were noted between the dielectric properties of the different types of tissue and their moisture contents. No correlation was evident between the dielectric properties and the SSC, which consists mainly of sugars, for the fruit and vegetable tissues. So, little promise was indicated for maturity detection by measurements at these microwave frequencies. A need for further frequency-dependent dielectric properties information, including lower frequencies, was suggested.

Short-circuited-waveguide permittivity measurements were taken on tissues of mature-green and full-ripe peaches at 2.45 GHz, and the dielectric properties were examined to see whether they might be useful in distinguishing the degree of maturity (Nelson, 1980). Permittivity measurements at the single frequency of 2.45 GHz did not offer promise for detecting differences in maturity. Measurements of the dielectric properties of a large variety of fruits and vegetables over a wide frequency range, 0.2–20 GHz, with an open-ended coaxial-line probe, revealed significant variation with frequency in both the dielectric constant and loss factor (Nelson et al., 1994b). Therefore, such measurements were taken on tree-ripened peaches over a range of maturity during the growing season to learn whether differences in microwave dielectric properties of the peaches might be useful in sensing the degree of maturity (Nelson et al., 1995). A permittivity-based maturity index was developed, based on differences in both components of the permittivity, the dielectric constants at the low end of the frequency range, and the loss factors at 10 GHz near the higher end of this frequency range. More research and developmental work was recommended for determining the potential for practical use of the technique, including measurements at frequencies lower than 200 MHz, because the curves for the dielectric constants of the two different maturities appeared to be diverging as they approached the lower end of the frequency range.

9.2.1 MELON STUDIES

Nondestructive techniques for sensing the quality of agricultural products are useful for growers, handlers and packers, marketers, and consumers of these products. For melons in general, there are no reliable methods for nondestructive determination of quality, the main attribute being the sweetness of the internal edible tissue. For watermelon, correlations have been reported between melon density and SSC, which were used in the sorting process (Kato, 1997), but other techniques, including near-infrared reflectance for SSC determination, are destructive. Sweetness can be rapidly assessed by taking a plug from a melon and measuring SSC with calibrated refractometers, but this leaves the melon vulnerable to rapid deterioration. Radio-frequency (RF) electric fields can penetrate melons well. Therefore, if the dielectric properties of the internal tissue of melons could be correlated with the SSC of those internal tissues, it might be possible to develop inexpensive instruments to sense those dielectric properties with electric fields and, thus, determine sweetness nondestructively.

The dielectric properties of cantaloupe were measured in the frequency range from 200 MHz to 20 GHz—along with many other fruits and vegetables—to provide some background data on those properties (Nelson et al., 1994a). Cantaloupe was also included in dielectric spectroscopy studies on several fruits and vegetables in the 10 MHz to 1.8 GHz range (Nelson, 2005). Efforts to find correlations between SSC and dielectric properties from 10 MHz to 1.8 GHz have been published for honeydew melons (Nelson et al., 2006) and watermelons (Nelson et al., 2007). An interesting correlation between the dielectric properties and SSC was reported for honeydew melons (Nelson et al., 2006), but attempts to use that correlation for predicting SSC were not successful (Guo et al., 2007b).

Further studies were conducted to determine whether useful correlations might be obtained between SSC and dielectric properties of cantaloupe, honeydew melons, and watermelons in the frequency range 200 MHz to 20 GHz (Nelson et al., 2008). Several cultivars of cantaloupe, honeydew melons, and watermelons were planted and harvested with a range of maturities for measurements of dielectric properties and determination of moisture content and SSC. Permittivities (dielectric constants and loss factors) were determined over the frequency range from 200 MHz to 20 GHz with an open-ended coaxial-line probe and network analyzer for both interior tissue and measurements on the surface of the melons. Permittivity data were presented graphically for all three types of melon. High correlations were noted between SSC and moisture content in the tissues of all three kinds of melon, with SSC increasing linearly with decreasing moisture content of the edible tissues. Dielectric properties determined by measurements on the external surface of the melons had lower values than those of the internal tissues. Dielectric properties were similar for all three types of melon, and they reflected the influence of the dielectric behavior of free water. No obvious correlations were noted between the dielectric properties and the SSC (sweetness) for sensing the quality of the melons.

9.2.2 APPLE STUDIES

The dielectric properties of fresh fruits can be rapidly sensed with suitable measurement instruments that employ RF electric fields for this purpose. Therefore, if adequate correlations can be found between the dielectric properties of such fruits and their quality factors, it may be possible to develop new instruments for rapid, nondestructive determination of quality. Dielectric properties of fresh apples were measured in several studies, including measurements from 300 to 900 MHz on immature and mature apples (Thompson and Zacharia, 1971), and measurements from 100 MHz to 12 GHz (Tran et al., 1984). Measurements on fresh apples, both on external surfaces and exposed flesh, were reported for the 150 MHz to 6.4 GHz range (Seaman and Seals, 1991). Dielectric properties of internal tissues of three apple cultivars over the range 200 MHz to 20 GHz were also reported (Nelson et al., 1994a,b). Dielectric properties of four apple cultivars were measured over the range 30 MHz to 3 GHz in connection with codling moth control studies (Ikediala et al., 2000). The temperature dependence of the dielectric properties was also included in the latter study, and apples were included in later studies on dielectric properties and temperature dependence at 2450 MHz (Sipahioglu and Barringer, 2003) and in the frequency range 10–1800 MHz (Nelson, 2003).

The dielectric properties of fresh apples of three cultivars were measured at 24°C shortly after harvest, and over 10 weeks in storage at 4°C, to determine whether these properties might be used to determine quality factors such as SSC, firmness, moisture content, and pH (Guo et al., 2007a). The dielectric constants and dielectric loss factors at 51 frequencies, from 10 to 1800 MHz, were determined for external surface and interior tissue measurements along with moisture content, firmness, and SSC, and the pH of juice expelled from the internal tissues.

The dielectric properties of fresh apples from 10 to 1800 MHz determined by open-ended coaxial-line probe measurements on the exterior surface and on exposed internal tissue were quite different. Surface measurements gave lower dielectric constants than did measurements of internal tissue, and much lower dielectric loss factors than for the internal tissue. Dielectric constants from both surface and interior tissue measurements showed regularly declining values with increasing frequency. Dielectric loss factors for internal tissues exhibited declining values with increasing frequency, characteristic of the dominance of ionic conduction at lower frequencies and that of dipolar relaxation at the higher frequencies. Loss factors for the surface measurements revealed an overriding dielectric relaxation around 20 MHz, most likely resulting from a complex combination of effects that might include Maxwell–Wagner, bound water, and ion-related phenomena. Correlations between the dielectric properties and the SSC, firmness, moisture content, and pH, were low at all frequencies. A high correlation was noted between the dielectric constant divided by SSC and the dielectric loss factor divided by SSC as plotted in the complex plane, but SSC was not predicted well by the dielectric properties because the individual dielectric properties were not well correlated with SSC. Dielectric properties remained relatively constant throughout the 10-week refrigerated storage period. SSC for all three apple cultivars was well correlated with moisture content of the tissues in an inverse relationship. Further research involving wider frequency ranges, and further analysis of dielectric properties, was believed to be necessary to satisfactorily assess the potential for sensing quality factors in apples by RF electric fields.

9.2.3 ONION STUDIES

Some research was conducted on utilizing the dielectric properties of onions for sensing moisture content, which is related to onion curing. Onions must be cured prior to storage. Curing is the process characterized by loss of moisture from the surface regions of the onion. Curing is an essential step that seals the moisture inside the onion. This forms a barrier to the intrusion of disease organisms during postharvest storage and shipping (Maw et al., 1997a). When done properly, curing involves sealing incisions that may exist in the outer ring, drying the outer skin, and sealing the neck and roots of the onions. Curing can be performed naturally by windrowing the onions in the field when climatic conditions are suitable, or artificially, by passing forced air heated to 40°C through the onions. Often, artificial curing, the more costly method, is used to supplement cheaper natural curing (Boyette et al., 1992). Currently, this process is controlled simply by time and human visual inspection (Maw et al., 2004). However, varietal, seasonal, and year-to-year differences can require different time periods for the curing procedure (Maw et al., 1997b). If a method for sensing the degree of curing, through sensing the onion skin's moisture content, could be developed, it would be very useful in determining when onions are properly cured. The curing process could then be improved by preventing excess weight loss from overcuring, detecting onions that have not completely cured, and saving energy costs associated with overcuring.

There is very little published research on the use of dielectric properties to predict the quality attributes of onions. Dielectric properties of three cultivars of onion were first measured at 2.45, 11.6, and 22 GHz in waveguide sample holders by the short-circuited-line technique (Nelson,

1992). Later, dielectric properties at a single frequency were measured for three varieties of onion, and variation was shown with moisture content and temperature (Abhayawick et al., 2002).

More recently, measurements were made on Vidalia onions at different moisture contents over a frequency range of 200 MHz to 20 GHz (McKeown et al., 2012). From these broad-band measurements, certain frequencies were investigated for their usefulness in predicting moisture content. In addition, a density-independent function, expressed in terms of the dielectric properties, was applied to the data to predict moisture content with a high level of accuracy up to 40% moisture content. Minced Vidalia onion samples were used for the measurements. The data were analyzed with respect to frequency and moisture dependence. Frequency analysis showed a nearly linear increase in the dielectric constant with rising moisture content at all frequencies. The dielectric loss factor exhibited similar behavior in the higher frequency range. Dielectric properties were plotted against moisture contents in the range 8.1–90.2%, showing that models could be developed for predicting moisture content from dielectric properties. By using a previously reported density-independent function of the dielectric properties, a model was developed that predicted moisture content up to 40%. The frequency dependence of these properties of Vidalia onions could potentially help in selecting an optimal frequency to use for moisture sensing. Based on the results of this study, the use of a density-independent function of the dielectric properties at higher frequencies, about 13 GHz, provided the best prediction of moisture content. These data on dielectric properties could be used to develop a sensor for use in the quality control of onions. Future work could investigate changes of the dielectric properties with density and temperature and determine correlations between dielectric properties of minced onions and those of the outer layer of intact onions at the same moisture content, which would be necessary for practical application.

9.2.4 SENSING THE MOISTURE CONTENT OF DATES

As with most agricultural products, the moisture content of dates (fruit of the date palm, *Phoenix dactylifera* L.) is a very important characteristic. At harvest, bunches of dates from the tree contain dates of widely ranging moisture content, and they need to be sorted promptly so that those of high moisture can be dried for safekeeping. Although moisture content is not a specific factor in the grading of dates, the different grades are highly correlated with the moisture content of date flesh.

Dates are generally sorted by hand into marketable and product grades. The marketable dates are judged suitable for packaging as whole or pitted dates. Product grades are exported or used for diced or ground-date products. Official USDA grades are determined subjectively based on color, uniformity of size, absence of defects, and character. Character, which accounts for 40% of the evaluation, involves subjective determination of development, ripening, and moisture content. The USDA standards define six grades: A, B, B (Dry), C, C (Dry), and Substandard—the first five of which have alternate designations of US Fancy, US Choice, US Choice(Dry), US Standard, and US Standard (Dry), respectively.

The US date industry uses four grades for marketing (Chesson et al., 1979): Natural, Waxy, No.1 Dry, and No. 2 Dry, in order of decreasing moisture content from about 23% (wet basis), or higher to less than 15%. No. 2 Dry must be rehydrated to moisture contents above 15% for marketing as fresh fruit. At harvest, dates may range in moisture content from about 12% to 30% (Davies, 1991). These dates grade all the way from Naturals that need to be dried for safe storage if moisture content is higher than about 23% to No. 2 Dry. Therefore, they must be sorted to separate those that need to be dried before they can be safely stored. The desirable moisture content for the widely grown Deglet Noor cultivar in California was reported to be 23–25% (Rygg, 1975; Nixon and Carpenter, 1978). At higher moisture contents, the dates are subject to molding and fermentation. Experience, using moisture contents determined by the Dried Fruit Moisture Tester (Type A Series, Dried Fruit Association (DFA) of California), places that desirable moisture range for Deglet Noor dates generally at 20–22%, although the allowable range varies significantly with growing conditions and the quality of the crop from year to year (Davies, 1992).

The moisture content of dates is also important in determining the permissible length of storage time without significant deterioration in quality (Nelson and Lawrence, 1992). Shelf life can be extended by storage at lower temperatures, but, in general, deterioration (darkening and loss of flavor) increases with increasing moisture content.

Hand sorters use at least three criteria in making grade determinations: elasticity (by feel), surface texture, and color (Chesson et al., 1979). In most packing houses, dates are hand sorted and graded as they move along on conveyors or oscillating tables. Although dates cannot be graded on the basis of moisture content alone, automatic sensing of individual dates for moisture content would be useful in separating high-moisture dates at harvest from those that do not require drying. It could also be used to separate the drier dates from those that require careful grading and thus reduce the quantity of material to be manually graded into marketable and product grades. Therefore, a practical rapid moisture sensor for individual fruit could reduce the amount of skilled labor required to sort and grade dates, thus being of significant benefit in this highly labor-intensive industry.

Methods for determining date moisture have also been reviewed (Nelson and Lawrence, 1992). When moisture contents are measured in the US date industry, most frequently the practical measurements are taken with an electrical resistance-type meter designed for the dried fruit industry (DFA moisture tester) that has been calibrated against vacuum-oven moisture determinations. Different oven temperatures and times for drying have been reported in the literature. The Official Methods of Analysis of the Association of Official Analytical Chemists (AOAC, 1984) includes a vacuum-oven method for dried fruits and the DFA Moisture Meter method for prunes and raisins. The vacuum-oven method specifies drying 5–10 g of ground or finely chopped sample for 6 h at 70°C under a vacuum of at least 100 mmHg with a slow current of air admitted after passing through H_2SO_4 for removal of moisture.

Because vacuum-oven methods are slow and tedious, and because the electrical meter methods still require grinding and are more suited to the laboratory, there is interest in a more rapid, nondestructive and less troublesome technique for moisture determination. Because of the need for online moisture sensing, work was initiated on ways to instantaneously sense the moisture content of whole individual dates nondestructively (Nelson and Lawrence, 1992). With laboratory equipment, RF impedance measurements at two frequencies, 1 and 5 MHz, on a parallel-plate capacitor of 5-cm diameter, holding a single date between the plates, were used to estimate moisture content of whole and pitted dates with standard errors of about 1% moisture content.

Previously developed calibration equations for estimating moisture content of individual whole dates and pitted dates, based on impedance measurements at 1 and 5 MHz on Deglet Noor dates from the 1988 and 1989 California crops, estimated moisture content (m.c.) of 1990 dates of the same cultivar from the same source with standard errors of comparable magnitude, about 1% m.c. (Nelson and Lawrence, 1994). The original calibration data—combined with the new verification data on dates from the 1990 crop—provided a new calibration equation more representative of California dates of the Deglet Noor cultivar for the moisture content range of 12−28%. The equation, based on a single variable model, required only capacitance measurements at 1 and 5 MHz on the parallel-plate electrode assembly. The accuracy and simplicity of this instantaneous, nondestructive measurement justified further exploration of this technique for potential application in the automatic sorting of dates to reduce the skilled-labor requirements in the manual processes of date sorting and grading.

REFERENCES

Abhayawick, L., Laguerre, J.C., Tauzin, V., Duquenoy, A., 2002. Pysical properties of three onion varieties as affected by the moisture content. J. Food Eng. 55 (3), 253−262.

AOAC, 1984. Sec. 22,013 Moisture in dried fruits, In: Official Methods of the Association of Official Analytical Chemists, Arlington, VA.

Boyette, M.D., Sanders, D.C., Estes, E.A., 1992. Postharvest cooling and handling of onions, AG-413-6, North Carolina Cooperative Extension Service. North Carolina State University, Raleigh, NC.

Chesson, J.H., Burkner, P.F., Perkins, R.M., 1979. An experimental vacuum separator for dates. Trans. ASAE 22 (1), 16−20.

Davies, J., 1991. Personal Communication. Dole Food Comopany, Thermal, CA.

Davies, J., 1992. Personal Communication. Dole Food Company, Thermal, CA.

Guo, W., Nelson, S.O., Trabelsi, S., Kays, S.J., 2007a. 10−1800-MHz dielectric properties of fresh apples during storage. J. Food Eng. 83, 562−569.

Guo, W.-C., Nelson, S.O., Trabelsi, S., Kays, S.J., 2007b. Dielectric properties of honeydew melons and correlation with quality. J. Microw. Power Electromagn. Energy 41 (2), 44−54.

Ikediala, J.N., Tang, J., Drake, S.R., Neven, L.G., 2000. Dielectric properties of apple cultivars and codling moth larvae. Trans. ASAE 43 (5), 1175−1184.

Kato, K., 1997. Electrical density sorting and estimation of soluble solids content of watermelon. J. Agric. Eng. Res. 67 (2), 161−170.

Maw, B.W., Smittle, D.A., Mullinix, B.C., 1997a. Artificially curing sweet onions. Appl. Eng. Agric. 13 (4), 517−520.

Maw, B.W., Smittle, D.A., Mullinix, B.C., 1997b. The influence of harvest maturity, curing, and storage conditions upon the storability of sweet onions. Appl. Eng. Agric. 13 (4), 511−515.

Maw, B.W., Butts, C.L., Purvis, A.C., Seebold, K., Mullinix, B.C., 2004. High-temperature continuous-flow curing of sweet onions. Appl. Eng. Agric. 20 (5), 657−663.

McKeown, M.S., Trabelsi, S., Tollner, E.W., Nelson, S.O., 2012. Dielectric spectroscopy measurements for moisture prediction in Vidalia onions. J. Food Eng. 111, 505−510.

Nelson, S.O., 1973. Electrical properties of agricultural products—a critical review. Trans. ASAE 16 (2), 384−400.

Nelson, S.O., 1980. Microwave dielectric properties of fresh fruits and vegetables. Trans. ASAE 23 (5), 1314−1317.

Nelson, S.O., 1983. Dielectric properties of some fresh fruits and vegetables at frequencies of 2.45 to 22 GHz. Trans. ASAE 26 (2), 613−616.

Nelson, S.O., 1992. Microwave dielectric properties of fresh onions. Trans. ASAE 35 (3), 963−966.

Nelson, S.O., 2003. Frequency- and temperature-dependent permittivities of fresh fruits and vegetables from 0.01 to 1.8 GHz. Trans. ASAE 46 (2), 567−574.

Nelson, S.O., 2005. Dielectric spectroscopy of fresh fruit and vegetable tissues from 10 to 1800 MHz. J. Microw. Power Electromagn. Energy 40 (1), 31–47.

Nelson, S.O., Lawrence, K.C., 1992. Sensing moisture content in dates by RF impedance measurement. Trans. ASAE 35 (2), 591–595.

Nelson, S.O., Lawrence, K.C., 1994. RF impedance sensing of moisture content in individual dates. Trans. ASAE 37 (3), 887–891.

Nelson, S.O., Forbus Jr., W.R., Lawrence, K.C., 1994a. Microwave permittivities of fresh fruits and vegetables from 0.2 to 20 GHz. Trans. ASAE 37 (1), 181–189.

Nelson, S.O., Forbus Jr., W.R., Lawrence, K.C., 1994b. Permittivities of fresh fruits and vegetables at 0.2 to 20 GHz. J. Microw. Power Electromagn. Energy 29 (2), 81–93.

Nelson, S.O., Forbus Jr., W.R., Lawrence, K.C., 1995. Assessment of microwave permittivity for sensing peach maturity. Trans. ASAE 38 (2), 579–585.

Nelson, S.O., Trabelsi, S., Kays, S.J., 2006. Dielectric spectroscopy of honeydew melons from 10 MHz to 1.8 GHz for quality sensing. Trans. ASABE 49 (6), 1977–1981.

Nelson, S.O., Guo, W., Trabelsi, S., Kays, S.J., 2007. Dielectric spectroscopy of watermelons for quality sensing. Meas. Sci. Technol. 18, 1887–1892.

Nelson, S.O., Trabelsi, S., Kays, S.J., 2008. Dielectric spectroscopy of melons for potential quality sensing. Trans. ASABE 51 (6), 2209–2214.

Nixon, R.W., Carpenter, J.B., 1978. Growing Dates in the United States, Agricultural Information Bulletin No. 207, US Department of Agriculture, Washington, DC.

Rygg, G.L., 1975. Date Development, Handling and Packing in the United States. Agriculture Handbook No. 482, US Department of Agriculture, Agricultural Research Service, Washington, DC.

Seaman, R., Seals, J., 1991. Fruit pulp and skin dielectric properties for 150 MHz to 6400 MHz. J. Microw. Power Electromagn. Energy 26 (2), 72–81.

Sipahioglu, O., Barringer, S.A., 2003. Dielectric properties of vegetables and fruits as a function of temperature, ash, and moisture content. J. Food Sci. 68 (1), 234–239.

Thompson, D.R., Zacharia, G.L., 1971. Dielectric theory and bioelectrical measurements (Part II. Experimental). Trans. ASAE 14 (2), 214–215.

Tran, V.N., Stuchly, S.S., Kraszewski, A., 1984. Dielectric properties of selected vegetables and fruits 0.1–10 GHz. J. Microw. Power 19 (4), 251–258.

MINING APPLICATIONS

The ability to measure dielectric properties over wide ranges of frequency, which was developed for agricultural applications, attracted interest from the mining industry, so efforts were made to provide assistance in solving some of the problems facing that industry. Thus, research findings on dielectric properties relating to mining interests are included in this chapter.

10.1 BACKGROUND INFORMATION—COAL

Coal has been an important energy source in the past and is expected to be essential for energy requirements many years into the future. However, environmental problems are associated with the use of coal, especially in the burning of coal with high sulfur content. Current coal beneficiation technology does not permit the separation of finely disseminated pyrite microcrystals from coal. Improved methods for removing pyrite would help in meeting air pollution control standards. One potential method involves finely grinding coal and using magnetic separation techniques. Heating coal can enhance the magnetic susceptibility of the pyrite (Marusak et al., 1976). Selective heating of the pyrite in coal by dielectric heating has been explored as a means of increasing the magnetic susceptibility of the pyrite with minimal heating of the coal (Bluhm et al., 1980).

10.2 DIELECTRIC PROPERTIES MEASUREMENTS ON COAL

For identification of the best frequency ranges for selective dielectric heating of the pyrite, information on the frequency dependence of the dielectric properties of pyrite and coal was needed. Suitable coal samples were obtained for measurements of dielectric properties over a range of frequencies. Materials were obtained as fresh face samples at the coal mines, processed through jaw crushers and roll mills, and air-dried before being pulverized to pass through a US Standard 60-mesh sieve. Coal from five coal beds or mines was selected for study. The pulverized samples were separated by heavy-media gravitational techniques to obtain materials with high and low pyrite contents, and the fractions were oven-dried before measurement of dielectric properties. Dense fractions containing most of the pyrite consisted of materials that sank in a liquid of 2.0 specific gravity (designated 2.0 Sink), whereas light fractions (designated 1.3 Float) floated in a liquid with a specific gravity of 1.3. The 12 samples selected for study included run-of-mine (ROM) samples and

Dielectric Properties of Agricultural Materials and Their Applications. DOI: http://dx.doi.org/10.1016/B978-0-12-802305-1.00010-5

the light and dense fractions of those samples. The dense fractions constituted a small part of the ROM material, generally less than 10%. Chemical compositions, heating values, moisture contents, and particle-size distributions were determined (Nelson et al., 1980).

The dielectric properties of the pulverized coal samples from identified sources were measured at 22°C over the frequency range from 1 MHz to 12 GHz, and results were presented graphically (Nelson et al., 1980). ROM samples, as well as dense pyrite-bearing fractions and light fractions, relatively free of pyrite, were included in the measurements. Dielectric constants decreased regularly with increasing frequency and were positively correlated with density. Dielectric loss factors of pyrite-bearing fractions of lower sulfur content were high at frequencies below about 50 MHz, and decreased with increasing frequency to low values at microwave frequencies. In contrast, the loss factor of a pyrite-bearing fraction with high sulfur content was low at the lower frequencies and increased with increasing frequency to high levels at microwave frequencies. Thus, selective dielectric heating of pyrite in coal to enhance its magnetic susceptibility for possible removal by magnetic separation can probably be best accomplished at microwave frequencies for some coals and at lower dielectric heating frequencies for others.

Similar measurements of the dielectric properties of 18 coal samples from different mines, separated into ROM and light and dense fractions, were made over the same frequency range (Nelson et al., 1981). Coal samples were prepared in the same way, but included some dense fractions that were separated in a liquid of 2.9 specific gravity (designated 2.9 Sink) to achieve a higher concentration of pyrite. The 2.9 Sink samples constituted a small part of the ROM material, generally less than 5%. In a further effort to obtain material of high pyrite content, coal pieces from one mine that contained large amounts of pyrite were selected, crushed, milled, and acid leached to remove minerals other than pyrite.

The dielectric properties of the 18 pulverized coal samples were measured over the frequency range from 1 MHz to 12 GHz at 22°C and presented graphically (Nelson et al., 1981). For Iowa coal, the dielectric loss factors of dense fractions containing most of the pyrite tended to increase noticeably as frequency increased. The loss factors of dense fractions of eastern coal (Kentucky and Ohio) tended to decrease with increasing frequency and then rose somewhat as frequency approached the microwave range. Frequency dependence of the dielectric constant and loss factor were generally similar to those noted earlier, but some differences were noted attributable to differences in density and moisture content. Data were also presented on the dielectric constant as a function of bulk density. Plots of the square root of the dielectric constant versus bulk density (Figure 10.1) resulted in straight lines that can be extrapolated with confidence to the coal particle density, thus providing estimates for the dielectric constant of the solid coal from measurements on pulverized samples.

10.3 DIELECTRIC HEATING OF COAL—PYRITE MIXTURES

Several studies have shown that the magnetic susceptibility of pyrite can be enhanced by heating (Marusak et al., 1976; Ergun and Bean, 1968). The advantage of this effect is the improved possibility of removing pyrite from coal by magnetic separation. However, the problem in heating pyrite in coal is that energy is wasted by also heating the coal. A possible solution is the preferential

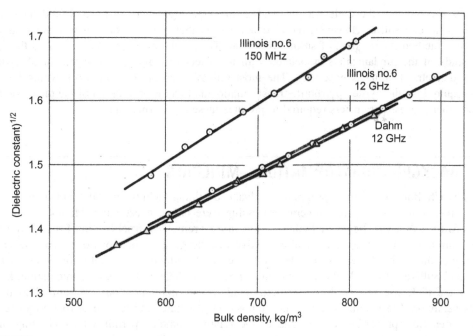

FIGURE 10.1

Linear relationships between the square root of the dielectric constant and bulk density for indicated run-of-mine coal samples at indicated frequencies at 22°C (Nelson et al., 1981).

dielectric heating of the pyrite if the dielectric properties of pyrite and coal are sufficiently different so that the pyrite will absorb more electromagnetic energy and heat faster than the coal.

Research was conducted on microwave heating of coal containing pyrite at 2.45 GHz to investigate possibilities for selectively heating the pyrite (Fanslow et al., 1980). ROM coal was separated by gravitational techniques into fractions designated as 1.30 float (clean coal) and 2.00 sink (pyrite and ash). Preferential dielectric heating ratios at 2.45 GHz were calculated, based on measured dielectric properties of the two coal fractions, for coal from four different sources, and predicted heating rates for the pyrite in coal ranged from 1.3 to 3.3 times faster than clean coal. Samples of the same coal fractions were then experimentally heated at 2.45 GHz. Results indicated that the pyrite and ash fraction was heated from 1.9 to 2.6 times faster than clean coal.

Additional research was conducted on selective heating of pyrite in coal and magnetic separation before and after 2.45-GHz microwave heating (Bluhm et al., 1986). Calculations, based on measured dielectric-constant and loss-factor values for coal and pyrite-bearing fractions from four different mines, predicted greater heating rates for pyrite than for the coal. Experimental dielectric heating of coal and pyrite mixtures confirmed that pyrite heated more rapidly than coal. The magnetic susceptibility of dielectrically heated coal increased; however, the increase in susceptibility was not large enough to enhance appreciably the removal of pyrite by magnetic separators.

The tests showed that the dielectrically heated, magnetically separated coals would have slight reductions in total sulfur. However, these reductions, in their present form, would not justify the use of this method for removal of sulfur from coal. The results of these tests suggested that additional study of the fundamentals associated with the selective magnetic enhancement of pyrite in coal by dielectric heating is necessary. The most important areas of future work would be treatments required for the pyrite−pyrrhotite transformation and the determination of the dielectric heating and other process parameters required to produce these transformations.

10.4 BACKGROUND INFORMATION—MINERALS

Under the US Bureau of Mines program of advanced mining technology, new and more efficient ways of rock fragmentation and mineral processing were being investigated. Methods were being developed for extracting the desired components and for remote internal heating for rock fragmentation (Thirumalai, 1970). Experimental extractive metallurgical applications of mineral separation make use of the large differences in dielectric properties of individual minerals (Jordan et al., 1980; Jordan and Sullivan, 1985). In the application of microwave heating for rock fragmentation, knowledge of the dielectric properties of the individual rock-forming minerals is required to predict heating behavior. The inhomogeneities in composition, structure, and texture of rock produce very complex electrical properties. Knowledge of the relative complex permittivity, consisting of the dielectric constant and loss factor, is required to determine the differential absorption of microwave energy in an aggregate of minerals (rock).

These properties are also essential in modeling electromagnetic wave propagation and interaction in layered earth problems (gas and oil exploration), remote sensing of ore deposits (Dobson and Ulaby, 1986), and detection of pollutants in ground water (Olhoeft, 1986). Data on dielectric properties of minerals have been reported and compiled (von Hippel, 1954; Parkhomenko, 1967; Iglesius and Westphal, 1967; Westphal and Sills, 1972; Westphal, 1977; Olhoeft et al., 1981; Webb and Church, 1986), but data at frequencies above 1 GHz were sparse. The measurements reported here were taken to obtain information on the dielectric properties of selected minerals as a function of frequency above 1 GHz.

10.5 MEASUREMENTS OF THE DIELECTRIC PROPERTIES OF MINERALS

Pulverized samples of 10 minerals, which are major or minor minerals common to rocks and ores being investigated for their response to microwave heating, were prepared for determination of dielectric properties (Nelson et al., 1989a,b). The rocks included gabbro and basalt, and the ores included magnetic taconite, nonmagnetic taconite, and manganese oxide ores. At least 100 g of the purified minerals were concentrated by handpicking under a stereoscopic microscope to reject other mineral contaminants, middlings, and weathered material.

Several samples were easily concentrated by hand at very coarse sizes, whereas others required gentle crushing and screening, followed by tedious and careful handpicking. Concentrates were carefully rechecked to remove any remaining middlings and tailings. Then they were gently crushed and screened to less than 104 μm. They were then divided into three aliquots for subsequent analysis: (i) light optical microscopy (LOM), X-ray diffraction, and chemical analysis, (ii) particle sizing, and (iii) a minimum of 5 cm^3 for dielectric properties measurement. Each aliquot for measurement of dielectric properties was dried in a forced-air oven at 105°C for 24 h and packaged in paraffin-sealed sample jars until the time for dielectric measurements.

Dielectric properties of the pulverized mineral samples were measured at 24°C and at frequencies of 1.0, 2.45, 5.5, 11.7, and 22.0 GHz employing the short-circuited waveguide method of Roberts and von Hippel (1946) with computation by the program described previously (Nelson et al., 1972, 1974). Measurement techniques are described in Section 2.1.5.1. At 1.0, 2.45, and 5.5 GHz, the Rhode and Schwarz[1] 21-mm coaxial airline short-circuited sample holder and systems described earlier (Nelson, 1973) were used. Rectangular-waveguide systems were used for the higher frequency measurements—an X-band system (Nelson, 1972) at 11.7 GHz, and a K-band system (Nelson, 1983b) at 22 GHz. Sample lengths for the various measurements ranged from about 15 to 35 mm. Measurements of the dielectric properties of each pulverized sample were taken at five or six different densities. Bulk densities of the samples were determined by dividing the weight of the samples by the volume they occupied in the sample holders. Densities for the solid materials were determined by measuring the volume occupied by the pulverized particles with a Beckman Model 930 air comparison pycnometer and dividing the sample weight by the volume measured. The bulk and particle densities were used in calculating the dielectric properties of the solid mineral materials from the measurements of dielectric properties at known bulk densities with Eq. (1.15) in accordance with the Landau and Lifshitz, Looyenga equation for dielectric mixtures (Landau and Lifshitz, 1960; Looyenga, 1965).

Dielectric properties for the ten minerals, amphibole (richterite), chlorite (clinochlore), feldspar (labradorite), mica (muscovite), mica (phlogopite), pyroxene (salite), goethite, hematite, ilmenite, and manganese oxide (hollandite) were published for five different bulk densities and the solid minerals, along with complete descriptions, chemical analyses, and particle-size distributions (Nelson et al., 1989a,b). Resulting permittivity values for the solid minerals are summarized in Table 10.1.

Measurements at 24°C at frequencies from 1 to 22 GHz of the complex relative permittivities of dry pulverized samples of ten minerals, six silicates, and four metal oxides, showed that the permittivities of one mineral from each class, pyroxene (salite) and goethite, were practically independent of frequency. The other metal oxides, hematite, ilmenite, and manganese oxide (hollandite), had high permittivities that decreased significantly with increasing frequency. The dielectric loss factor of manganese oxide (hollandite) revealed a dielectric relaxation in the 1- to 22-GHz range with losses peaking between 2.45 and 11.7 GHz. The other silicates, amphibole (richterite), chlorite (clinochlore), feldspar (labradorite), and muscovite and phlogopite micas, had lower permittivities that showed moderate reductions with increasing frequency. Their loss factors showed relatively little frequency dependence, but moderate reductions with increasing frequency were evident for amphibole (richterite), feldspar (labradorite), and pyroxene (salite).

---

[1]Mention of trade names or commercial products in this publication is solely for the purpose of providing specific information and does not imply recommendation or endorsement by the US Department of Agriculture.

Table 10.1 Dielectric Properties of Solid Minerals Determined from Measurements on Dry Pulverized Samples of Selected Minerals at 24°C (Nelson et al., 1989a,b)

Mineral	Density, g/cm^3	Dielectric Constants, ε', and Loss Factors, ε'', at Specified Frequencies, GHz									
		1.0		2.45		5.5		11.7		22.0	
		ε'	ε''	ε'	ε''	ε'	ε''	ε'	ε''	ε'	ε''
Amphibole	3.07	7.53	0.032	7.37	0.026	7.29	0.019	7.12	0.011	7.03	0.027
Chlorite	2.76	7.20	0.126	7.06	0.137	7.01	0.121	6.96	0.128	6.94	0.146
Feldspar	2.72	6.03	0.112	6.01	0.090	5.99	0.066	5.91	0.034	5.67	0.049
Mica (muscovite)	2.74	8.85	0.111	8.69	0.091	8.48	0.103	8.48	0.066	8.29	0.091
Mica (phlogopite)	2.96	10.59	0.21	9.77	0.14	9.15	0.16	8.91	0.18	8.27	0.19
Pyroxene	3.30	7.18	0.22	7.18	0.17	7.18	0.14	7.18	0.11	7.18	0.10
Geothite	4.04	13.6	0.34	13.6	0.38	13.6	0.33	13.6	0.30	13.6	0.35
Hematite	5.03	18.9	2.88	18.3	2.23	17.2	1.68	14.4	1.12	10.9	0.53
Ilmenite	4.75	30.7	10.2	23.6	11.2	19.6	7.0	16.4	3.9	14.3	1.7
Manganese oxide	5.05	68.3	6.5	61.9	10.7	51.3	16.8	41.2	7.9	38.8	6.8

10.6 COAL AND LIMESTONE MEASUREMENTS

Interest in the permittivities or dielectric properties of coals and minerals arises from their influence on electromagnetic wave propagation in exploration (Singh et al., 1979), for control of mining processes (Balanis et al., 1976, 1978, 1980), and possible dielectric heating applications for modifying coal characteristics (Nelson et al., 1980, 1981; Bluhm et al., 1986) or in rock fragmentation (Nelson et al., 1989a,b). Permittivity measurements on pulverized samples of relatively pure materials have been used with dielectric mixture equations to estimate the permittivities of the solid materials (Nelson et al., 1989a; Nelson and You, 1990). Of several well-known dielectric mixture equations tried, the Landau and Lifshitz, Looyenga equation provided the best results for relatively low-permittivity and low-loss materials. Federal regulations (CFR 30, 1995) require rock dusting in coal mines to insure that "the incombustible content of the combined coal dust, rock dust, and other dust shall not be less than 65 percentum" to limit explosion hazards.

Measurements of permittivities were made on pulverized samples of Pittsburgh coal, rock dust (limestone), and a 35−65% mixture of coal and limestone (Nelson, 1996) to learn how different the dielectric properties of these materials might be and whether there might be potential for developing techniques of rapidly sensing coal and rock dust concentrations.

Pulverized samples of Pittsburgh coal and limestone (more than 90% calcium carbonate), with most of the particle diameters ranging from 5 to 100 μm, were furnished for the measurements by the Pittsburgh Research Center, US Bureau of Mines. Moisture content, determined by drying at 105°C for 24 h in an air oven, was 1.4% for the coal and 0.14% for the limestone. A mixed sample,

35% coal and 65% limestone by weight, was also included for the dielectric measurements. Permittivity measurements were made at 11.7 GHz with an X-band measurement system (Nelson, 1972), and the short-circuited waveguide method (Roberts and von Hippel, 1946; Nelson et al., 1974). The samples were weighed before they were placed into a 5-cm long, WR-90-waveguide, short-circuited sample holder for the measurements. Sample length was determined for each of a sequence of measurements at successively increasing sample bulk densities that were determined from sample weight, sample length, and waveguide cross-sectional area.

Particle densities ρ_s, which correspond to solid material densities, were calculated from sample weights of 15−25 g and corresponding particle volumes that were determined by measurements in a Beckman Model 930 air comparison pycnometer (Nelson et al., 1989b). Initial pycnometer measurements on pulverized coal samples revealed that the coal was compressible, as indicated by continual drift of the pressure null indicator for several minutes after the sample was placed under 2 atmospheres of air pressure. Therefore, the 1-1/2-1 atmosphere mode of operation was used so that the air displacement measurement could be determined at a pressure of 1 atmosphere. Pycnometer measurements on limestone samples were made in the 1−2 atmosphere compression mode, because both methods gave nearly the same volume values, and the 2-atmosphere mode has better sensitivity for the null pressure determination. Pycnometer measurements on the coal−limestone mixture were made with the 1-1/2-1 atmosphere mode because of the compressibility of the coal.

Results of the permittivity measurements and sample bulk density determinations on the coal and limestone samples are shown in Tables 10.2 and 10.3, where the results of analyses are also summarized.

It was shown earlier (Nelson, 1983a; Nelson, 1992) that the linearity with bulk density of the cube root of the dielectric constant of the air−particle mixture, ε', the real part of the relative complex permittivity, $\varepsilon = \varepsilon' - j\varepsilon''$, where ε'' is the dielectric loss factor, is consistent with the Landau and Lifshitz, Looyenga dielectric mixture equation, which can be stated as follows for a two-phase mixture:

$$(\varepsilon)^{1/3} = v_1(\varepsilon_1)^{1/3} + v_2(\varepsilon_2)^{1/3} \tag{10.1}$$

where subscripts 1 and 2 refer to the air and the solid particulate material, respectively, and v represents the volume fraction occupied by a component of the mixture. For the two-phase (air−particle) mixture, $v_1 + v_2 = 1$ and the permittivity of air is $1 - j0$. Solving Eq. (10.1) for ε_2 in terms of ε (relative complex permittivity of the mixture), and considering that $v_2 = v_s$, the volume fraction occupied by the solid material provides an expression from which the permittivities of the particulate material can be calculated:

$$\varepsilon_s = \varepsilon_2 = \left[\frac{\varepsilon^{1/3} + v_2 - 1}{v_2} \right] = \left[\frac{\varepsilon^{1/3} + v_s - 1}{v_s} \right] \tag{10.2}$$

The necessary value for v_s can be obtained if the bulk density ρ of the mixture and the density ρ_s of the solid particulate material are known, since $v_s = \rho/\rho_s$.

The cube roots of the dielectric constants of the three pulverized samples (see Tables 10.1 and 10.2) are shown as functions of sample bulk density in Figure 10.2.

Linear regression analyses showed very high coefficients of determination (r^2 values), and the intercepts are very close to the theoretical value of 1, which, for zero bulk density, is the dielectric constant of air. These regression constants are given in Table 10.4.

Table 10.2 Measured Permittivities, $\varepsilon = \varepsilon' - j\varepsilon''$, and Bulk Densities of Pulverized Pittsburgh Coal at 11.7 GHz and 20°C and Permittivities of Solid Coal Estimated by the Landau and Lifshitz, Looyenga Dielectric Mixture Equation (Nelson, 1996)

| Measured Values | | Cube Root Dielectric Constant | Volume Fraction | Estimated Value |
| Air–Particle Permittivity | Bulk Density | | | Solid Particle Permittivity |
ε	ρ, g/cm^3	$(\varepsilon')^{1/3}$	v_s	ε_s
1.894−j0.035	0.565	1.237	0.382	4.262−j0.157
1.948−j0.037	0.598	1.249	0.404	4.220−j0.153
1.983−j0.037	0.619	1.256	0.418	4.195−j0.146
2.012−j0.039	0.632	1.262	0.427	4.208−j0.149
2.044−j0.041	0.648	1.269	0.438	4.208−j0.152
2.089−j0.043	0.671	1.278	0.453	4.203−j0.151
2.113−j0.046	0.682	1.283	0.461	4.208−j0.158
2.181−j0.051	0.716	1.297	0.484	4.200−j0.163
2.203−j0.051	0.724	1.301	0.489	4.217−j0.161
2.229−j0.054	0.736	1.306	0.497	4.218−j0.166

Table 10.3 Measured Permittivities, $\varepsilon = \varepsilon' - j\varepsilon''$, and Bulk Densities of Pulverized Limestone at 11.7 GHz and 20°C and Permittivities of Solid Limestone Estimated by the Landau and Lifshitz, Looyenga Dielectric Mixture Equation (Nelson, 1996)

| Measured Values | | Cube Root Dielectric Constant | Volume Fraction | Estimated Value |
| Air–Particle Permittivity | Bulk Density | | | Solid Particle Permittivity |
ε	ρ, g/cm^3	$(\varepsilon')^{1/3}$	v_s	ε_s
2.363−j0.011	0.972	1.332	0.353	7.292−j0.066
2.415−j0.011	1.008	1.342	0.367	7.212−j0.062
2.519−j0.011	1.064	1.361	0.387	7.213−j0.057
2.565−j0.012	1.088	1.369	0.396	7.215−j0.060
2.847−j0.013	1.228	1.417	0.447	7.240−j0.054
2.961−j0.014	1.278	1.436	0.465	7.280−j0.055
3.098−j0.015	1.332	1.458	0.484	7.359−j0.055
3.258−j0.017	1.395	1.482	0.507	7.427−j0.058
3.462−j0.020	1.476	1.513	0.537	7.477−j0.062
3.772−j0.024	1.587	1.557	0.577	7.582−j0.066
3.930−j0.026	1.642	1.578	0.597	7.624−j0.068
4.154−j0.031	1.715	1.608	0.624	7.094−j0.075
4.221−j0.031	1.739	1.616	0.632	7.695−j0.073

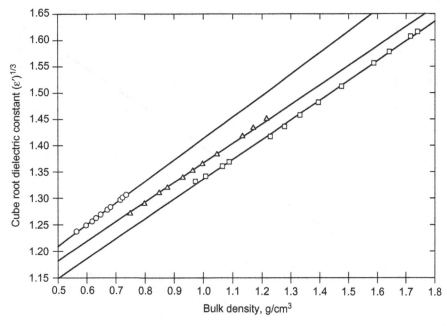

FIGURE 10.2

Linear relationships between the cube roots of the dielectric constants of pulverized samples and their bulk densities at 20°C and 11. 7 GHz. Symbols: circle—Pittsburgh coal, triangle—35−65% coal−limestone mixture, square—limestone.

Table 10.4 Linear Regression Statistics for the Cube Root of the Dielectric Constant of Pulverized Coal and Limestone Samples as a Function of Sample Bulk Density, $(\varepsilon')^{1/3} = a + b\rho$

	Regression without Point (0,1)			Regression with Point (0,1) Included		
	Intercept	Slope	Coefficient of Determination	Intercept	Slope	Coefficient of Determination
Material	*a*	*b*	r^2	*a*	*b*	r^2
Coal	1.0049	0.4083	0.9987	1.0003	0.4152	0.9999
Limestone	0.9612	0.3750	0.9990	0.9877	0.3561	0.9982
35−65% Coal−limestone mixture	0.9823	0.3846	0.9997	0.9965	0.3703	0.9995

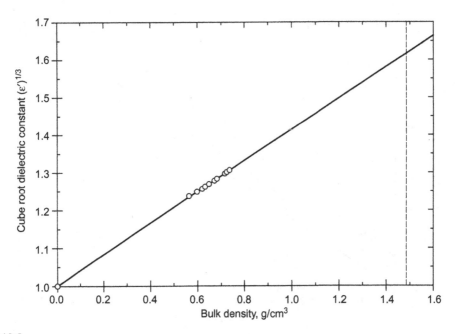

FIGURE 10.3

Linear regression of the cube root of the dielectric constant $(\varepsilon')^{1/3}$ of pulverized Pittsburgh coal on bulk density (ρ) with point (0, 1) included in the regression calculation, showing intersection of regression line with $\rho_s = 1.48$ line at 1.615, $\varepsilon'_s = (1.615)^3 = 4.21$.

Since the point $(\rho = 0, \; \varepsilon' = 1)$ is a valid reference point, it can be included in the regression calculation, and this brings the intercept even closer to $\varepsilon' = 1$ at zero bulk density, as illustrated in Figure 10.3.

When the linear regression of the cube root of the dielectric constant on bulk density, provides an r^2 value so nearly 1, and the zero-bulk-density intercept is so close to the value of 1, the Landau and Lifshitz, Looyenga dielectric mixture equation can be used with confidence to estimate the permittivity of the solid material (see Tables 10.2 and 10.3). The linear extrapolation of $(\varepsilon')^{1/3}$ to the density of the solid material is illustrated for the coal measurements in Figure 10.3 and for the limestone measurements in Figure 10.4.

Values provided by the regression equations for the dielectric constants of the three material samples are given in Table 10.5.

The mean values of the solid material permittivities calculated with the Landau and Lifshitz, Looyenga mixture equation for the permittivity measurements at each bulk density, as illustrated for coal in Table 10.2 and for limestone in Table 10.3, are also included in Table 10.5.

Permittivity values for the pulverized coal in Table 10.2 agree reasonably well with those reported for other measurements on coal (Nelson et al., 1980, 1981; Balanis et al., 1976, 1978, 1980; Klein, 1981) when differences in frequency, moisture content, and density are taken into account. Values obtained check extremely well with those reported earlier for Pittsburgh No. 8 ROM coal at the same frequency and similar densities (Nelson et al., 1980).

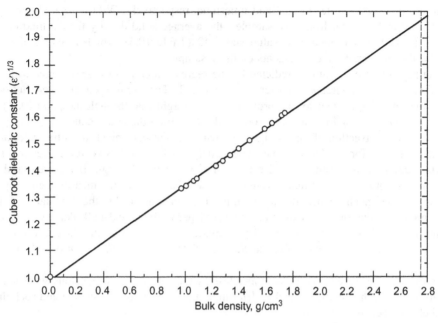

FIGURE 10.4

Linear regression of the cube root of the dielectric constant $(\varepsilon')^{1/3}$ of pulverized limestone on bulk density (ρ) with point (0, 1) included in the regression calculation, showing intersection of regression line with $\rho_s = 2.75$ line at 1.967, $\varepsilon'_s = 1.967^3 = 7.61$.

Table 10.5 Estimated Permittivities of Solid Materials, $\varepsilon_s = \varepsilon'_s - j\varepsilon''_s$, from Measurements on Pulverized Samples at 20°C and 11.7 GHz

Material	Density, g/cm³	ε'_s Predicted by Linear Regression (Point (0,1) Included)	ε_s by Mixture Equation (Mean Values)
Coal	1.48	4.21	$4.21 - j0.156$
Limestone	2.75	7.61	$7.41 - j0.063$
35−65% Coal−limestone mixture	2.13	5.69	$5.64 - j0.108$

It is interesting to check the ρ_s value measured with the air comparison pycnometer for the 35−65% (by weight) coal−limestone mixture by calculating the expected average solid material density from the pycnometer determinations of ρ_s for the pulverized coal and limestone samples independently. The expected ρ_s for the mixture can be obtained from the appropriate relationship between these solid densities:

$$\frac{1}{\rho_s} = \frac{0.35}{\rho_{sc}} + \frac{0.65}{\rho_{sl}} \tag{10.3}$$

where subscripts sc and sl refer to coal and limestone, respectively. With measured density values of 1.48 for coal and 2.75 for limestone samples, the average solid density for the mixture is given as 2.11 by Eq. (10.3). The measured value was 2.13 (Table 10.5), which is within the expected accuracy for the pycnometer measurements on these samples.

One can note that the ε'_s values predicted by the regression equations agree better with the mixture equation when the intercept is closer to the value 1. The mixture equation prediction of the dielectric constant is equivalent to that provided by a straight line through the point (0, 1) and the selected single point defined by the cube root of the measured dielectric constant at any particular bulk density. The intersection of that straight line with the vertical line at the solid material density ρ_s gives the estimate for ε'_s. Thus, if the measured (ρ, ε') point is above the regression line (see Figures 10.3 and 10.4, for example), the estimated ε'_s value will be high. If the measured point is below the regression line, the estimated value for ε'_s will be low. For the measurements reported on these samples, the mean values of the permittivities calculated by the Landau and Lifshitz, Looyenga mixture equation, taken over all measured permittivity and bulk density points, should provide the most reliable estimates. And they provide values for both the dielectric constant and the loss factor, which is of greater interest than the dielectric constant for most dielectric heating applications.

Values of both the dielectric constants and loss factors of coal and limestone were sufficiently different to justify further studies aimed at determining rock dust content in coal and rock dust mixtures by dielectric sensing techniques.

10.7 SENSING PULVERIZED MATERIAL MIXTURE PROPORTIONS

Rock dusting in coal mines is required to reduce explosion hazards by providing at least 65% noncombustible content in the combined coal dust and other dust in the mines (CFR 30, 1995). Currently, the noncombustible content is determined by heating processes that burn the coal without decomposing the rock dust. Techniques are needed for rapid reliable determinations of the rock dust content in samples of dust from coal mines. Measurements of microwave permittivities of pulverized coal and pulverized limestone samples have been used to obtain the permittivities of the solid materials through the use of Landau and Lifshitz, Looyenga dielectric mixture equation (Nelson, 1996). This mixture equation implies the linearity of the cube root of the dielectric constant with bulk density of an air–particle mixture, and the usefulness of the relationship has been demonstrated with a number of materials (Nelson and You, 1990; Nelson, 1992, 1988).

At 11.7 GHz and 20°C, the dielectric constants of dry coal and limestone are about 4.2 and 7.6, respectively, while the loss factors are about 0.16 and 0.06, respectively (Nelson, 1996; Table 10.5). Thus, both components of the relative complex permittivity of coal and limestone rock have significant differences that should be detectable by suitable measurements. Broad-frequency-range measurements of coal and limestone permittivity showed reasonably small variations in dielectric constants between 200 MHz and 20 GHz with both decreasing somewhat at higher frequencies (Nelson and Bartley, 1997). Thus, it appeared reasonable to determine proportions of coal and rock dust by measurements of the permittivities of dust samples. Therefore, resonant cavity measurements were explored for this purpose.

10.7.1 PRINCIPLES OF RESONANT CAVITY MEASUREMENT

Resonant cavity measurement techniques are convenient for measuring the relative dielectric complex permittivity $\varepsilon = \varepsilon' - j\varepsilon''$ of materials at single microwave frequencies. It follows from resonant cavity perturbation theory that when a dielectric object of low loss, $(\varepsilon')^2 \gg (\varepsilon'')^2$, is inserted into the cavity, the change in the resonant frequency Δf and the change in the cavity transmission factor ΔT can be expressed as follows (Kraszewski and Nelson, 1995):

$$\Delta f \approx 2(\varepsilon' - 1)f_0 K \left(\frac{v_s}{v_0}\right) \tag{10.4}$$

$$\Delta T \approx 4Q_0 \varepsilon'' K^2 \left(\frac{v_s}{v_0}\right) \tag{10.5}$$

where f_0 is the resonant frequency of the empty cavity, Q_0 is the Q-factor of the empty cavity, K is the shape factor for the dielectric object, v_s is the volume of the sample (object), and v_0 is the volume of the empty cavity. For the sample configuration used in this study, the shape factor is approximately one. Use of these equations and the $\Delta f / \Delta T$ ratio for measurements on low-loss dielectric objects has shown that desired permittivity-related characteristics of objects can be determined independent of object mass (Kraszewski et al., 1990; Kraszewski and Nelson, 1996). Thus, the technique can be expected to sense permittivities of powdered samples relatively independent of the bulk density of the powdered materials. This is an important advantage, because bulk densities of powdered materials vary greatly, depending on settling or packing, and carefully controlling the degree of packing is troublesome for a practical measurement.

10.7.2 MEASURING MIXTURE PROPORTIONS

Pulverized samples of Pittsburgh coal and limestone (greater than 90% calcium carbonate), with most particle diameters ranging from 5 to 100 μm, were furnished by the Pittsburgh Research Center, National Institute of Occupational Safety and Health (formerly US Bureau of Mines). Moisture content, determined by drying samples for 24 h at 105°C in an air oven, was 1.4% for the coal and 0.5% for the limestone. Samples of pure coal and pure limestone and mixtures ranging from 10.00% to 80.00% coal, by weight, were prepared for the microwave resonant cavity measurements. The microwave cavity and sample holder are shown in Figure 10.5, and the instruments and procedures used for the measurements have been described in detail elsewhere (Nelson and Kraszewski, 1998).

The resonant frequency of the empty rectangular-waveguide cavity was 2.473 GHz.

A series of resonant cavity measurements was taken to provide a calibration curve. Powdered coal−limestone mixture samples of 0%, 10%, 20%, 40%, 60%, 80%, and 100% coal were used to develop the calibration curve as shown in Figure 10.6.

The performance of the calibration was checked with additional coal−limestone mixture samples prepared with 10% increments for the 0−60% coal range. The tests showed that the proportion of coal in the mixtures was determined with a standard error of performance of 1.8%. Thus, the resonant cavity measurement technique, based on the resonant frequency shift

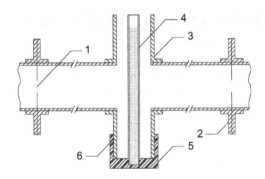

FIGURE 10.5

Cross section of rectangular-waveguide resonant cavity and sample holder, showing coupling irises (1), waveguide flanges (2), aluminum sleeves (3), 12-mm diameter glass tube (4), and Delrin cap (5) with O-ring seal (6) (Nelson and Kraszewski, 1998).

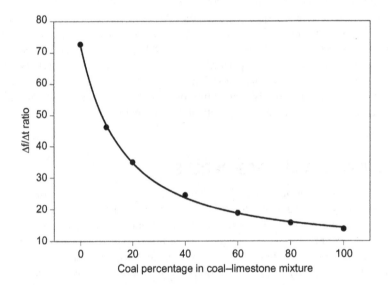

FIGURE 10.6

Relationship between $\Delta f / \Delta T(y)$ and coal percentage in coal−limestone powder mixture (x) for calibration data set: $y = a\, e^{b/(c+x)}$, where $a = 6.949$, $b = 102.8$, and $c = 43.82$ (Nelson and Kraszewski, 1998).

and the change in the transmission factor when mixed coal and limestone samples are inserted into the cavity, provides a means for rapidly estimating the percentage of coal in the mixture. The technique might be developed for rapid tests of dust in coal mines to determine whether the 65% noncombustible content requirement is being met for prevention of coal mine explosions.

ic
REFERENCES

Balanis, C.A., Rice, W.S., Smith, N.S., 1976. Microwave measurements of coal. Radio Sci. 11 (4), 413−418.

Balanis, C.A., Jeffrey, J.L., Yoon, Y.K., 1978. Electrical properties of eastern bituminous coal as a function of frequency, polarization and direction of the electromagnetic wave and temperature of the sample. IEEE Trans. Geosci. Electron. GE-16 (4), 316−323.

Balanis, C.A., Shepard, P.W., Ting, F.T.C., Kardosh, W.F., 1980. Anisotropic electrical properties of coal. IEEE Trans. Geosci. Remote Sens. GE-18 (3), 250−256.

Bluhm, D.D., Fanslow, G.E., Beck-Montgomery, S.R., Nelson, S.O., 1980. Selective magnetic enhancement of pyrite in coal by dielectric heating. In: Proceedings of the Conference on Chemistry and Physics of Coal Utilization, Morgantown, WV.

Bluhm, D.D., Fanslow, G.E., Nelson, S.O., 1986. Enhanced magnetic separation of pyrite from coal after microwave heating. IEEE Trans. Magn. MAG-22 (6), 1887−1890.

CFR 30, 1995. US Code of Federal Regulations, Mineral Resources 30, Part 75, Par. 75.403, p. 492.

Dobson, M.C., Ulaby, F.T., 1986. Active microwave soil moisture research. IEEE Trans. Geosci. Remote Control 24 (1), 23−36.

Ergun, S., Bean, E.H., 1968. Magnetic Separation of Pyrite from Coals. US Bureau of Mines, Report of Investigations 7181.

Fanslow, G.E., Bluhm, D.D., Nelson, S.O., 1980. Dielectric heating of mixtures containing coal and pyrite. J. Microw. Power 15 (3), 187−191.

Iglesius, J., Westphal, W.B., 1967. Supplementary Dielectric Constant and Loss Measurements on High Temperature Materials. Tech. Report 203, Laboratory for Insulation Research, Massachusetts Institute of Technology, Cambridge, MA.

Jordan, C.E., Sullivan, G.V., 1985. Dielectric Separation of Minerals. US Bureau of Mines Bulletin 685: 17.

Jordan, C.E., Sullivan, G.V., Davis, B.E., Weaver, C.P., 1980. A Continuous Dielectric Separator for Mineral Beneficiation. US Bureau of Mines, Report of Investigation 8437.

Klein, A., 1981. Microwave determination of moisture in coal—comparison of attenuation and phase measurement. J. Microw. Power 16 (3−4), 289−304.

Kraszewski, A.W., Nelson, S.O., 1995. Contactless mass determination of arbitrarily shaped dielectric objects. Meas. Sci. Technol. 6, 1598−1604.

Kraszewski, A.W., Nelson, S.O., 1996. Resonant cavity perturbation—some new applications of an old measuring technique. J. Microw. Power Electromagn. Energy 31 (3), 178−187.

Kraszewski, A.W., Nelson, S.O., You, T.-S., 1990. Use of a microwave cavity for sensing dielectric properties of arbitrarily shaped biological objects. IEEE Trans. Microw. Theory Tech. 38 (7), 858−863.

Landau, L.D., Lifshitz, E.M., 1960. Electrodynamics of Continuous Media. Pergamon Press Inc., Oxford.

Looyenga, H., 1965. Dielectric constants of heterogeneous mixtures. Physica 31, 401−406.

Marusak, L.A., Walker, P.L.J., Mulay, L.N., 1976. The magneto-kinetics of the oxidation of pyrite. IEEE Trans. Magn. MAG-12 (6), 889−891.

Nelson, S.O., 1972. A system for measuring dielectric properties at frequencies from 8.2 to 12.4 GHz. Trans. ASAE 15 (6), 1094−1098.

Nelson, S.O., 1973. Microwave dielectric properties of grain and seed. Trans. ASAE 16 (5), 902−905.

Nelson, S.O., 1983a. Observations on the density dependence of the dielectric properties of particulate materials. J. Microw. Power 18 (2), 143−152.

Nelson, S.O., 1983b. Dielectric properties of some fresh fruits and vegetables at frequencies of 2.45 to 22 GHz. Trans. ASAE 26 (2), 613−616.

Nelson, S.O., 1988. Estimating the permittivity of solids from measurements on granular or pulverized materials. In: Sutton, W.H., Brooks, M.H., Chabinsky, I.J. (Eds.), Microwave Processing of Materials, vol. 124. Materials Research Society, Pittsburgh, PA, pp. 149−154.

Nelson, S.O., 1992. Estimation of permittivities of solids from measurements on pulverized or granular materials. In: Priou, A. (Ed.), Dielectric Properties of Heterogeneous Materials, vol. PIER 6. Elsevier, New York, Amsterdam, London, Tokyo, pp. 231–271.

Nelson, S.O., 1996. Determining dielectric properties of coal and limestone by measurements on pulverized samples. J. Microw. Power Electromagn. Energy 31 (4), 215–220.

Nelson, S.O., Bartley, P.G., Jr., 1997. Estimating properties of solids from permittivity measurements on pulverized samples. In: Proceedings of the Conference on Microwave and High Frequency Heating, Fermo, Italy, 488–491.

Nelson, S.O., Kraszewski, A., 1998. Sensing pulverized material mixture proportions by resonant cavity measurements. IEEE Trans. Instrum. Meas. 47 (5), 1201–1204.

Nelson, S.O., You, T.-S., 1990. Relationships between microwave permittivities of solid and pulverised plastics. J. Phys. D Appl. Phys. 23, 346–353.

Nelson, S.O., Stetson, L.E., Schlaphoff, C.W., 1974. A general computer program for precise calculation of dielectric properties from short-circuited waveguide measurements. IEEE Trans. Instrum. Meas. 23 (4), 455–460.

Nelson, S.O., Fanslow, G.E., Bluhm, D.D., 1980. Frequency dependence of the dielectric properties of coal. J. Microw. Power 15 (4), 277–282.

Nelson, S.O., Beck-Montgomery, S.R., Fanslow, G.E., Bluhm, D.D., 1981. Frequency dependence of the dielectric properties of coal—Part II. J. Microw. Power 16 (3–4), 319–326.

Nelson, S.O., Lindroth, D.P., Blake, R.L., 1989a. Dielectric properties of selected minerals at 1 to 22 GHz. Geophysics 54 (10), 1344–1349.

Nelson, S.O., Lindroth, D.P., Blake, R.L., 1989b. Dielectric properties of selected and purified minerals at 1 to 22 GHz. J. Microw. Power Electromagn. Energy 24 (4), 213–220.

Nelson, S.O., Stetson, L.E. and Schlaphoff, C.W., 1972. Computer Program for Calculating Dielectric Properties of Low- or High-Loss Materials from Short-Circuited Waveguide Measurements. ARS-NC-4, Agricultural Research Service, US Department of Agriculture.

Olhoeft, G.R., 1981. Electrical properties of rocks. In: Touloukian, Y.S., Ho, C.Y. (Eds.), Physical Properties of Rocks and Minerals. McGraw-Hill, Inc., New York, NY (Chapter 9)

Olhoeft, G.R., 1986. Direct detection of hydrocarbons and organic chemicals with ground penetrating radar and complex resistivity, National Well Water Association Conference on Peteroleum Hyrdocarbons and Organic Chemicals in Ground Water, Houston, TX.

Parkhomenko, E.I., 1967. Electrical Properties of Rocks (Trans. from Russian). Plenum Press, New York, NY.

Roberts, S., von Hippel, A., 1946. A new method for measuring dielectric constant and loss in the range of centimeter waves. J. Appl. Phys. 17 (7), 610–616.

Singh, R., Singh, K.P., Singh, R.N., 1979. Microwave measurements on some Indian coal samples. Proc. Indian Natl. Sci. Acad. 45A, 397–405.

Thirumalai, K., 1970. Rock Fragmentation by Creating a Thermal Inclusion with Dielectric Heating. US Bureau of Mines RI 7424: 33.

von Hippel, A.R., 1954. Dielectric Materials and Applications. The Technology Press of M.I.T. and John Wiley & Sons, New York, NY.

Webb, W.E., Church, R.H., 1986. Measurement of Dielectric Properties of Minerals at Microwave Frequencies. US Bureau of Mines, Report of Investigation 9035: 8.

Westphal, W.B., 1977. Dielectric Constant and Loss Data. Techn. Report AFML-TR-74-250, III, Air Force Systems Command, Wright-Patterson Air Force Base, OH.

Westphal, W.B., Sills, A., 1972. Dielectric Constant and Loss Data. Tech Report AFML-TR-72-39. Air Force Systems Command, Wright-Patterson Air Force Base, OH.

DIELECTRIC PROPERTIES OF SELECTED FOOD MATERIALS

The dielectric properties of foods are important in understanding and modeling their behavior in radio-frequency (RF) and microwave processing, heating, and cooking of food materials (Mudgett et al., 1995; Nelson and Datta, 2001). They are important because they influence the absorption of energy from the high-frequency electric fields and the conversion to heat. Thus, they are also important in the design of RF and microwave processing equipment and in the design of foods and meals intended for microwave preparation. Data on the dielectric properties of foods have been compiled for reference in several publications (Nelson, 1973; Tinga and Nelson, 1973; Kent, 1987; Datta et al., 1995). However, the variety in food materials is so diverse that the dielectric properties of interest must often be determined by measurement for particular applications. Materials such as foods generally have significant loss factors, and their dielectric properties are therefore dependent on both temperature and frequency as well as the physical and chemical characteristics of the material that include composition and density. Although a single frequency is generally used for heating or processing, knowledge of the frequency dependence of the permittivity of the food material is important in predicting its behavior under RF or microwave heating. Because the dielectric properties of materials are dependent on temperature, the measurement of the dielectric properties as a function of temperature, as well as frequency, is often advisable.

11.1 MEASUREMENT OF THE DIELECTRIC PROPERTIES OF SOME FOOD MATERIALS

An open-ended coaxial-line probe was used with network and impedance analyzers and a sample temperature control assembly, designed for use with the probe, to measure the dielectric properties of some food materials as a function of frequency and temperature (Nelson and Bartley, 2002a,b). The food materials included a macaroni and cheese dinner preparation, whey protein gel, ground whole-wheat flour, and apple juice.

Results of measurements on samples of these foods are shown in Figures 11.1−11.5. Values for the dielectric constant and loss factor of the macaroni and cheese material at three ISM frequencies (frequencies allocated for industrial, scientific, and medical applications), 27, 40, and 915 MHz, and at 1800 MHz, the upper frequency limit of the analyzer used, are shown in Figure 11.1.

The gradual decrease in the dielectric constant with increasing temperature appears reasonable for a food material containing 66% water, wet basis (w.b.), although less temperature dependence

Dielectric Properties of Agricultural Materials and Their Applications. DOI: http://dx.doi.org/10.1016/B978-0-12-802305-1.00011-7

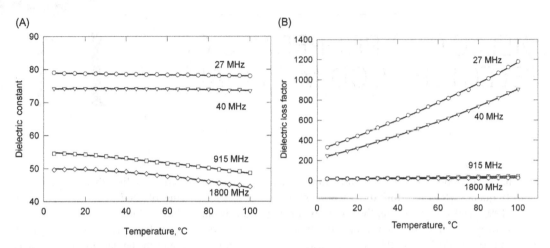

FIGURE 11.1

Temperature dependence of the dielectric properties of homogenized macaroni and cheese of 66% moisture content at indicated frequencies (Nelson and Bartley, 2002a).

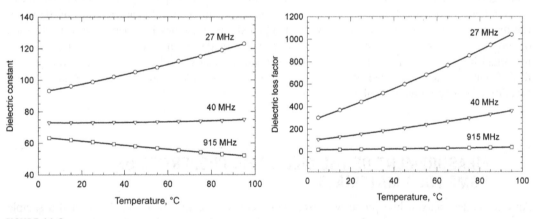

FIGURE 11.2

Temperature dependence of the dielectric properties of a whey protein gel of 74% moisture content and density of 1.05 g/cm^3 at indicated frequencies (Nelson and Bartley, 2002a).

was noted at 27 and 40 MHz than at the microwave frequencies. The increase in the loss factor, as temperature increases, is mainly attributable to the increase in ionic conduction with increasing temperature at 27 and 40 MHz. For dielectric heating applications, this means that the rate of energy absorption from high-frequency electric fields can be expected to increase along with the increase in temperature. The loss factor, which has a greater influence on heating rate for RF and microwave dielectric heating than the dielectric constant, shows much less temperature dependence

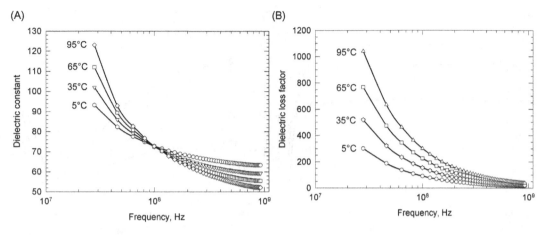

FIGURE 11.3

Frequency dependence of the permittivity of a whey protein gel of 74% moisture content and density of 1.05 g/cm^3 at indicated temperatures (Nelson and Bartley, 2002a).

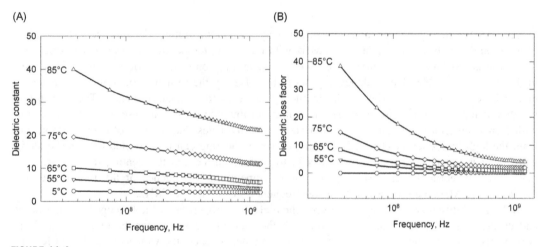

FIGURE 11.4

Frequency dependence of the dielectric properties of ground wheat (whole-wheat flour) of 12% moisture content at indicated temperatures (Nelson and Bartley, 2002a). Sample density = 0.78 g/cm^3.

at the microwave frequencies than it does at the lower frequencies of 27 and 40 MHz. In addition, the curves for the loss factor at 915 and 1800 MHz are nearly identical and indicate that they should also well represent the behavior of the loss factor at the common ISM frequency of 2450 MHz. The dielectric constant at 2450 MHz can also be estimated by extrapolation of data from the curves for 915 and 1800 MHz. Values for the dielectric constant and loss factor of whey protein gel at three ISM frequencies (27, 40, and 915 MHz) are shown in Figure 11.2.

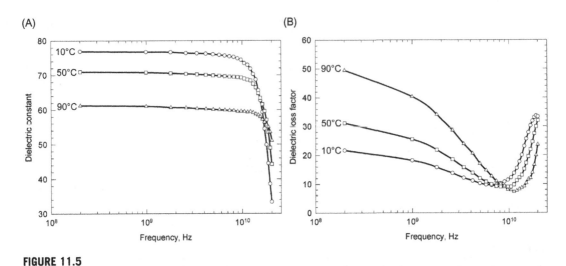

FIGURE 11.5

Frequency dependence of the dielectric properties of apple juice at indicated temperatures (Nelson and Bartley, 2002a) (88.5% water).

At 40 MHz, the dielectric constant shows very little temperature dependence. The gradual decrease in the dielectric constant at 915 MHz with increasing temperature appears reasonable for a food material containing this much water (74%) outside a region of dielectric relaxation. However, in the data for 27 MHz, that tendency appears to be offset by the influence of some low-frequency dielectric relaxation process (below the frequency range of these measurements). The increase in the loss factor, as temperature increases, is mainly attributable to the increase in ionic conduction with increasing temperature at 27 and 40 MHz. It also indicates that the rate of energy absorption from the high-frequency electric fields during dielectric heating can be expected to increase along with the increase in temperature. The loss factor, which influences the heating rate for RF and microwave dielectric heating, shows much less temperature dependence at the microwave frequency of 915 MHz than it does at the lower frequencies of 27 and 40 MHz.

Substantial differences are noted between the dielectric properties of the homogenized macaroni and cheese and the whey protein gel. Although the dielectric constant for the two at 40 MHz is very nearly the same and shows very little temperature dependence, the values of the dielectric constant of the whey protein gel are greater than those of the macaroni and cheese at both 27 and 915 MHz. And although the temperature dependence of the dielectric constants for the two are similar at 915 MHz, the temperature dependence at 27 MHz is much greater for the whey protein gel. The loss factors for the two different foods show similar values and similar temperature-dependent behavior at 27 and 915 MHz, but the value at 40 MHz is considerably greater for the macaroni and cheese than for the whey protein gel. It is interesting to examine the frequency dependence of the whey protein gel at different temperatures, as shown in Figure 11.3.

The influence of ionic conductivity is evident at the lower frequencies in the data for the dielectric loss factor, where it combines with the influence of bound (rotationally hindered) water. At frequencies above about 100 MHz, the dielectric constant decreases with increasing temperature, as might be expected for materials in the absence of dielectric relaxation. However, at frequencies below 100 MHz, the dielectric constant increases with increasing temperature, probably in response to a low-frequency dielectric relaxation shifting to higher frequencies as temperature rises. This would also contribute to the increase in the loss factor as temperature rises, along with the increased ionic conduction due to increasing temperature.

Dependence of the permittivity of ground whole wheat on frequency and temperature is illustrated in Figure 11.4.

Because moisture content is much lower (12%, w.b.), both the dielectric constant and loss factor are relatively low, but they increase substantially with increasing temperature. Because both parts of the permittivity increase with increasing temperature, these changes are likely due to relaxation mechanisms operating at frequencies below the frequency range illustrated. The increases in the loss factor at very low frequencies are also related to increased ionic or protonic conductivity (Funk, 2001) as a result of higher temperatures. Data for the permittivity of apple juice, shown in Figure 11.5, clearly show the influence of liquid water, with a relaxation frequency below 20 GHz at 10°C, which shifts to higher frequencies above 20 GHz at temperatures of 50°C and 90°C.

Below the range of the liquid water relaxation, the dielectric constant is reduced by increasing temperature in much the same way as it is with water. The increase in the loss factor with temperature at the lower frequencies is most likely attributable to the influence of ionic conduction.

To explore further the influence of water content on the dielectric properties of food materials, measurements with the open-ended coaxial-line probe were taken with the same equipment for a wider range of water contents. To best illustrate the dielectric behavior of various food materials with respect to dependence on frequency and temperature as influenced by moisture content, data are presented graphically. In Figures 11.6 and 11.7, the dielectric constants and loss factors of hard red winter wheat are presented for two different moisture contents, 11.2% and 25.0%, w.b., from 10 MHz to 1.8 GHz.

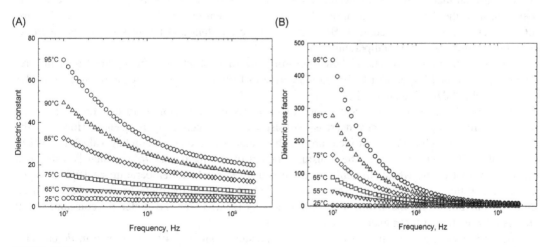

FIGURE 11.6

Frequency dependence of the dielectric properties of ground hard red winter wheat of 11.2% moisture content at indicated temperatures over the frequency range from 10 MHz to 1.8 GHz (Nelson and Trabelsi, 2006).

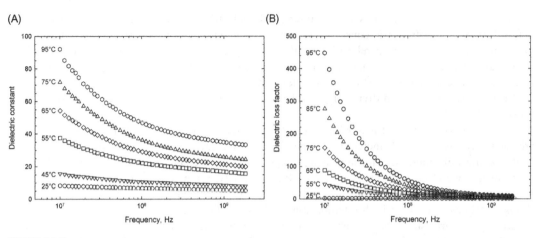

FIGURE 11.7

Frequency dependence of the dielectric properties of ground hard red winter wheat of 25.0% moisture content at indicated temperatures over the frequency range from 10 MHz to 1.8 GHz (Nelson and Trabelsi, 2006).

Because of the relatively large dimensions of wheat kernels (ca. 6 by 3 mm) compared to the 3-mm diameter open-ended coaxial-line probe used for the permittivity measurements, samples were ground to provide a more homogeneous sample (Nelson and Trabelsi, 2006). However, dielectric properties of whole-kernel and ground wheat of the same bulk density are very similar (Nelson, 1984), so the results shown are illustrative of wheat permittivity at a bulk density of about 0.8 g/cm³. Trends in the behavior of the wheat dielectric properties at the two different moisture contents are similar in that both decrease consistently with increasing frequency. However, the values of both the dielectric constant and loss factor are much greater for the wheat at 25% moisture than at 11.2% moisture content. Both the dielectric constant and the loss factor also increase consistently with increasing temperature.

Results of the measurements of dielectric properties of fresh chicken breast meat over the frequency range from 10 MHz to 1.8 GHz at temperatures from 5°C to 65°C are shown in Figure 11.8 (Nelson et al., 2007b; Zhuang et al., 2007).

The behavior of both the dielectric constant and loss factor is similar to that for wheat, with respect to frequency, in that they decrease consistently with increasing frequency. However, the properties for fresh chicken breast meat are much greater than the corresponding values for wheat, even at the 25% moisture level. The moisture content of the meat was 76%, which is also much greater than that of the wheat. In addition, there is a reversal of the temperature coefficient of the dielectric constant as frequency increases, which was not noted in the dielectric behavior of wheat. Below about 200 MHz, the temperature coefficient is positive for the dielectric constant, but it changes to negative at the higher frequencies.

Dielectric properties of a fresh fruit of similar moisture content, 74%, are shown in Figure 11.9 for banana over the same frequency range at temperatures from 5°C to 65°C (Nelson, 2003).

The behavior of the dielectric properties of banana is similar to that of the chicken breast meat, with dielectric constants of about the same magnitude. However, the dielectric loss factor values

(A)

(B)

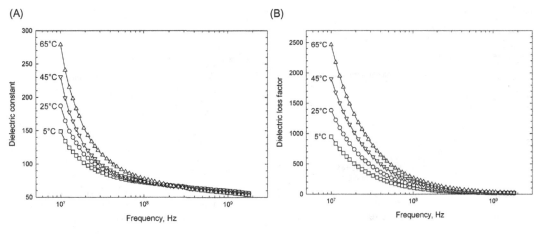

FIGURE 11.8

Frequency and temperature dependence of the dielectric properties of fresh chicken breast meat, moisture content: 76% (Nelson et al., 2007b).

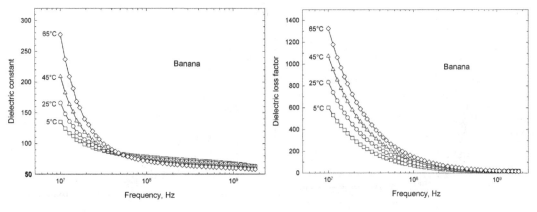

FIGURE 11.9

Frequency and temperature dependence of fresh banana permittivity, moisture content: 74% (Nelson, 2003).

for banana are considerably less than those of the chicken breast meat. Also, the frequency where temperature dependence disappears for the dielectric constant is about 50 MHz compared to 200 MHz for the meat.

For fresh fruit tissue of apples, with 85% moisture content, the dielectric properties over the same frequency and temperature ranges are shown in Figure 11.10 (Nelson, 2003). Here, both the dielectric constant and loss factor have lower values than corresponding values for banana or chicken breast meat. The point of change in the sign of the temperature coefficient for the dielectric constant is a little above 20 MHz.

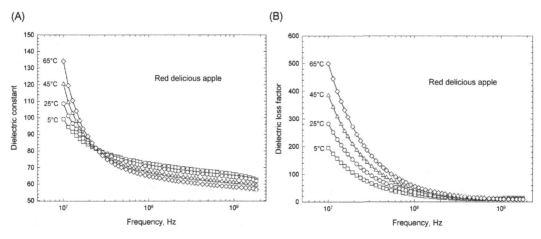

FIGURE 11.10

Frequency and temperature dependence of fresh apple permittivity, moisture content: 85% (Nelson, 2003).

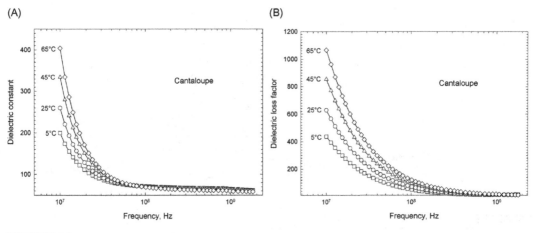

FIGURE 11.11

Frequency and temperature dependence of fresh cantaloupe permittivity, moisture content: 87% (Nelson, 2003; Nelson and Trabelsi, 2009).

Dielectric properties of fresh cantaloupe tissue of 87% moisture content over the same frequency and temperature ranges are presented in Figure 11.11 (Nelson, 2003). Here, the dielectric constants are greater than either those of apple or banana, but the loss factor values for the cantaloupe are intermediate between those of apple and banana at corresponding frequencies and temperatures. The temperature coefficient for the dielectric constant changes sign at about 70 MHz.

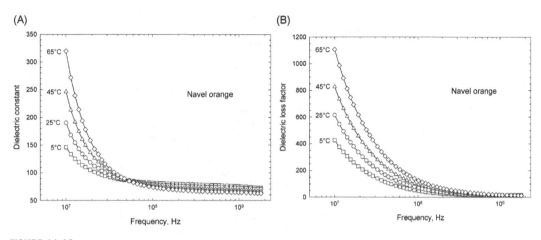

FIGURE 11.12

Frequency and temperature dependence of fresh navel orange permittivity, moisture content: 89% (Nelson, 2003).

For navel orange tissue of 89% moisture content, dielectric constants and loss factors over the same frequency and temperature ranges are shown in Figure 11.12 (Nelson, 2003; Nelson and Trabelsi, 2009).

Dielectric constants of fresh orange tissue are somewhat less than those of cantaloupe but greater than those of apple and only a little larger than those of banana tissue. The frequency of zero temperature dependence for the orange tissue dielectric constant is also about 50 MHz, like that of the banana tissue.

Dielectric properties of cantaloupe of 87% moisture content at 24°C, over the frequency range 200 MHz to 20 GHz, are shown in Figure 11.13 (Nelson et al., 2008), where scales are such that the dielectric constant and loss factor can be presented on the same graph. Here, the dielectric constant decreases consistently with increasing frequency, but the loss factor decreases with frequency to a minimum between 1 and 2 GHz, and then increases to a maximum somewhat below 20 GHz.

The dielectric properties of honeydew melon and watermelon at 24°C, both of 90% moisture content, over the same 200-MHz to 20-GHz frequency range, are presented in Figures 11.14 and 11.15 (Nelson et al., 2008).

Both the dielectric constants and loss factors of the three types of melon (Figures 11.13−11.15) are quite similar. They have similar values for the three types of melon and exhibit similar frequency dependence.

The sugar content of the various fresh fruit tissues was determined by measurement of the total soluble solids content of juices expressed from the tissues, but correlations between dielectric properties and soluble solids content were very low (Nelson, 2003, 2005; Guo et al., 2007; Nelson et al., 2007a). Some of these studies were made to learn whether a correlation between dielectric properties and soluble solids content might exist that could be useful for nondestructive sensing of sweetness in melons. Even though soluble solids content, which is mostly sugars in fruits, ranged from about 4% to about 14%, no useful correlations were identified, and the influence of these amounts of sugar on the dielectric relaxation observed was negligible (Nelson et al., 2008; Guo et al., 2007).

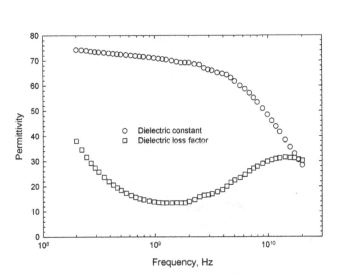

FIGURE 11.13

Frequency dependence of the permittivity of fresh cantaloupe at 24°C, moisture content: 87% (Nelson and Trabelsi, 2009).

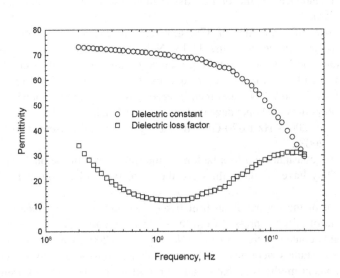

FIGURE 11.14

Frequency dependence of fresh honeydew melon permittivity at 24°C, moisture content: 90% (Nelson and Trabelsi, 2009).

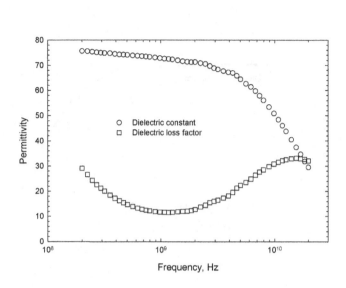

FIGURE 11.15

Frequency dependence of fresh watermelon permittivity at 24°C, moisture content: 90% (Nelson and Trabelsi, 2009).

Considering the dielectric properties data presented and the behavior of these properties with respect to moisture content of the different food products, several general observations can be noted. The dielectric constant always decreases with increasing frequency. For wheat below 25% moisture content, the dielectric constant and loss factor both increase with increasing temperature in the frequency range 10−1800 MHz. For other food materials, such as fresh chicken breast meat and fresh fruits, with moisture contents above about 70%, the dielectric constant increases with temperature at frequencies below about 20 MHz, but at some point in the range between 20 and 200 MHz, the temperature coefficient of the dielectric constant changes sign and the dielectric constant decreases with increasing temperature at higher frequencies. At frequencies below about 1 GHz, the dielectric loss factor appears to increase consistently with increasing temperature.

With respect to the reversal of the dielectric-constant temperature coefficient in the frequency range between 20 and 200 MHz, this phenomenon can be explained by the dominance of one of two dielectric loss mechanisms operating in these food materials. At lower frequencies, the large values for the dielectric properties are accounted for by the polarization and loss associated with ionic mechanisms. These mechanisms diminish as frequency increases, and above the point of temperature independence for the dielectric constant, the polarization and loss associated with dipole orientation becomes dominant. The dielectric behavior of pure liquid water at 25°C, as shown in Figure 11.16 (Nelson and Trabelsi, 2008, 2009), is illustrative of dielectric dipole relaxation, whereas the influence of ionic conduction on dielectric behavior is evident at the lower frequencies in all of the other graphic data illustrated for food materials.

The influence of free liquid water on the dielectric behavior of the fresh melon tissues is also clearly shown in comparing that of pure liquid water (Figure 11.16) with those of the melons in Figures 11.13−11.15.

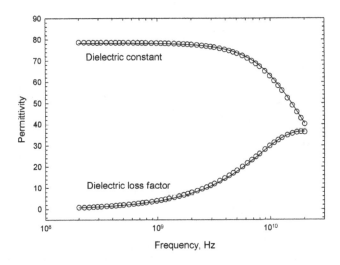

FIGURE 11.16

Frequency dependence of the dielectric properties of pure liquid water at 25°C. Circles represent measured points (Nelson and Trabelsi, 2009). Lines are provided by the Debye equation values calculated with single relaxation parameters published by Kaatze(1989).

Free water at 25°C exhibits a dipole relaxation centered at 19.24 GHz (Kaatze, 1989). The dielectric behavior of the melon tissues is also influenced by chemically bound water and ions in solution at the lower frequencies. Whereas the losses for pure water decrease monotonically at both frequencies above and below the relaxation frequency (19.24 GHz for pure water at 25°C) as shown in Figure 11.16, for frequencies below 19 GHz, the loss factors for the melon tissues decrease below the relaxation frequency to a minimum between 1 and 2 GHz and then increase due to ionic conduction as frequency continues to decrease.

The influence of free water is also evident in the dielectric behavior of apple juice as shown in Figure 11.11. In this instance, the peak of the relaxation curve just below 20 GHz is clearly noted in the dielectric loss factor at 10°C. At higher temperatures, the relaxation frequency shifts to higher frequencies beyond the 20-GHz limit shown in Figure 11.11, with consequent decreases in both the dielectric constant and loss factor with increasing temperature. At lower frequencies, the dielectric constant decreases with increasing temperature, as does the dielectric constant of pure water (Kaatze, 1989), and the loss factor increases with increasing temperature owing to the influence of temperature on ionic conduction. In Figure 11.10 for fresh apple tissue and in Figure 11.12 for fresh orange tissue, there is a hint of temperature coefficient sign reversal for the dielectric loss factor at frequencies a bit below 1 GHz. This might also be explained by the approach to the dielectric relaxation region as frequencies continue to increase.

Thus, the dielectric behavior of the food materials illustrated in this chapter, and other foods as well, is highly influenced by their water content. For those foods with very high water content, the influence of free liquid water is dominant, but bound water also exerts some influence. For lower moisture content food materials, the influence of bound water is important in addition to the composition of the materials, and ions in solution have an important role in determining the dielectric behavior of food materials, particularly at lower frequencies.

The open-ended coaxial-line probe, when used with appropriate sample temperature control equipment, can be useful in determining the temperature-dependent as well as frequency-dependent dielectric properties of food materials. Data presented for a macaroni and cheese dinner preparation, whey protein gel, ground whole wheat flour, apple juice, fresh chicken breast meat, and fresh fruit and melon tissues show the diverse frequency- and temperature-dependent behavior of food materials and illustrate the necessity for permittivity measurements when reliable data on dielectric properties are required. They further indicate the potential usefulness of permittivity in explaining the influence of different forms of water (e.g., free and rotationally hindered) and dielectric relaxation mechanisms when measurements are taken over sufficiently wide ranges of frequency and temperature.

11.2 MEASUREMENTS ON HYDROCOLLOID FOOD INGREDIENTS

The permittivities, that is, dielectric constants and loss factors, of five powdered hydrocolloids were measured at 2.45 GHz at different moisture contents over the temperature range 20–100°C (Nelson et al., 1991; Prakash et al., 1992). The hydrocolloids were potato starch, locust bean gum, gum arabic, carageenan, and carboxymethylcellulose. The moisture ranges were those of equilibrium moisture contents when sublots were conditioned at 25°C at known relative humidities up to 80%.

The dielectric constant and loss factor of powdered potato starch, locust bean gum, gum arabic, and carageenan, at 2.45 GHz, increased regularly with increasing moisture content in the range from 0% to 20%, w.b., and the same kind of behavior was noted for carboxymethylcelluose in the moisture range from 0% to 30% (Nelson et al., 1991). These dielectric properties of all five hydrocolloids also increased with increasing temperature over the range 20–100°C, and the temperature dependence increased markedly as moisture content increased. At 1% moisture content, the dielectric constant and loss factor of gum arabic, carageenan, and carboxymethylcellulose exhibited practically no temperature dependence, but between 5% and 10% that dependence became substantial and increased markedly with increasing moisture content throughout the moisture range. The dielectric properties of potato starch and locust bean gum showed considerable temperature dependence at 1% moisture content, and that temperature dependence increased much more slowly with increasing moisture content than observed for the other three hydrocolloids. Loss factors of these powdered hydrocolloids were not greatly different, but locust bean gum and carageenan had somewhat greater loss factors than the other three at high moisture contents and high temperatures. The dielectric constants and loss factors of the solid material of these hydrocolloids are naturally much greater than the corresponding properties of the powdered materials, and they exhibit similar relationships with moisture content and temperature.

11.3 DIELECTRIC PROPERTIES OF CHICKEN MEAT FOR QUALITY SENSING

The quality attributes of chicken meat are of interest to both producers and consumers. They are often assessed through measurement of physical properties, including color, pH, water-holding capacity (WHC), drip loss, cook yield, and texture (Warner−Bratzler shear force value).

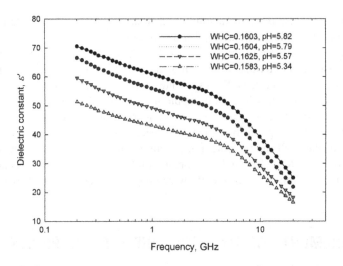

FIGURE 11.17

Dielectric constant of chicken breast meat samples as a function of frequency at 24°C for samples with water holding capacity of 16% and different pH levels (Trabelsi and Nelson, 2009).

The procedures for determining these properties are tedious and time consuming, and they require various types of laboratory equipment. Chicken meat is composed of 70% water, 20% proteins, and 5% lipids, on average. Because water is a major component in chicken meat, dielectric spectroscopy might be useful for rapid assessment of chicken meat quality (Zhuang et al., 2007; Kent et al., 2000; Dashner and Knochel, 2003; Trabelsi and Nelson, 2009; Kent and Anderson, 1996).

Because dielectric properties can be rapidly sensed, those properties of fresh chicken meat were measured and examined for correlations between dielectric properties and some of the quality attributes. Dielectric properties of chicken breast meat were measured with an open-ended coaxial-line probe between 200 MHz and 20 GHz at temperatures ranging from −20°C to 25°C (Trabelsi and Nelson, 2009; Trabelsi, 2012).

The variation of the dielectric constant and dielectric loss factor with frequency at 24°C for fresh chicken breast meat of similar WHC and variable pH values are shown in Figures 11.17 and 11.18.

They reveal that the dielectric properties are greater for higher pH, and they show clearly the effect of ionic conduction, which is dominant at lower frequencies. The properties also reveal a dipolar-type relaxation at higher frequencies, which is attributable to the water present in the chicken breast tissue (Figure 11.18). For all samples, the dielectric constant decreases with increasing frequency with a slope change at about 4 GHz. This may signal the presence of two distinct relaxations, with one in the lower frequency region and another above 4 GHz. It is customary to look into the frequency dependence of the dielectric loss factor to identify these relaxations. However, in this instance, the dielectric loss factor is dominated by ionic conduction in the lower frequency range, thus totally masking any relaxation there, and only the second relaxation at about 10 GHz is observed. This relaxation, which is of dipolar nature, occurs at a frequency lower than

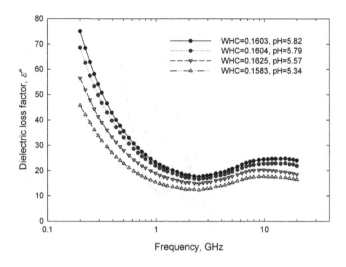

FIGURE 11.18

Dielectric loss factor of chicken breast meat as a function of frequency at 24°C for samples with water holding capacity of 16% and different pH levels (Trabelsi and Nelson, 2009).

that of liquid water at the same temperature and reflects the effect of water binding in the chicken meat. However, the relaxation appears to be broader than that of liquid water, indicating a range of relaxation frequencies rather than a single relaxation frequency of 19.2 GHz characteristic of liquid water at 25°C (Kaatze, 1989; Hasted, 1973). This is confirmed by the complex-plane representation of the dielectric properties (Cole−Cole plot) shown in Figure 11.19.

For each chicken breast sample, the plot consists of two distinct sections: a linear section corresponding to the lower frequencies, where the dielectric response is dominated by ionic conduction, and a circular arc, with its center below the ε'-axis, which represents the broad relaxation taking place at higher frequencies and is characteristic of the dielectric response of bound water (Trabelsi and Nelson, 2006). This is to be expected because water in the chicken breast meat is of bound form with different degrees of binding.

The variation of dielectric constant and dielectric loss factor with temperature at 2.936 GHz for chicken breast meat samples and distilled water are shown in Figures 11.20 and 11.21.

At low temperatures, dielectric properties of the chicken breast meat samples behave in a similar fashion to those of distilled water, with a sharp increase around 0°C. This is characteristic of the phase change from solid to liquid water. The temperatures shown on the graphs are those in the wall of the stainless steel sample holder. The temperature of samples at the tip of the coaxial-line probe are obviously several degrees higher than those shown on the axis. At temperatures above 0°C, dielectric constants of the different chicken breast samples are distinct and follow a trend similar to that of distilled water. However, their values are less than those of water. In contrast, values of the dielectric loss factor of distilled water decrease more rapidly than those of the chicken breast meat as temperature increases above 0°C.

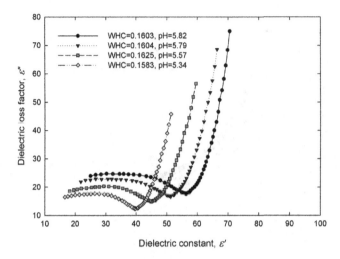

FIGURE 11.19

Complex-plane representation of the dielectric properties of chicken breast meat samples at 24°C at frequencies ranging from 200 MHz to 20 GHz for samples with water holding capacity of 16% and different pH levels (Trabelsi and Nelson, 2009).

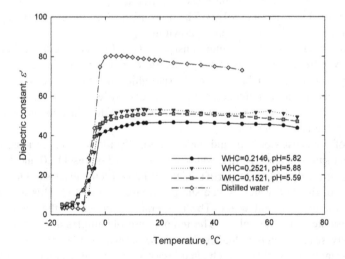

FIGURE 11.20

Dielectric constant of chicken breast meat samples with indicated water holding capacity and pH, as a function of sample-holder temperature at 2.936 GHz (Trabelsi and Nelson, 2009).

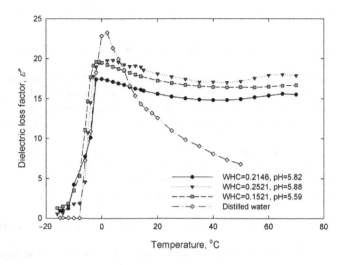

FIGURE 11.21

Dielectric loss factor of chicken breast meat samples with indicated water holding capacity and pH, as a function of sample-holder temperature at 2.936 GHz (Trabelsi and Nelson, 2009).

Additional measurements of the dielectric properties of chicken meat were made in the 200-MHz to 50-GHz range to study potential correlations between those properties and quality attributes with respect to aging (Trabelsi, 2013), comparison of intact and ground broiler breast meat (Samuel and Trabelsi, 2012), and for predicting meat quality (Samuel et al., 2012). The dielectric constant decreased with storage duration at all frequencies while the loss factor increased with storage duration at frequencies below 4 GHz, due to changes in ionic conduction, and at frequencies above that, the loss factor remained constant (Trabelsi, 2013). Both dielectric constant and loss factor decreased with water loss (Trabelsi, 2013). Whole intact muscle had slightly greater dielectric constants and loss factors than did ground muscle (Samuel and Trabelsi, 2012), but the small differences could be accounted for by unknown sample densities at the tip of the coaxial probe as well as amounts of free and bound water and the influence of grinding the muscle. Low correlations were noted between both the dielectric constant and loss factor of chicken breast meat and their quality factors, but the loss tangent showed some correlation with pH and color of the samples (Samuel and Trabelsi, 2012).

Measurements of the dielectric properties of chicken breast tissue over broad frequency ranges have revealed that it is possible to characterize tissue samples according to WHC and pH. Also, the complex-plane representations of the dielectric properties of chicken breast meat samples indicate a broad relaxation that is typical of bound water with different degrees of binding. Although indications that dielectric spectroscopy has potential for determining quality attributes of chicken breast meat from measurement of their dielectric properties, much further research is necessary to determine whether such techniques might provide any practically useful methods for quality measurements.

REFERENCES

Dashner, F., Knochel, R., 2003. Dielectric microwave sensors with multivariate calibration. Adv. Radio Sci. 1, 9–13.

Datta, A.K., Sun, E., Solis, A., 1995. Food dielectric property data and their compostion-based prediction. In: Rao, M.A., Rizvi, S.S.H. (Eds.), Engineering Properties of Foods. Marcel Dekker, Inc., New York, NY.

Funk, D.B., 2001. An Investigation of the Nature of the Radio-Frequency Dielectric Response in Cereal Grains and Oilseeds with Engineering Implications for Grain Moisture Meters. University of Missouri-Kansas City, MO.

Guo, W.-C., Nelson, S.O., Trabelsi, S., Kays, S.J., 2007. Dielectric properties of honeydew melons and correlation with quality. J. Microw. Power Electromagn. Energy 41 (2), 44–54.

Hasted, J.B., 1973. Aqueous Dielectrics. Chapman and Hall, London.

Kaatze, U., 1989. Complex permittivity of water as a function of frequency and temperature. J. Chem. Eng. Data 34, 371–374.

Kent, M., 1987. Electrical and Dielectric Properties of Food Materials. Science and Technology Publishers, Hornchurch, Essex, England.

Kent, M., Anderson, D., 1996. Dielectric studies of added water in poultry meat and scallops. J. Food Eng. 28, 239–259.

Kent, M., Knochel, R., Dashner, F., Beerger, U.-K., 2000. Composition of foods using microwave dielectric spectra. Eur. Food Res. Technol. 210, 359–366.

Mudgett, R.E., 1995. Electrical properties of foods. In: Rao, M.A., Rizvi, S.S.H. (Eds.), Engineering Properties of Foods. Marcel Dekker, Inc., New York, Basel, Hong Kong.

Nelson, S.O., 1973. Electrical properties of agricultural products—a critical review. Trans. ASAE 16 (2), 384–400.

Nelson, S.O., 1984. Density dependence of the dielectric properties of wheat and whole-wheat flour. J. Microw. Power 19 (1), 55–64.

Nelson, S.O., 2003. Frequency- and temperature-dependent permittivities of fresh fruits and vegetables from 0.01 to 1.8 GHz. Trans. ASAE 46 (2), 567–574.

Nelson, S.O., 2005. Dielectric spectroscopy of fresh fruit and vegetable tissues from 10 to 1800 MHz. J. Microw. Power Electromagn. Energy 40 (1), 31–47.

Nelson, S.O., Bartley Jr., P.G., 2002a. Frequency and temperature dependence of the dielectric properties of food materials. Trans. ASAE 45 (4), 1223–1227.

Nelson, S.O., Bartley Jr., P.G., 2002b. Measuring frequency- and temperature-dependent permittivities of food materials. IEEE Trans. Instrum. Meas. 51 (4), 589–592.

Nelson, S.O., Datta, A.K., 2001. Dielectric properties of food materials and electric field interactions. In: Datta, A.K., Anantheswaran, R.C. (Eds.), Handbook of Microwave Technology for Food Applications. Marcel Dekker, Inc., New York, NY.

Nelson, S.O., Trabelsi, S., 2006. Dielectric spectroscopy of wheat from 10 MHz to 1.8 GHz. Meas. Sci. Technol. 17, 2294–2298.

Nelson, S.O., Trabelsi, S., 2008. Dielectric spectroscopy measurements on fruit, meat, and grain. Trans. ASABE 51 (5), 1829–1834.

Nelson, S.O., Trabelsi, S., 2009. Influence of water content on RF and microwave dielectric behavior of foods. J. Microw. Power Electromagn. Energy 43 (2), 13–22.

Nelson, S.O., Prakash, A., Lawrence, K.C., 1991. Moisture and temperature dependence of the permittivities of some hydrocolloids at 2.45 GHz. J. Microw. Power Electromagn. Energy 26 (3), 178–185.

Nelson, S.O., Guo, W., Trabelsi, S., Kays, S.J., 2007a. Dielectric spectroscopy of watermelons for quality sensing. Meas. Sci. Technol. 18, 1887–1892.

Nelson, S.O., Trabelsi, S., Zhuang, H., 2007b. Dielectric Spectroscopy of Fresh Chicken Breast Meat. ASABE Paper No. 073095, American Society of Agricultural and Biological Engineers, St. Joseph, MI.

Nelson, S.O., Trabelsi, S., Kays, S.J., 2008. Dielectric spectroscopy of melons for potential quality sensing. Trans. ASABE 51 (6), 2209–2214.

Prakash, A., Nelson, S.O., Mangino, M.E., Hansen, P.M.T., 1992. Variation of microwave dielectric properties of hydrocolloids with moisture content, temperature and stoichiometric charge. Food Hydrocolloids 6 (3), 315–322.

Samuel, D., Trabelsi, S., 2012. Measurement of dielectric properties of intact and ground broiler breast meat over the frequency range from 500 MHz to 50 GHz. Int. J. Poult. Sci. 11 (3), 172–176.

Samuel, D., Trabelsi, S., Karnwaugh, A.B., Anthony, N.B., Aggrey, S.E., 2012. The use of dielectric spectroscopy as a tool for predicting meat quality in poultry. Int. J. Poult. Sci. 11 (9), 551–555.

Tinga, W.R., Nelson, S.O., 1973. Dielectric properties of materials for microwave processing—tabulated. J. Microw. Power 8 (1), 23–65.

Trabelsi, S., 2012. Frequency and temperature dependence of dielectric properties of chicken meat. In: Proceedings of the 2012 IEEE International Instrumentation and Technology Conference. Institute of Electrical and Electronics Engineers (IEEE), Piscataway, NJ, pp. 1515–1518.

Trabelsi, S., 2013. Investigating effects of aging on radiofrequency dielectric properties of chicken meat. In: Proceedings of the 2013 IEEE Instrumentation and Meaasurement Technology Conference. Institute of Electrical and Electronics Engineers (IEEE), Piscataway, NJ, pp. 1685–1688.

Trabelsi, S., Nelson, S.O., 2006. Temperature-dependent behaviour of dielectric properties of bound water in grain at microwave frequencies. Meas. Sci. Technol. 17, 2289–2293.

Trabelsi, S., Nelson, S.O., 2009. Use of Dielectric Spectroscopy for Determining Quality Attributes of Meat. ASABE Paper No. 097035, American Society of Agricultural and Biological Engineers, St. Joseph, MI.

Zhuang, H., Nelson, S.O., Trabelsi, S., Savage, E.M., 2007. Dielectric properties of uncooked chicken breast muscles from ten to one thousand eight hundred megahertz. Poult. Sci. 86, 2433–2440.

SENSING MOISTURE AND DENSITY OF SOLID BIOFUELS

Biofuels are an alternative source of energy with the advantage of renewability as compared to fossil fuels. Moisture content is an important factor in pricing, optimizing combustion, and storing solid biofuels. Methods for moisture determination have been reviewed and compared (Nystrom and Dahlquist, 2004; Samuelsson et al., 2006). They are often categorized as direct methods, such as chemical titration and oven-drying, and indirect methods such as near-infrared, nuclear, and dielectric properties-based techniques, which rely on correlating a measured parameter with moisture content. Standard techniques are destructive, time and labor consuming, and are impractical for in-process, real-time measurement of moisture content. In contrast, indirect methods are mostly nondestructive, and they provide information on moisture content rapidly, making them more suitable for on-line applications.

Dielectric properties-based sensors use electric fields for sensing the dielectric properties of the material, which are intrinsic electrical properties that characterize the wave–material interaction (Kraszewski, 1996). In general, dielectric properties of lossy materials are dependent on frequency, temperature, density, and the composition of the material. It is this dependence that permits the use of dielectric properties for sensing applications (Kraszewski, 1996; Kupfer, 2005). For instance, at microwave frequencies, dielectric properties of moist materials are strongly affected by the amount of water in the materials because of the polar nature of the water molecules (Hasted, 1973). Therefore, measurement of dielectric properties has been useful in measuring moisture content in these materials.

12.1 PINE PELLETS (PELLETED SAWDUST)

The microwave sensing technique described in Section 7.4 was used to determine simultaneously bulk density and moisture content of pine pellets from measurement of their dielectric properties, the dielectric constant ε' and loss factor ε''. Among techniques for measuring dielectric properties (Bussey, 1967; von Hippel, 1954), free-space methods are nondestructive and require no special sample preparation, making them suitable for both static laboratory measurements and on-line industrial applications. Dielectric properties of pine-pellet samples of different moisture contents and different bulk densities were measured with a free-space-transmission technique (Trabelsi and Nelson, 2003) at 23°C and 10 GHz. Moisture content was predicted from ε' and ε'' with a density-independent moisture calibration function (Trabelsi et al., 1998a, 2001a).

Dielectric Properties of Agricultural Materials and Their Applications. DOI: http://dx.doi.org/10.1016/B978-0-12-802305-1.00012-9

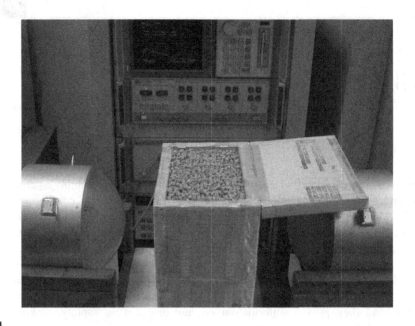

FIGURE 12.1

Free-space measurement arrangement (Trabelsi and Nelson, 2009).

The measurement technique used for determining the dielectric properties of pine pellets was initially developed for characterizing cereal grains and oilseeds (Trabelsi et al., 1998b). It consisted of a free-space-transmission measurement technique, in which ε' and ε'' were determined from measurements of the wave attenuation and phase shift caused by the pine pellets with an 8510C Hewlett-Packard[1] vector network analyzer (VNA). The free-space-transmission measurement arrangement, shown in Figure 12.1, consisted of two linearly polarized horn-lens antennas connected to the VNA with high-quality coaxial cables.

The horn-lens antennas collimated the electromagnetic energy in a relatively narrow beam and provided a plane wave at a short distance from the transmitting antenna. The distance between the two antennas was 37 cm. The VNA was calibrated in transmission mode (response-type calibration) in the frequency range 2–18 GHz with an empty sample holder between the two antennas. The sample holder was a box of rectangular cross section, made of Styrofoam sheet, a material with dielectric properties close to those of air. After calibrating the VNA, each pine-pellet sample was poured into the sample holder, which was then placed between the two antennas. From measurement of the

---

[1]Mention of trade names or commercial products in this publication is solely for the purpose of providing specific information and does not imply recommendation or endorsement by the US Department of Agriculture.

modulus $|S_{21}|$ and argument φ of the scattering transmission coefficient S_{21}, the attenuation and phase shift were calculated as follows:

$$A = 20 \log|S_{21}| \tag{12.1}$$

$$\phi = \varphi - 2\pi n \tag{12.2}$$

where n is an integer that can be determined (Trabelsi et al., 2000).

The dielectric properties of each sample were calculated as follows, assuming that a plane wave was propagating through a low-loss material:

$$\varepsilon' = \left[1 - \frac{(\varphi - 360n)}{360d} \frac{c}{f} \right]^2 \tag{12.3}$$

$$\varepsilon'' = \frac{-20 \log|S_{21}|}{8.686\pi d} \frac{c}{f} \sqrt{\varepsilon'} \tag{12.4}$$

where c is the speed of light in m/s, f is the frequency in Hz, and d is the thickness of the layer of material in meters. The dielectric properties, as given by Eqs (12.3) and (12.4), are the average values for the air−material mixture.

12.1.1 BULK DENSITY DETERMINATION FOR PINE PELLETS FROM COMPLEX-PLANE REPRESENTATION

Figure 12.2 shows the complex-plane representation (Trabelsi et al., 2001b) of the dielectric properties, divided by bulk density, for pine pellets measured at 10 GHz and 23°C.

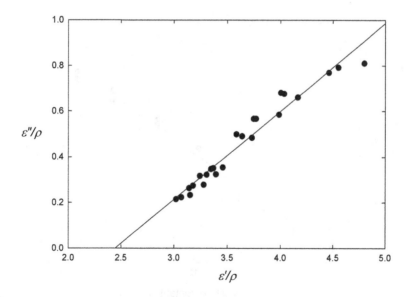

FIGURE 12.2

Complex-plane representation of dielectric properties divided by bulk density for pine pellets at 10 GHz and 23°C (Trabelsi and Nelson, 2009).

All the data points fall along a straight line, which can be fitted with linear regression calculation as:

$$\frac{\varepsilon''}{\rho} = a_f\left(\frac{\varepsilon'}{\rho} - k\right)$$

(12.5)

where a_f is the slope in the complex plane and k is the ε'/ρ-axis intercept. At 10 GHz, a_f was 0.385, k was 2.44 and the coefficient of determination r^2 was 0.95. The bulk density of each pine-pellet sample can be calculated by solving Eq. (12.5) for ρ which is independent of moisture content and temperature (see Eq. (7.1)). The standard error of calibration corresponding to use of Eq. (12.5) for predicting bulk density was 0.026 g/cm^3.

12.1.2 MOISTURE CONTENT DETERMINATION WITH DENSITY-INDEPENDENT CALIBRATION FUNCTION

A density-independent moisture calibration function ψ expressed in terms of the dielectric properties was used to determine moisture content in pine pellets independent of density. This calibration function is defined as follows (Trabelsi et al., 1998a):

$$\psi = \sqrt{\frac{\varepsilon''}{\varepsilon'(a_f\varepsilon' - \varepsilon'')}}$$

(12.6)

Figure 12.3 shows that calibration function ψ increases linearly with moisture content.

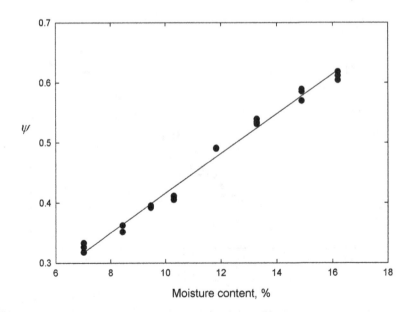

FIGURE 12.3

Variation of density-independent calibration function ψ with moisture content of pine pellets at 10 GHz and 23°C (Trabelsi and Nelson, 2009).

A linear regression provides the following relationship between ψ and moisture content M in percent (%):

$$\psi = 0.033M + 0.086 \quad r^2 = 0.99 \tag{12.7}$$

Moisture content is determined without knowledge of bulk density by solving Eq. (12.7) for M. The standard error of calibration corresponding to use of Eq. (12.7) for predicting moisture content from measurement of the dielectric properties was 0.35% moisture content.

Bulk density and moisture content of pine pellets can be determined simultaneously and independently from measurement of their dielectric properties. Bulk density was determined from a complex-plane representation of dielectric constant and dielectric loss factor without need for moisture content and temperature information. Moisture content was determined independent of bulk density with the density-independent moisture calibration function expressed in terms of the dielectric properties (Eq. (12.6)). Standard errors of calibration for bulk density and moisture content were predicted from measurement of the dielectric properties with standard errors of calibration of 0.026 g/cm^3 and 0.35%, respectively. The method can be implemented with microwave sensors for on-line determination of bulk density and moisture content in pine pellets from measurement of their dielectric properties at a single microwave frequency.

12.2 PEANUT-HULL PELLETS

Each year, about 500,000 t of peanut hulls are generated when the peanut crop is shelled in the Southeastern states where peanuts are grown and processed (Fasina, 2008). The peanut hulls are pelleted to improve efficiency in handling and transportation. Thus, peanut-hull pellets can be used as a biofuel in several energy conversion processes, including direct combustion, gasification, and pyrolysis (Berndes et al., 2003). Efficient use of peanut-hull pellets on an industrial scale requires their characterization in real time. Among their physical characteristics, moisture content and bulk density are the most important. Moisture content is important in pricing, quality control, optimizing conversion processes, and in storage management. Bulk density is a factor influencing the quality of the peanut-hull pellets and their flow dynamics. An indirect, nondestructive method was developed for rapid determination of bulk density and moisture content in peanut-hull pellets following the same procedures described in this chapter for pine pellets (Trabelsi et al., 2010). A free-space-transmission technique was used for measurement of the dielectric properties of the peanut-hull pellets. Results of measurements at 5, 10, and 15 GHz are shown in Figure 12.4, where the complex plane representation is used with Eqs (12.5) and (12.6) and appropriate linear regressions to obtain the bulk density and moisture content of peanut-hull pellets (Trabelsi et al., 2010, 2013).

The regression calculations provide the values for the regression constants a and b in the regression equation for ψ, the density-independent moisture calibration function,

$$\psi = aM + b \tag{12.8}$$

and, thus, the moisture content is obtained as:

$$M = \frac{\psi - b}{a} \tag{12.9}$$

The linear plots for ψ at 5, 10, and 15 GHz are shown in Figure 12.5.

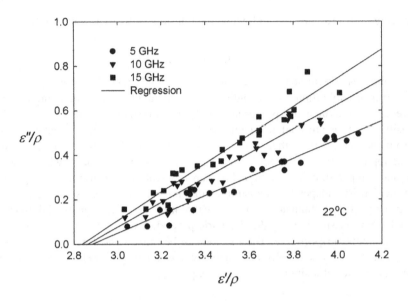

FIGURE 12.4

Complex-plane representation of the dielectric properties, divided by bulk density, of peanut-hull pellets at indicated frequencies and 22°C (Trabelsi et al., 2013).

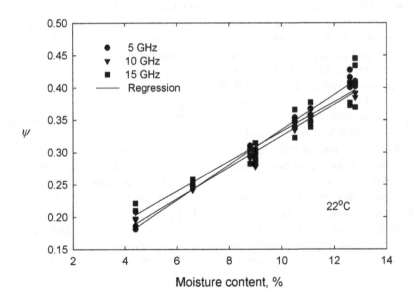

FIGURE 12.5

Variation of density-independent calibration function ψ with moisture content in peanut-hull pellets at indicated frequencies and 22°C (Trabelsi et al., 2013).

Coefficients of determination for the regressions (Eq. (12.8)) were, respectively, 0.988, 0.971, and 0.913, and the standard errors of calibration were 0.30%, 0.49%, and 0.87% moisture content, respectively. Corresponding standard errors of calibration for the bulk density determinations were 0.016, 0.018, and 0.016 g/cm^3, at 5, 10, and 15 GHz, respectively.

Therefore, the microwave technique for simultaneous sensing of moisture content, independent of bulk density, and bulk density, independent of moisture content and temperature, was successfully tested for peanut-hull pellets. The method provides suitable accuracy for on-line applications in the handling of peanut-hull pellets.

REFERENCES

Berndes, G., Hoogwijk, M., van den Broek, R., 2003. The contribution of biomass in the future global energy supply: a review of 17 studies. Biomass Bioenergy 25 (1), 1–28.

Bussey, H.E., 1967. Measurement of RF properties of materials—a survey. Proc. IEEE 55 (6), 1046–1053.

Fasina, O., 2008. Physical properties of peanut hull pellets. Bioresour. Technol. 99 (5), 1259–1266.

Hasted, J.B., 1973. Aqueous Dielectrics. Chapman and Hall, London.

Kraszewski, A., 1996. Microwave Aquametry. IEEE Press, Piscataway, NJ.

Kupfer, K., 2005. Electromagnetic Aquametry: Electromagnetic Wave Interaction with Water and Moist Substances. Springer Verlag Berlin Heidelberg, New York, NY.

Nystrom, J., Dahlquist, E., 2004. Methods for determination of moisture in woodchips for power plants—a review. Fuel 83, 773–779.

Samuelsson, R., Burvall, J., Jirjis, R., 2006. Comparison of different methods for the determination of moisture content in biomass. Biomass Bioenergy 30, 929–934.

Trabelsi, S., Nelson, S.O., 2003. Free-space measurement of dielectric properties of cereal grain and oilseed at microwave frequencies. Meas. Sci. Technol. 14, 589–600.

Trabelsi, S., Nelson, S.O., 2009. Sensing moisture content and bulk density of pelleted biofuels by microwave permittivity measurements. In: Proceedings of the 8th International Conference on Electromagnetic Wave Interaction with Water and Moist Substances (ISEMA 2009), Espoo, Finland, pp. 256–262.

Trabelsi, S., Kraszewski, A., Nelson, S.O., 1998a. New density-independent calibration function for microwave sensing of moisture content in particulate materials. IEEE Trans. Instrum. Meas. 47 (3), 613–622.

Trabelsi, S., Kraszewski, A., Nelson, S.O., 1998b. Nondestructive microwave characterization for determining the bulk density and moisture content of shelled corn. Meas. Sci. Technol. 9, 1548–1556.

Trabelsi, S., Kraszewski, A.W., Nelson, S.O., 2000. Phase-shift ambiguity in microwave dielectric properties measurements. IEEE Trans. Instrum. Meas. 49 (1), 56–60.

Trabelsi, S., Kraszewski, A., Nelson, S.O., 2001a. New calibration technique for microwave moisture sensors. IEEE Trans. Instrum. Meas. 50 (877–881), .

Trabelsi, S., Kraszewski, A., Nelson, S., 2001b. Microwave dielectric sensing of bulk density of granular materials. Meas. Sci. Technol. 12, 2192–2197.

Trabelsi, S., Paz, A.M., Nelson, S.O., 2010. Dielectric-based method for determining moisture content and bulk density of peanut-hull pellets. In: Proceedings of the 24th Annual Microwave Power Symposium (IMPI). International Microwave Power Institute, Mechanicsville, Virginia, pp. 190–195.

Trabelsi, S., Paz, A.M., Nelson, S.O., 2013. Microwave dielectric method for the rapid, non-destructive determination of bulk density and moisture content of peanut hull pellets. Biosyst. Eng. 115, 332–338.

von Hippel, A.R., 1954. Dielectric Materials and Applications. The Technology Press of M.I.T. and John Wiley & Sons, New York, NY.

DIELECTRIC PROPERTIES MODELS FOR GRAIN AND SEED

13

When sufficient information is available on the dielectric properties of a material and the variables that influence the values of the dielectric properties, models can be developed that are useful in calculating estimates of those values. The dielectric properties of grain are of particular importance because of their value in the electrical measurement of grain moisture content (Nelson, 1977). They have also been needed in the exploration of dielectric heating applications (Nelson, 1985c) and in studies on controlling stored-grain insects by high-frequency and microwave dielectric heating (Nelson and Charity, 1972; Nelson, 1996). The various factors that influence the dielectric properties of grain have been studied, and frequency of the applied electric field, moisture content of the grain, bulk density of the grain, and temperature have been identified as the most important variables (Nelson, 1982). The dielectric constant was identified as the best single property for use in predicting moisture content.

Establishment of linear relationships between functions of the dielectric properties of particulate dielectrics and their bulk densities (Nelson, 1983b) led to the development of a simple mathematical model for estimating the dielectric constant of shelled field corn (Nelson, 1984b). Application of these same principles enabled the development of a similar model for calculating the dielectric constant of hard red winter wheat (Nelson, 1985a). For purposes of illustrating model development for grain, the procedures followed in developing this model are included here.

13.1 MODEL DEVELOPMENT FOR WHEAT

Extensive measurements of the dielectric properties of seven cultivars of hard red winter wheat, *Triticum aestivum* L., were summarized previously (Nelson, 1982; Nelson and Stetson, 1976). These measurements spanned the frequency range from 250 Hz to 12.1 GHz and included moisture contents ranging from 2.7% to 23.8% (wet basis). The dielectric-constant data for the averages of these cultivars at 15 moisture contents spanning that range were used in the analysis. Data were obtained at 18 frequencies over the range from 250 Hz to 12.1 GHz with seven measurement systems described previously (Nelson and Stetson, 1976).

The dielectric constant ε' is the real part of the complex relative permittivity, $\varepsilon = \varepsilon' - j\varepsilon''$, and ε'' is the dielectric loss factor (Nelson, 1973). It was shown that, for pulverized and granular materials, $(\varepsilon')^{1/2}$ and $(\varepsilon')^{1/3}$ were essentially linear functions of the bulk density ρ of the particulate material (Nelson, 1983b). The quadratic relationship predicted measured values to

Dielectric Properties of Agricultural Materials and Their Applications. DOI: http://dx.doi.org/10.1016/B978-0-12-802305-1.00013-0

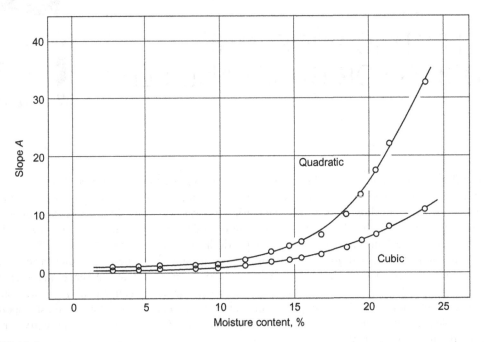

FIGURE 13.1

Dependence of slope A, of the $(\varepsilon')^{1/2}$ versus ρ and $(\varepsilon')^{1/3}$ versus ρ regression lines, on moisture content of hard red winter wheat at 500 Hz and 24°C (Nelson, 1985a).

within 1% and 4%, respectively, for pulverized coal and whole-kernel wheat, whereas the corresponding accuracies for the cubic relationship were 0.1% and 2%, respectively.

Because the dielectric constant of air–particle mixtures is 1 at zero density, that is, the dielectric constant of air, these linear relationships may be written as follows:

$$(\varepsilon')^{1/2} = 1 + A_2\rho \tag{13.1}$$

$$(\varepsilon')^{1/3} = 1 + A_3\rho \tag{13.2}$$

where A_2 and A_3 are the slopes of the straight lines. Thus, the values of A_2 and A_3 can be calculated from the value of ε', which is known at any given frequency and moisture content, if the bulk density is known.

$$A_2 = ((\varepsilon')^{1/2} - 1)/\rho \tag{13.3}$$

$$A_3 = ((\varepsilon')^{1/3} - 1)/\rho \tag{13.4}$$

Corresponding values of slope A were calculated for both the quadratic and cubic cases for each of the 15×18 (moisture content $\times$ frequency) points at which dielectric constants were available for hard red winter wheat. The slope values, A, were then plotted against moisture content at each of the 18 frequencies. These are illustrated for five of those frequencies in Figures 13.1–13.5.

Slope A was also plotted against the $\log_{10}$ of frequency, in MHz, for each of the 15 moisture contents. An illustrative example is shown in Figure 13.6.

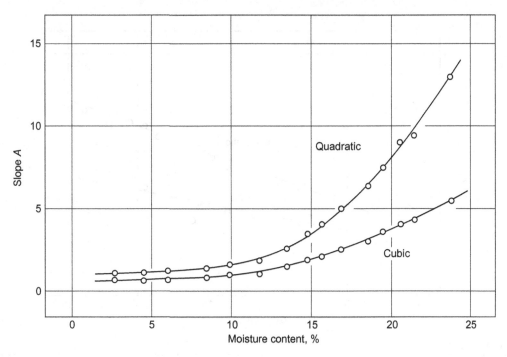

FIGURE 13.2

Dependence of slope A, of the $(\varepsilon')^{1/2}$ versus ρ and $(\varepsilon')^{1/3}$ versus ρ regression lines, on moisture content of hard red winter wheat at 5 kHz and 24°C (Nelson, 1985a).

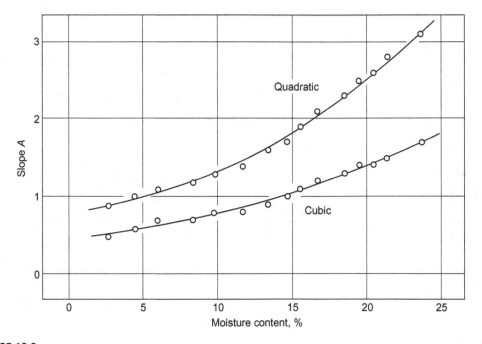

FIGURE 13.3

Dependence of slope A, of the $(\varepsilon')^{1/2}$ versus ρ and $(\varepsilon')^{1/3}$ versus ρ regression lines, on moisture content of hard red winter wheat at 500 kHz and 24°C (Nelson, 1985a).

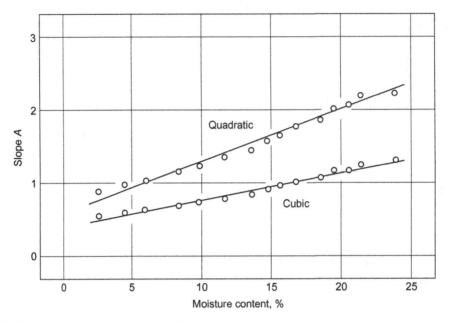

FIGURE 13.4

Dependence of slope A, of the $(\varepsilon')^{1/2}$ versus ρ and $(\varepsilon')^{1/3}$ versus ρ regression lines, on moisture content of hard red winter wheat at 5.1 MHz and 24°C (Nelson, 1985a).

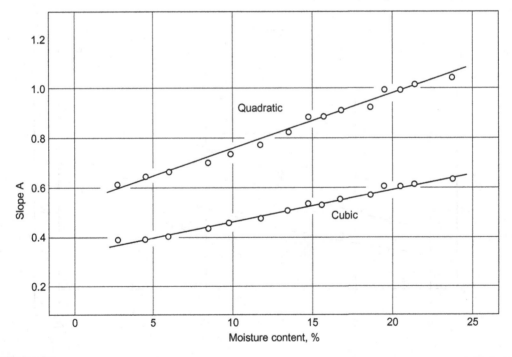

FIGURE 13.5

Dependence of slope A, of the $(\varepsilon')^{1/2}$ versus ρ and $(\varepsilon')^{1/3}$ versus ρ regression lines, on moisture content of hard red winter wheat at 12.1 GHz and 24°C (Nelson, 1985a).

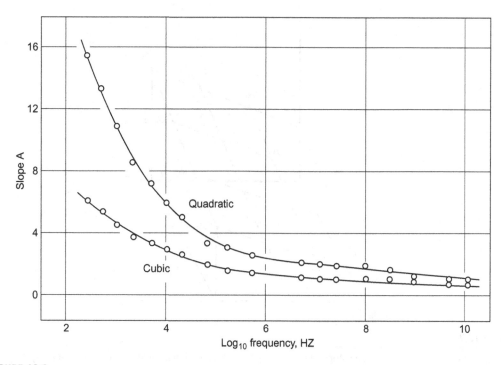

FIGURE 13.6

Dependence of slope A, of the $(\varepsilon')^{1/2}$ versus ρ and $(\varepsilon')^{1/3}$ versus ρ regression lines, on the log of frequency for hard red winter wheat at 19.5% moisture content at 24°C (Nelson, 1985a).

The relationships between A and both moisture content and log frequency were found to be very nearly linear at frequencies of 5.1 MHz and greater. Therefore, the development of the linear model for calculating dielectric constants was restricted to the frequency range from 5.1 MHz to 12.1 GHz. At any given frequency, slope A is a linear function of moisture content, m, and at any given moisture content, slope A is a linear function of $\log_{10}$ frequency, $\log f$.

Therefore, values of A define a surface in three-dimensional space as illustrated in Figure 13.7.

Although this surface is not necessarily a plane, its characteristics are such that planes parallel to the $A - m$ plane and planes parallel to the $A - \log f$ plane intersect it in straight lines (Nelson, 1985a). Lines on this surface parallel to the $A - m$ plane represent the linear dependence of A on m at particular frequencies, f_1, f_2, f_3 etc., whereas lines on the surface parallel to the $A - \log f$ plane represent the linear dependence of A on $\log f$ at particular moisture contents, m_1, m_2, m_3, etc. These two independent relationships can be expressed as follows:

$$A = B_0 + B_m \tag{13.5}$$

$$A = C_0 + C_{\log f} \tag{13.6}$$

For a given frequency, Eq. (13.5), with the proper values for B_0 and B, provides the dependence of A on m. For any given moisture content, Eq. (13.6) provides the dependence of A on $\log f$, when

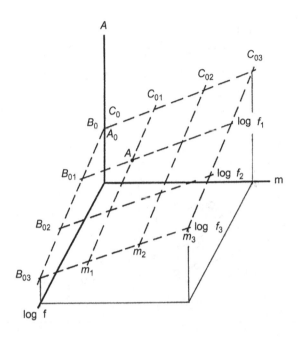

FIGURE 13.7

Representation of the regression-line slope, $A = F(m, \log f)$, as a three-dimensional surface with A linearly dependent on both moisture content, m, and $\log_{10}$ of frequency, $\log f$ (Nelson, 1985a; Nelson and Noh, 1992).

the proper values for C_0 and C are known. According to Eqs (13.5) and (13.6), when $m = \log f = 0$, B_0 will equal C_0, and we will call this value A_0. Referring to Figure 13.7, we can now write an expression for the value of point A on the surface of interest by progressing from point A_0 along the line of intersection of the surface and the $A - m$ plane to point C_{01}, and thence to point A along the line on the surface resulting from the intersection with that surface of the plane $m = m_1$. This expression may be written as follows, where slope B is a function of $\log f$, and slope C is a function of m:

$$A = A_0 + B(\log f = 0)m + C(m)\log f \qquad (13.7)$$

Alternatively, one might progress from A_0 to B_{01} in the $A - \log f$ plane and thence to point A on the line formed by the intersection of the surface with the $\log f = \log f_1$ plane. Then we would write:

$$A = A_0 + C(m = 0)\log f + B(\log f)m \qquad (13.8)$$

Equations (13.7) and (13.8) provide equivalent expressions for any point A on the surface as functions of m and $\log f$. Once the value of A is obtained for any desired moisture content and frequency, the dielectric constant ε', can be calculated by solving Eqs (13.3) or (13.4), as appropriate, for ε'.

Upon noting the linearity of the plots of slope A (the slopes of the $(\varepsilon')^{1/2}$ versus ρ and $(\varepsilon')^{1/3}$ versus ρ relationships of Eqs (13.1) and (13.2)) versus m and A versus $\log f$, linear regression-analysis calculations were performed to obtain the intercept and slope values for Eqs (13.5) and (13.6). Regression analyses were calculated at each of the eight frequencies (5.1, 13.5, 28, 100, and 300 MHz, and 1, 5.5, and 12.1 GHz) over the 18 moisture contents from 2.7% to 23.8%, and these analyses provided the B_0 intercept and slope-B values of Eq. (13.5) for each frequency (Nelson, 1985a). From the resulting regression lines, values for A were obtained at moisture contents of 5%, 10%, 15%, 20%, and 25%, and linear regressions were calculated with these data as a function of $\log f$ to obtain the C_0 intercept and slope-C values of Eq. (13.6) at each moisture level (Nelson, 1985a). (The reader is referred to the original reference for detailed presentation of the remaining analysis (Nelson, 1985a).) The following two equations were obtained, for the quadratic and cubic cases, respectively, as mathematical models for the dielectric constant as a function of moisture content, frequency, and bulk density at 24°C:

$$\varepsilon_2' = [1 + (0.6202 + 0.0815m - 0.0378 \log f - 0.0143m \log f)\rho]^2 \qquad (13.9)$$

$$\varepsilon_3' = [1 + (0.4034 + 0.0343m - 0.0378 \log f - 0.0265m \log f)\rho]^3 \qquad (13.10)$$

where m is moisture content in percent, wet basis, f is frequency in Hz, and ρ is bulk density in g/cm^3.

The accuracy of the predictive Eqs. (13.9) and (13.10) was tested by calculating the dielectric constants at each of the eight frequencies and 15 moisture contents and comparing the results to the originally measured values for the hard red winter wheat. The quadratic and cubic models predicted very nearly the same values, and the differences between measured and predicted values varied over the range of frequencies and moisture contents. The average error of prediction over all data was 4.1% for the quadratic model and 4.0% for the cubic model. The models provide a practical mathematical method for estimating the dielectric constant of hard red winter wheat as a function of frequency, moisture content, and bulk density over wide ranges of moisture content (9.9–23.8%) and frequency (5 MHz to 12 GHz) at 24°C.

13.2 MODELS FOR CORN

An extensive series of measurements was taken on shelled samples of 21 yellow-dent field corn, *Zea mays* L., hybrids as they dried from moisture contents in the field as high as 35–40% down to about 10% over a period of several weeks (Nelson, 1979). Dielectric properties were measured with three different systems at 20, 300, and 2450 MHz in coaxial sample holders of identical cross-sectional dimensions (Nelson, 1979). Other measurements on the samples included moisture content, bulk density, kernel dimensions, kernel density, crude protein and fat acidity. Models were developed for the dielectric constant as a function of moisture content, bulk density, and temperature for the three selected frequencies as follows (Nelson, 1979):

$$\varepsilon_{20}' = 3.51 + 0.132(m - 10) + (\rho_b - \overline{\rho}_b)(0.839 - 0.086m + 0.027m^2) + 0.012(T - 24) \qquad (13.11)$$

$$\varepsilon_{300}' = 2.89 + 0.098(m - 10) + (\rho_b - \overline{\rho}_b)(0.460m - 2.16) + 0.015(T - 24) \qquad (13.12)$$

$$\varepsilon_{2450}' = 2.48 + 0.099(m - 10) + (\rho_b - \overline{\rho}_b)(0.387m - 3.22) + 0.013(T - 24) \qquad (13.13)$$

where m is moisture content in percent, wet basis (range 10–35%), ρ_b is bulk density in g/cm^3, T is the temperature of the grain in °C, and

$$\overline{\rho}_b = 0.6829 + 0.01422m - 0.000979m^2 + 0.0000153m^3 \tag{13.14}$$

Later, the same data that were used for the development of the models for the dielectric constant at 20, 300, and 2450 MHz were supplemented with new measurements relating the dielectric constant and bulk density (Nelson, 1984b). They showed that the square roots and cube roots of the dielectric constant were essentially linear with bulk density, and a new model for the dielectric constant was developed following the same procedures used in development of the model for the dielectric constant of hard red winter wheat described in Section 13.1. The following models resulted for the dielectric constant of shelled, yellow-dent field corn based on the quadratic and cubic relationships, respectively (Nelson, 1984b):

$$\varepsilon_2' = [1 + (0.685 + 0.0674m - 0.121 \log f - 0.0158m \log f)\rho]^2 \tag{13.15}$$

$$\varepsilon_3' = [1 + (0.466 + 0.0342m - 0.077 \log f - 0.0023m \log f)\rho]^3 \tag{13.16}$$

where m is moisture content in percent, wet basis, f is frequency in Hz, and ρ is bulk density in g/cm^3.

These models predicted dielectric constant values at 24°C for shelled corn with average errors of 2.9% and 2.7%, respectively, over the frequency range from 20 to 2450 MHz and the moisture content range from 10% to 33% (Nelson, 1984b).

13.3 MODELS FOR SOYBEANS

Dielectric properties of soybeans, *Glycine max* (L.) Merrill, representing 10 cultivars, were measured at 24°C over the moisture range from 6% to 24% (wet basis), at frequencies of 20, 300, and 2450 MHz (Nelson, 1985b). Models for the dielectric constant of soybeans were developed following the same procedures used for development of similar models for hard red winter wheat. These models for soybeans for the quadratic and cubic relationships, respectively, were:

$$\varepsilon_2' = [1 + (0.138 + 0.1211m + 0.074 \log f - 0.0228m \log f)\rho]^2 \tag{13.17}$$

$$\varepsilon_3' = [1 + (0.172 + 0.0626m + 0.026 \log f - 0.0112m \log f)\rho]^3 \tag{13.18}$$

The models predicted the dielectric constant over the wide frequency range from 20 to 2450 MHz and the 6−24% moisture range with average errors of 2.9% for the quadratic model and 2.8% for the cubic model (Nelson, 1985b).

13.4 MODELS FOR BARLEY

Dielectric properties of Certified seed of two hardy cultivars from Nebraska and two semihardy cultivars from Georgia of winter barley, *Hordeum vulgare* L., were measured at 24°C and frequencies of 20, 300, and 2450 MHz over a range of moisture contents from 8% to 25% (Nelson, 1986b). The resulting data were used for development of models for the dielectric constant of winter barley following the same procedures used in model development for wheat described in Section 13.1.

The resulting models, based on quadratic and cubic relationships, respectively, for the hardy winter barley were:

$$\varepsilon'_2 = [1 + (0.539 + 0.0771m - 0.0361 \log f - 0.0103m \log f)\rho]^2 \tag{13.19}$$

$$\varepsilon'_3 = [1 + (0.382 + 0.040m - 0.0297 \log f - 0.0050m \log f)\rho]^3 \tag{13.20}$$

For the semihardy winter barley, the quadratic and cubic models were, respectively:

$$\varepsilon'_2 = [1 + (0.436 + 0.0878m - 0.0076 \log f - 0.0136m \log f)\rho]^2 \tag{13.21}$$

$$\varepsilon'_3 = [1 + (0.329 + 0.0469m - 0.0150 \log f - 0.0067m \log f)\rho]^3 \tag{13.22}$$

Errors in moisture prediction when dielectric constants calculated with these models were compared to measured values were about 4% moisture content for the quadratic models and 3% for the cubic models (Nelson, 1986b).

Similar measurements on Certified seed of four spring barley cultivars were used to develop the following quadratic and cubic models for the dielectric constant as follows:

$$\varepsilon'_2 = [1 + (0.317 + 0.0946m + 0.0410 \log f - 0.0164m \log f)\rho]^2 \tag{13.23}$$

$$\varepsilon'_3 = [1 + (0.270 + 0.0498m + 0.0101 \log f - 0.0082m \log f)\rho]^3 \tag{13.24}$$

Average prediction accuracy of these equations for the dielectric constant of spring barley was about 4% over the moisture range from 8% to 25% and the frequency range from 20 to 2450 MHz (Nelson, 1986a).

13.5 MODELS FOR RICE

Measurements used for the development of dielectric properties models were taken on grain lots of two rice, *Oryza sativa* L., cultivars, a long-grain Indica variety, "Lebonnet", and a medium-grain Japonica variety, "Brazos", which were described in more detail previously (Noh and Nelson, 1989). Methods of conditioning the rice samples to avoid checking, determining moisture content by standard oven-drying procedures, and the procedures and methods for measuring the dielectric properties were also described (Noh and Nelson, 1989). Dielectric properties and bulk densities were measured at several moisture levels of practical interest for rough, brown, and white rice samples of each cultivar at frequencies from 50 Hz to 12 GHz. Measurements of the dielectric properties were taken three times and averaged at each moisture content and frequency. These mean values of ε' and ε'' at each frequency and moisture level were the data used for model development.

The resulting models, based on quadratic and cubic relationships, respectively, were as follows (Nelson and Noh, 1992):

$$\varepsilon' = [1 + (a_2 - b_2 \log f + c_2 m - d_2 m \log f)\rho]^2 \tag{13.25}$$

$$\varepsilon' = [1 + (a_3 - b_3 \log f + c_3 m - d_3 m \log f)\rho]^3 \tag{13.26}$$

where a, b, c, and d are constants for a particular kind of rice as listed in Table 13.1.

Table 13.1 Constants of Eqs (13.25) and (13.26) for Estimating Dielectric Constants of Rough (Paddy), Brown (Hulled), and White (Milled) Rice at 24°C (Nelson and Noh, 1992)

Cultivar and Form of Rice	Quadratic Model				Cubic Model				Moisture Range, %	Average Error, % (Quadratic)
	a_2	b_2	c_2	d_2	a_3	b_3	c_3	d_3		
Lebonnet (Long-Grain)										
Rough rice	0.4460	0.0230	0.0798	0.01044	0.3406	0.0236	0.0414	0.0050	11–17	3.0
Brown rice	0.4758	0.0556	0.0878	0.01154	0.3498	0.0419	0.0447	0.0053	11–16	4.8
White rice	0.6458	0.0950	0.0828	0.01052	0.4453	0.0622	0.0412	0.0047	11–16	4.3
Brazos (Medium Grain)										
Rough rice	0.3532	0.0022	0.0891	0.01208	0.2913	0.1250	0.0463	0.00585	11–18	3.9
Brown rice	0.5592	0.0690	0.0853	0.01004	0.3990	0.0484	0.0424	0.00430	12–16	3.2
White rice	0.7145	0.1383	0.0749	0.00751	0.4854	0.0685	0.0370	0.00311	12–16	3.1

Equations (13.25) and (13.26) were used to calculate the dielectric constant at each frequency and moisture level for each of the rice lots, and calculated values were compared with the measured values. The absolute value of the differences was termed "error" and the mean error over all frequencies and moisture contents is listed in Table 13.1 for the quadratic model. Errors for the cubic model were nearly equivalent to those for the quadratic model (Nelson and Noh, 1992).

13.6 COMPOSITE MODEL FOR CEREAL GRAIN

A model was developed for the complex permittivity, $\varepsilon = \varepsilon' - j\varepsilon''$, that is, both the dielectric constant ε' and loss factor ε'', based on physical principles and measured values of the dielectric properties of several cereal grains of different moisture contents and bulk densities at frequencies from 5 to 5000 MHz, and, thus, it was considered a composite model for the complex permittivity of cereal grains (Kraszewski and Nelson, 1989). Included in the study were data for yellow-dent field corn, hard and soft red winter wheat, rye, barley, and oats.

Dielectric properties of grain depend upon the density and moisture content of the grain as well as upon the frequency of measurement. The effects of all three variables were considered.

13.6.1 DENSITY DEPENDENCE

Dielectric properties of bulk grain were determined earlier for several densities by packing and compressing the grain between successive measurements (Nelson, 1984a). Second-order polynomial, or quadratic, curves of the dielectric constant and the loss factor as functions of bulk density fit the experimental data well. Kent (1977) showed that both the dielectric constant and loss factor should be quadratic functions of the density of a particulate material. Klein (1981) noted in work on granular coal that the square root of the dielectric constant was a linear function of density. These observations are consistent (Nelson, 1983a) with the dielectric mixture formula for two-phase mixtures (Kraszewski et al., 1976; Kraszewski, 1977) in the following form:

$$\varepsilon = v_1\sqrt{\varepsilon_1} + v_2\sqrt{\varepsilon_2} \qquad (13.27)$$

where ε, ε_1, and ε_2 are the permittivities of the air–kernel mixture (bulk grain), host medium (air), and inclusions (grain kernels), respectively, and v_1 and v_2 are the volume fractions for components 1 and 2, with $v_1 + v_2 = 1$. It can be shown that Eq. (13.27) has the same form for both the dielectric constant and the loss factor if $\tan^2 \delta \ll 1$ for both components of the mixture (Kraszewski and Nelson, 1989).

For bulk grain, which can be considered a mixture of grain kernels and air, assume that component 1 in Eq. (13.27) is air. Because ε' for air is 1 and ε'' is zero, $\varepsilon_1 = 1$. For the kernels, $\varepsilon_2 = \varepsilon'_k - j\varepsilon''_k$ and $v_2 = \rho/\rho_k$, where ρ is the grain bulk density and ρ_k is the kernel density. Thus, it follows that, for the real and imaginary components of the bulk grain permittivity (Kraszewski and Nelson, 1989):

$$\frac{\sqrt{\varepsilon'} - 1}{\rho} = \frac{\sqrt{\varepsilon'_k} - 1}{\rho_k} = A \qquad (13.28)$$

$$\frac{\varepsilon''}{\rho^2} = \frac{\varepsilon_k}{\rho_k{}^2} = B \tag{13.29}$$

where A and B are both functions of the frequency and moisture content of the grain. Equations (13.28) and (13.29) imply that the model for grain might be developed on the basis of the dielectric properties of bulk grain or on those of grain kernels. More data are available on the dielectric properties of grain than on the kernels. Therefore, the model development proceeded with available data on bulk grain.

13.6.2 FREQUENCY DEPENDENCE

It is well known that the dielectric properties of most biological materials exhibit a pronounced dispersive behavior with sequences of dielectric relaxations related to molecular, macromolecular, subcellular and cellular relaxation phenomena (Pethig and Kell, 1987). Therefore, similar relaxations can be expected in grain kernels, and consequently in bulk grain. The complex structure of grain kernels (changing with stage of maturity and storage), the dependence of the dielectric properties upon moisture content, bulk density and temperature, and, finally, partially unknown character of the dielectric relaxations, all contribute to the complexity of the dielectric behavior of grain. The permittivity of an ideal Debye-type dispersive material (Pethig and Kell, 1987) can be described as:

$$\varepsilon = \varepsilon_\infty - j\frac{\sigma_0}{\omega\varepsilon_0} + \sum_{i=1}^{n}\frac{\Delta_i}{1 + j(\omega/\omega_i)} \tag{13.30}$$

where ε_∞ is the limiting value of ε' as $\omega \to \infty$, σ_0 is the limit of the conductivity value as $\omega \to 0$ (often called the dc conductivity), ε_s is the limit of ε' as $\omega \to 0$, and $\Delta_i = \varepsilon_s - \varepsilon_\infty$ is the dielectric increment expressing the change in the relative permittivity due to the dispersion associated with ω_i which is the relaxation frequency of the ith dispersion region. Separating Eq. (13.30) into its real and imaginary components yields:

$$\varepsilon' = \varepsilon_\infty - j\frac{\sigma_0}{\omega\varepsilon_0} + \sum_{i=1}^{n}\frac{\Delta_i}{1 + j(f/f_i)^2} \tag{13.31}$$

$$\varepsilon'' = \sum_{i=1}^{n}\frac{\Delta_i f/f_i}{1 + (f/f_i)^2} + \frac{\sigma_0}{2\pi f\varepsilon_0} \tag{13.32}$$

Thus, knowledge of ε_s, ε_∞, and f_i; for every dispersion region is necessary for calculation of the values of ε' and ε'' of grain in this region with Eqs (13.31) and (13.32).

13.6.3 MOISTURE DEPENDENCE

Relationships between the parameters A and B, Eqs (13.28) and (13.29), and moisture content in grain arise from a number of distinct phenomena, each with its own frequency dependence, and these effects are difficult to separate from each other. For example, the effect of dielectric relaxation of bound water at some frequencies may be overshadowed by ionic losses or free water relaxation when moisture content increases. With particulate materials such as grain,

Maxwell–Wagner interfacial polarization might also be expected to occur (Daniel, 1967), but most likely at quite low frequencies. However, their contributions can be accounted for with the appropriate values substituted in Eq. (13.30).

The moisture dependence of A and B was examined for experimental data on several kinds of grain (Kraszewski and Nelson, 1989), and the relationship between moisture content m and density noted in the definition of moisture content as:

$$m = \frac{w_w}{w_w + w_d} = \frac{w_w/V}{w_w/V + w_d/V} \tag{13.33}$$

where w_w is the mass of water, w_d is the mass of dry material, and V is the volume of the material. The density of the wet material is $\rho = w_w/V + w_d/V$ and the amount of water contained in the material (water concentration in g/cm^3) is simply $w_w/V = m\rho$. In many instances, the water concentration has greater physical meaning than moisture content m, which is most often used in practice and expressed in percentage, $M = 100m$.

13.6.4 MODEL DEVELOPMENT

For model development, about 300 experimentally determined values were used for the dielectric constant and loss factor of barely, corn, rye, and wheat, taken at different densities in the frequency range from 5 to 5000 MHz and for moisture contents from 8% to 26%, wet basis (Kraszewski and Nelson, 1989). The numerical values of A and B, Eqs (13.28) and (13.29), were calculated for all data points at respective densities. Then a multivariable linear regression procedure was used to find the combination of variables that provided the maximum value of the coefficient of determination for a given number of variables. From several combinations, the following expressions were selected for their simplicity and basic physical significance (Kraszewski and Nelson, 1989):

$$1/A = 0.0358 + 1.8904M^{-1/2} + 1.8622M^{-1} \log f \tag{13.34}$$

$$B = 0.145 + 0.00148M^2 \log f - 0.004615M^2 + 0.008044M^2/\log f \tag{13.35}$$

Combining Eqs (13.28) and (13.34) and Eqs (13.29) and (13.35), with minor adjustments, the final form of the expressions for the bulk grain model were obtained as follows:

$$\varepsilon' = \left(1 + \frac{0.504M\rho}{\sqrt{M + \log f}}\right)^2 \tag{13.36}$$

$$\varepsilon'' = 0.146\rho^2 + 0.004615M^2\rho^2(0.32 \log f + 1.743/\log f - 1) \tag{13.37}$$

where M is the moisture content in percent, ρ is the grain bulk density in g/cm^3, and f is the frequency in MHz. Equations (13.36) and (13.37) were proposed as a general model for all cereal grains in the frequency range from 5 to 5000 MHz and for moisture contents from 8% to 26%, wet basis at temperatures of about 24°C (Kraszewski and Nelson, 1989).

Evaluation of the composite model for cereal grains was done by comparing values calculated from the model equations and measured values of many samples of corn, wheat, rye, barley, and oats, and the average accuracies were 5% for the dielectric constant and 10% for the loss factor over the specified frequency and moisture ranges at about 24°C.

13.7 MODELS AT MICROWAVE FREQUENCIES

Models for the dielectric constant and loss factor at microwave frequencies from 5 to 15 GHz were developed for several cereal grain and oilseed crops over wide ranges of moisture content at 23°C (Nelson and Trabelsi, 2011). Dielectric constant and loss factor data reported previously (Trabelsi and Nelson, 2004, 2005) for hard red winter wheat, *T. aestivum* L., yellow-dent field corn, *Z. mays*, L., spring barley, *H. vulgare* L., winter oats, *Avena sativa* L., grain sorghum, *Sorghum bicolor* (L.) Moench, soybeans, *G. max* (L.) Merr., canola, *Brassica napus* L., and Runner type peanuts, *Arachis hypogaea* L., were examined in their relationships with frequency and moisture content as shown for example in Figures 13.8–13.12.

The nonlinearity shown in Figure 13.8 was avoided by taking the log of frequency as shown in Figure 13.9. Then appropriate multiple linear regression calculations were performed to fit the data to the following models for the dielectric constant ε' and loss factor ε'':

$$\varepsilon' = a + b \log f + cM \qquad (13.38)$$

$$\varepsilon'' = d + ef + gM \qquad (13.39)$$

where a, b, c, d, e, and g are constants determined by the regression analyses, f is frequency in GHz, and M is moisture content in percent.

Values for the regression constants are listed for the cereal grain and oilseed crops in Table 13.1.

The models provide reasonably good estimates for the dielectric properties of these commodities at 23°C over the ranges of frequencies and moisture contents investigated, which are shown in Table 13.2.

Standard errors for the dielectric constant were on the order of 1–2%, and those for the loss factor were generally a few percent, but for small loss factors at the extremes of the data they

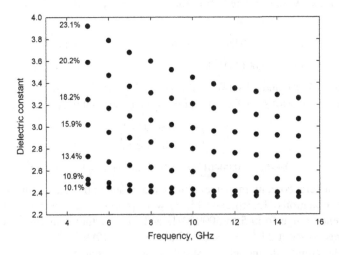

FIGURE 13.8

Frequency dependence of the dielectric constant of grain sorghum at indicated moisture contents (Nelson and Trabelsi, 2011).

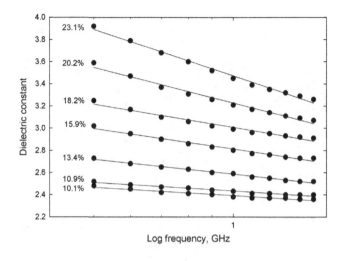

FIGURE 13.9

Linearity of the dependence of the dielectric constant of grain sorghum on log of frequency at indicated moisture contents (Nelson and Trabelsi, 2011).

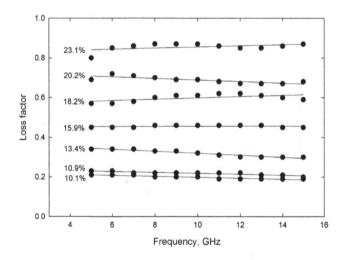

FIGURE 13.10

Frequency dependence of the dielectric loss factor of grain sorghum at indicated moisture contents (Nelson and Trabelsi, 2011).

can naturally be much larger. Although the models developed do not include bulk density as an independent variable, well-tested models are available for correcting dielectric properties for variations in bulk density (Nelson, 1984a; Trabelsi and Nelson, 2012) as discussed in Section 1.2.3.

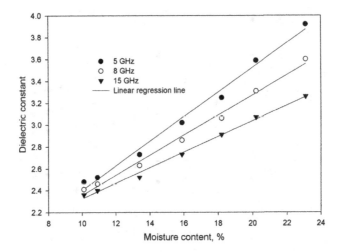

Moisture dependence of the dielectric constant of grain sorghum at indicated frequencies (Nelson and Trabelsi, 2011).

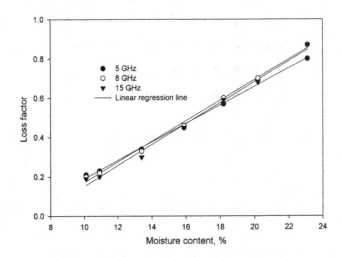

Moisture dependence of the dielectric loss factor of grain sorghum at indicated frequencies (Nelson and Trabelsi, 2011).

Table 13.2 Constants of the Following Equations for Estimating the Dielectric Constant ε′ and Loss Factor ε″ of Indicated Cereal Grain or Oilseed at 23°C, Where f is Frequency in GHz and M is Moisture Content in Percent, Wet Basis (ASABE, 2013):

$$\varepsilon' = a + b \log f + cM \qquad \varepsilon'' = d + ef + gM$$

Commodity	Constants a	b	c	d	e	g	Frequency Range, GHz	Moisture Range, %	Bulk Density, kg/m³
Hard red winter wheat, (*Triticum aestivum* L.)	2.5075	−0.4198	0.0406	−0.1330	−0.0021	0.0360	5–15	10.2–18.0	746–822
Shelled yellow-dent field corn (*Zea mays* L.)	2.5546	−0.6035	0.0598	−0.4488	0.0146	0.0579	5–15	10.8–19.6	710–805
Spring barley (*Hordeum vulgare* L.)	2.2685	−0.4549	0.0316	−0.1984	0.0042	0.0285	5–15	10.8–24.8	567–686
Winter oats (*Avena sativa* L.)	1.6547	−0.2277	0.0279	−0.0942	0.0011	0.0171	5–15	9.1–18.3	456–479
Grain sorghum (*Sorghum bicolor* (L.) Moench)	2.1388	−0.6753	0.0866	−0.3338	−0.0011	0.0514	5–15	10.1–23.1	746–776
Soybeans (*Glycine max* (L.) Merr.)	2.0411	−0.5087	0.0877	−0.5344	0.0071	0.0659	5–15	8.5–20.3	728–741
Canola (*Brassica napus* L.)	2.0824	−0.5938	0.0837	−0.2174	−0.0052	0.0499	5–15	6.6–18.8	662–678
Shelled peanuts (*Arachis hypogaea* L.)	2.0029	−0.3966	0.0764	−0.4524	0.0099	0.0691	5–12	5.1–17.2	604–639
Unshelled peanuts (*A. hypogaea* L.)	1.7246	−0.6661	0.0580	−0.9290	0.0454	0.0666	5–12	7.7–21.7	328–374

REFERENCES

ASABE, 2013. ASABE, D293.4, Dielectric Properties of Grain and Seed, in ASABE Standards 2013 American Society of Agricultural and Biological Engineers, St. Joseph, MI.

Daniel, V.V., 1967. Dielectric Relaxation. Academic Press, London, New York, NY.

Kent, M., 1977. Complex permittivity of fish meal: a general discussion of temperature, density, and moisture dependence. J. Microw. Power 12 (4), 341–345.

Klein, A., 1981. Microwave determination of moisture in coal—comparison of attenuation and phase measurement. J. Microw. Power 16 (3–4), 289–304.

Kraszewski, A., 1977. Prediction of the dielectric properties of two-phase mixtures. J. Microw. Power 12 (3), 216–222.

Kraszewski, A.W., Nelson, S.O., 1989. Composite model of the complex permittivity of cereal grain. J. Agric. Eng. Res. 43, 211–219.

Kraszewski, A., Kulinski, S., Matuszewski, M., 1976. Dielectric properties of a model of biphase water suspension at 9.4 GHz. J. Appl. Phys. 47 (4), 1275–1277.

Nelson, S.O., 1973. Electrical properties of agricultural products—a critical review. Trans. ASAE 16 (2), 384–400.

Nelson, S.O., 1977. Use of electrical properties for grain moisture measurement. J. Microw. Power 12 (1), 67–72.

Nelson, S.O., 1979. RF and microwave dielectric properties of shelled, yellow-dent field corn. Trans. ASAE 22 (6), 1451–1457.

Nelson, S.O., 1982. Factors affecting the dielectric properties of grain. Trans. ASAE 25 (4), 1045–1049, 1056.

Nelson, S.O., 1983a. Observations on the density dependence of the dielectric properties of particulate materials. J. Microw. Power 18 (2), 143–152.

Nelson, S.O., 1983b. Density dependence of the dielectric properties of particulate materials. Trans. ASAE 26 (6), 1823–1825.

Nelson, S.O., 1984a. Density dependence of the dielectric properties of wheat and whole-wheat flour. J. Microw. Power 19 (1), 55–64.

Nelson, S.O., 1984b. Moisture, frequency, and density dependence of the dielectric constant of shelled, yellow-dent field corn. Trans. ASAE 27 (5), 1573–1578, 1585.

Nelson, S.O., 1985a. A mathematical model for estimating the dielectric constant of hard red winter wheat. Trans. ASAE 28 (1), 234–238.

Nelson, S.O., 1985b. A model for estimating the dielectric constant of soybeans. Trans. ASAE 28 (6), 2047–2050.

Nelson, S.O., 1985c. Potential agricultural applications for RF and microwave energy. Trans. ASAE 30 (3), 818–822, 831.

Nelson, S.O., 1986a. Mathematical models for the dielectric constants of spring barley and oats. Trans. ASAE 29 (2), 607–610.

Nelson, S.O., 1986b. Models for estimating the dielectric constant of winter barley. Int. Agrophys. 2 (3), 189–200.

Nelson, S.O., 1996. Review and assessment of radio-frequency and microwave energy for stored-grain insect control. Trans. ASAE 39 (4), 1475–1484.

Nelson, S.O., Charity, L.F., 1972. Frequency dependence of energy absorption by insects and grain in electric fields. Trans. ASAE 15 (6), 1099–1102.

Nelson, S.O., Noh, S.H., 1992. Mathematical models for the dielectric constants of rice. Trans. ASAE 35 (5), 1533–1536.

Nelson, S.O., Stetson, L.E., 1976. Frequency and moisture dependence of the dielectric properties of hard red winter wheat. J. Agric. Eng. Res. 21, 181−192.

Nelson, S.O., Trabelsi, S., 2011. Models for the microwave dielectric properties of grain and seed. Trans. ASABE 54 (2), 549−553.

Noh, S.H., Nelson, S.O., 1989. Dielectric properties of rice at frequencies from 50 Hz to 12 GHz. Trans. ASAE 32 (3), 991−998.

Pethig, R., Kell, D.B., 1987. The passive electrical properties of biological systems: their significance in physiology, biophysics, and biotechnology. Phys. Med. Biol. 32 (8), 933−970.

Trabelsi, S., Nelson, S., 2004. Microwave dielectric properties of shelled and unshelled peanuts. Trans. ASAE 47 (4), 1215−1222.

Trabelsi, S. and Nelson, S., 2005. Microwave Dielectric Properties of Cereal Grain and Oilseed. ASAE Paper No. 056165, American Society of Agricultural Engineers, St. Joseph, MI.

Trabelsi, S., Nelson, S.O., 2012. Microwave dielectric properties of cereal grains. Trans. ASABE 55 (5), 1989−1996.

DEVELOPMENT OF MICROWAVE MOISTURE-SENSING INSTRUMENTATION

14

The successful use of microwave frequencies for sensing the moisture content in grain and seed, as discussed in Section 7.4, has stimulated interest in the practical use of this information for solving problems that hamper some current agricultural practices. Two such problems have been addressed with efforts to develop instrumentation applicable to the solution of the problems. One is the development of an instrument to provide the kernel moisture content from microwave measurements on the unshelled peanuts, or peanut pods—a microwave peanut kernel moisture meter. The other is for control of peanut driers by providing kernel moisture content from nondestructive monitoring of kernel moisture in unshelled peanut pods during the drying process. Both are practical applications in the peanut industry, but extension to the grain and seed (and other) industries is only natural for the potential benefits to be derived.

14.1 PEANUT KERNEL MOISTURE METER

14.1.1 BACKGROUND

Water content, or moisture content, is often used to assess the quality of agricultural products and to determine their optimum handling and processing conditions. In agriculture, moisture content of crops is the single most important factor determining the proper time for harvest and requirements for safe storage, and it is used for price determination. For peanuts, rapid methods for moisture assessment are essential in minimizing opportunities for mold growth and development of aflatoxin during storage, maintaining kernel quality, and improving profitability for peanut growers and high-quality safe products for consumers.

When farmers bring their peanut crop to the buying points for sale, all peanuts must be sampled and graded to determine condition and value. Peanut grading is a labor-intensive process during which inspectors evaluate samples according to five grading parameters: foreign material, size of pods, meat content, damaged kernels, and kernel moisture content. At the peanut buying stations, it is only after sorting, cleaning, and shelling that kernel moisture content is determined and a decision is made as to whether the peanuts meet the standard moisture level (10.49%, or lower, in the United States) for sale. Excessive moisture content is a common basis for rejection of peanut lots. If kernel moisture content is greater than 10.49%, the peanut loads are returned to the drying

Dielectric Properties of Agricultural Materials and Their Applications. DOI: http://dx.doi.org/10.1016/B978-0-12-802305-1.00014-2

facilities for further reduction of moisture content before the sale can be completed. Therefore, considerable time and labor costs could be saved if kernel moisture content could be determined by measurements on unshelled peanut pods at the beginning of the grading process.

14.1.2 DIELECTRIC PROPERTIES

Microwave dielectric properties of unshelled (pod) peanuts and kernels (shelled peanuts) were reported for moisture contents between 6% and 22% at frequencies between 6 and 18 GHz at 23°C and different bulk densities (Trabelsi and Nelson, 2004). The free-space measurements were made carefully observing several precautions to obtain reliable reference data for peanuts. Dielectric properties at different bulk densities were corrected to normally encountered densities by use of the Landau and Lifshitz, Looyenga dielectric mixture equation (Nelson, 1990, 1992). The resulting values of dielectric constant and loss factor were linearly dependent on the bulk density of shelled and unshelled peanuts. For the dependence of the dielectric constant and loss factor on moisture content, both increased in value with increasing moisture content.

Additional measurements of the dielectric properties of shelled and unshelled peanuts were studied for their usefulness in sensing moisture contents and bulk densities of these materials following the same principles presented in Section 7.4. Measurements were taken with laboratory equipment, including a vector network analyzer, horn/lens antennas, and tunnel-shaped radiation absorbing enclosure for the sample holder to limit interference from surroundings, following procedures that insured collection of reliable data for the dielectric properties at 8 GHz and 24°C (Trabelsi and Nelson, 2006). Calibration measurements were performed on peanuts from the 2003 harvest, and peanuts from the 2004 harvest were used for the validation measurements. Results of the studies revealed standard errors of calibration of 0.34% moisture content for both unshelled and shelled peanuts in the determination of moisture contents in the range from about 5% to 18% moisture content. Figure 14.1 shows the predicted moisture contents based on the calibration.

Validation measurements provided moisture content predictions with standard errors of performance of less than 0.7% moisture content over the same moisture range. The technique for determining kernel moisture contents from microwave measurements on the unshelled pod peanuts, as explained in Section 7.4, was also presented (Trabelsi and Nelson, 2006).

Further measurements of dielectric properties of shelled and unshelled peanuts over a range of temperatures from 0.5°C to 58°C revealed that kernel moisture content from measurements on unshelled pod peanuts could still be predicted with standard errors of less than 0.8% moisture content (Trabelsi et al., 2009).

14.1.3 PRACTICAL METER DEVELOPMENT

For the development of a practical, low-cost microwave moisture meter using the principles established successfully with the laboratory instruments, a prototype was assembled with available commercial components (Trabelsi and Nelson, 2007). It is shown schematically in Figure 14.2 and pictured in Figure 14.3.

It consisted of a voltage-controlled oscillator operated at 5.8 GHz, isolator, power splitter, two high-gain (19 dBi) 5.8-GHz microstrip patch antennas facing each other, followed by an I/Q demodulator (attenuation and phase detector), which was connected to a laptop computer through a

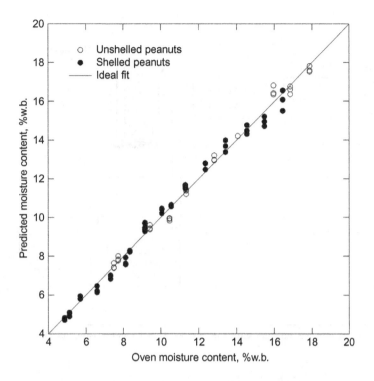

FIGURE 14.1

Predicted moisture content versus oven standard reference moisture content in unshelled and shelled peanuts. Solid line corresponds to ideal fit (Trabelsi and Nelson, 2006).

multichannel single-ended BNC connector box. The grain or seed sample to be characterized was poured into a Styrofoam container of rectangular cross-section and placed between the microstrip patch antennas. As the electromagnetic wave propagates through the sample it experiences energy loss and decreased velocity, which quantitatively are measured as attenuation and phase shift by comparing the differences between the reference levels (values without the sample) and those after the sample is poured into the Styrofoam sample holder. Two voltages V_1 and V_2 (Figure 14.2) are measured first with an empty sample holder between the two antennas and then again with the sample in the sample holder.

From these voltages, the attenuation and phase shift caused by the sample are determined, and the two components of the complex permittivity, ε' and ε'', are calculated assuming plane wave propagation through a low-loss material ($\varepsilon'' \ll \varepsilon'$):

$$\varepsilon' = \left[1 - \frac{\phi}{360d}\frac{c}{f}\right]^2 \tag{14.1}$$

$$\varepsilon'' = \frac{-A}{8.686\pi d}\frac{c}{f}\sqrt{\varepsilon'} \tag{14.2}$$

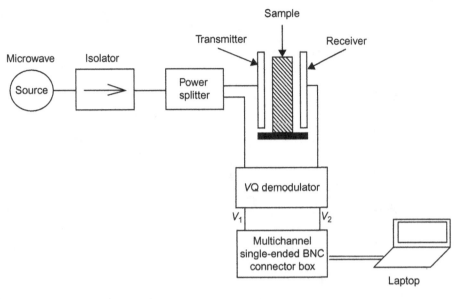

FIGURE 14.2

Schematic diagram of microwave moisture sensor prototype (Trabelsi and Nelson, 2007).

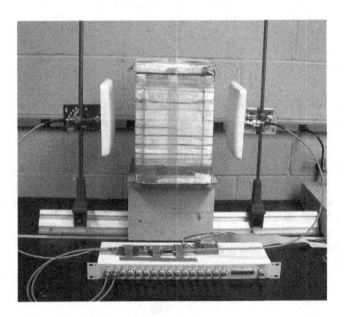

FIGURE 14.3

Original prototype microwave moisture sensor showing assembled microwave circuit components in foreground and peanut sample holder between two 5.8-GHz patch antennas.

FIGURE 14.4

Microwave moisture meter and polycarbonate sample holder bucket filled with peanut pods (Trabelsi and Nelson, 2010a,b).

where $\phi = \varphi - 360n$, φ is the measured phase angle in degrees, n is an integer to be determined, c is the speed of light in meters per second, d is the sample thickness in meters and f is the frequency in hertz. The attenuation A is expressed in decibels. For determination of n, a routine that selects the proper n value (Trabelsi et al., 2000) was embedded in the algorithm for ε' and ε'' computation. The moisture content of the peanut sample was determined from the measured dielectric properties (Trabelsi and Nelson, 2006; Trabelsi et al., 2008) as outlined in Section 7.4.

A new prototype microwave moisture meter for peanuts was built into a cabinet and provided features suitable for collecting data at peanut buying points and grading stations. Figure 14.4 shows the microwave moisture meter and polycarbonate (Lexan) sample holder.

Microwave components and related accessories were housed in an 18- by 12- by 16-in high aluminum cabinet of rectangular cross-section composed of two parts, a base cabinet and a cabinet cover. The base cabinet (Figure 14.5)—which is a closed rectangular box about 18 in wide, 12 in deep, and 5 in high—housed the microwave circuit components, fused power switch, power supply, cooling fan, and analog-to-digital interface, and it served as a mounting platform for the transmitting and receiving antennas, which faced each other in alignment with L-shaped mounting brackets spaced 11-1/2 in apart above the base cabinet.

A ½-in thick PVC top for the cabinet cover has an opening to receive the sample holder registering the sample in proper position between the antennas for measurements. The entire interior of the antenna chamber was lined with 3/4-in microwave absorber to prevent any reflections in that chamber which might interfere with the measurements. At this stage of development, the output from the

FIGURE 14.5

Base cabinet housing microwave circuit components with patch antennas mounted on top (Trabelsi and Nelson, 2010a,b).

prototype meter was connected by cable to a laptop computer for calculation of moisture content and records storage.

14.1.3.1 Moisture determination

From measurements of the dielectric properties of peanut samples, the moisture content was determined from the dielectric properties through use of the density-independent moisture calibration function ψ (Trabelsi et al., 1998, 2001):

$$\psi = \sqrt{\frac{\varepsilon''}{\varepsilon'(a_f\varepsilon' - \varepsilon'')}} \tag{14.3}$$

where a_f is the slope of a line determined from the complex-plane representation (see Figure 14.6) of the dielectric properties, each divided by bulk density (Trabelsi et al., 1998, 2011).

For granular and particulate materials, the function ψ varied linearly with moisture content (Trabelsi et al., 1999; Volgyi et al., 2002).

14.1.3.2 Laboratory calibration

Samples of peanut kernels and peanut pods, "Georgia Green" Runner type, of different moisture contents and temperatures were used to generate peanut moisture calibration equations for kernels and pods. Table 14.1 shows the ranges for moisture content, temperature, and bulk density.

The moisture content in each sample was determined on the wet basis by the standard oven-drying method (ASAE, 2002). The samples were dried at 130°C for 6 h. The temperature of each

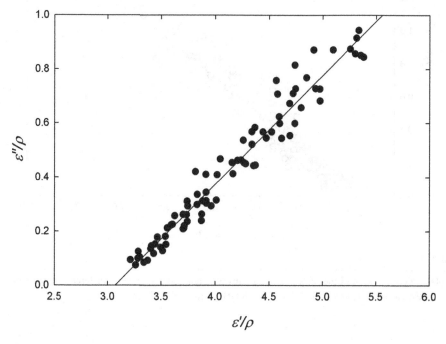

FIGURE 14.6

Complex-plane representation of the dielectric properties of peanut kernel samples of different moisture contents and temperatures, each divided by bulk density, at 5.8 GHz (Trabelsi et al., 2011).

Table 14.1 Physical Characteristics of Peanut Samples

Material	Moisture Content, %	Temperature, °C	Bulk Density, g/cm³
Peanut kernels	6.8−16.5	8.2−40.6	0.558−0.674
Peanut Pods	6.2−21.0	7.2−35.6	0.277−0.365

sample was determined with a digital thermometer at the time of microwave measurement. The bulk density of each sample was determined gravimetrically, dividing the weight of the sample by the sample holder volume. Figures 14.6 and 14.7 show the complex-representations of the dielectric properties of samples of peanut kernels and samples of peanut pods of different moisture contents and different temperatures at 5.8 GHz.

The slope a_f required in Eq. (14.3) was determined for each material from these representations. For peanut kernels, a_f was 0.402 and for peanut pods, a_f was 0.454.

Figures 14.8 and 14.9 show the dependence of the density-independent moisture calibration function ψ on moisture content and temperature of peanut kernels and peanut pods, respectively.

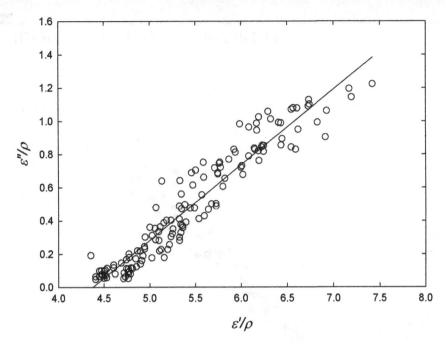

FIGURE 14.7

Complex-plane representation of the dielectric properties of peanut pod samples of different moisture contents and temperatures, each divided by bulk density, at 5.8 GHz (Trabelsi et al., 2011).

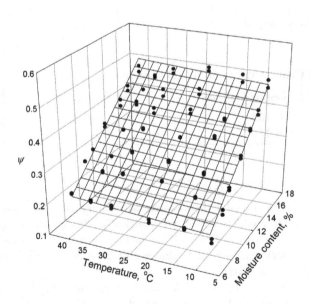

FIGURE 14.8

Three-dimensional representation of the variation of density-independent calibration function ψ with moisture content and temperature for peanut kernels at 5.8 GHz (Trabelsi et al., 2011).

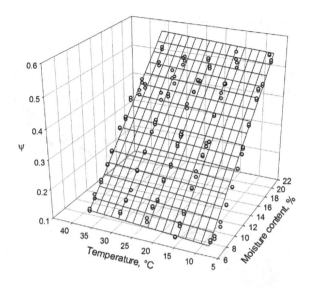

FIGURE 14.9

Three-dimensional representation of the variation of density-independent calibration function ψ with moisture content and temperature for peanut pods at 5.8 GHz (Trabelsi et al., 2011).

For each material, the data describe a plane for which the equation can be determined by multiple linear regression. Resulting ψ functions for peanut kernels ψ_k and peanut pods ψ_p were as follows:

$$\psi_k = 0.0293M_k + 0.0007T - 0.0040 \quad r^2 = 0.96 \tag{14.4}$$

$$\psi_p = 0.0284M_p + 0.0009T - 0.0516 \quad r^2 = 0.96 \tag{14.5}$$

Equating ψ_k and ψ_p, and solving for kernel moisture content M_k, as explained in Section 7.4, we have:

$$M_k = \frac{\psi_p - 0.0007T + 0.004}{0.0293} \tag{14.6}$$

Equation (14.6) was embedded in the measurement algorithm for the microwave moisture meter and provides kernel moisture determination from dielectric properties measurements on pods.

14.1.3.3 Field testing of the microwave moisture meter

After the laboratory calibration was completed, the microwave moisture meter was field tested at peanut buying points in Georgia, USA, where the inspectors used it at the beginning of the grading process. A user-friendly interface on the laptop computer guided the operator through step-by-step instructions to complete the measurements. Results obtained with the microwave moisture meter, for two consecutive harvest seasons, 2008 and 2009, compared well with those obtained with the

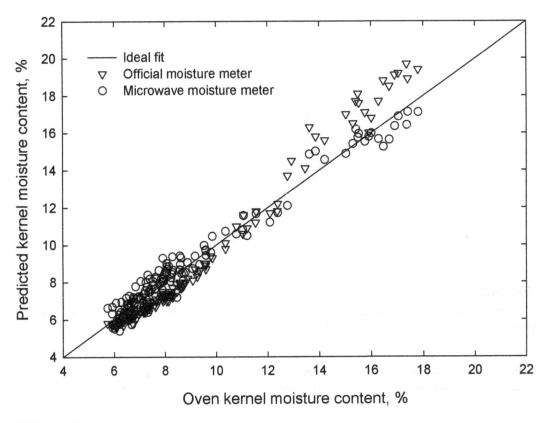

FIGURE 14.10

Kernel moisture content predicted with microwave moisture meter from measurements on peanut pods and with the official moisture meter reading for kernels plotted against kernel moisture content obtained with the standard oven-drying method (Trabelsi et al., 2011).

official moisture meter (a commercial capacitance-type meter operating at a frequency below 20 MHz). However, a more objective evaluation of the microwave moisture meter performance requires comparison of results obtained with this meter and the official meter to those obtained with the standard oven-drying technique. Those results, compared in Figure 14.10, show clearly that the microwave moisture meter had a much better performance than the official meter across the moisture range. The official meter underestimated the kernel moisture content below 10% and overestimated it for moisture levels higher than 13%. For data presented in Figure 14.10, the standard errors of performance were 0.51% and 0.31% moisture content for the official meter and the microwave moisture meter, respectively (Trabelsi et al., 2011).

Physical properties other than kernel moisture content, such as pod bulk density and pod moisture content, can be determined with this microwave moisture meter from the same measurement of pod dielectric properties. However, the principal advantage offered is the reliable determination of kernel moisture content from measurements on the unshelled peanuts that allows decisions on

FIGURE 14.11

Microwave moisture meter shown with LCD and keyboard (Lewis and Trabelsi, 2012).

suitability for sale or need for further drying before unnecessary time and effort are spent on grading and handling.

14.1.3.4 The addition of a microcontroller

To eliminate the necessity for a laptop computer to control the microwave moisture meter and to provide a complete instrument with easy portability, an embedded system employing a microcontroller was added to the moisture meter (Lewis and Trabelsi, 2012).

The embedded system included a graphical 144×32 pixel LCD and 16-button keypad to facilitate user interaction. The embedded system provides the following functionalities: user interface (input/output), event execution, process control, data acquisition, and data storage. Testing showed that the moisture meter with the new embedded system maintained the performance and accuracy observed for the original meter with personal computer or laptop computer control. Standard deviations for the attenuation and phase measurements, from which moisture contents were determined, for measurements with the embedded system, were only 60% as large as those taken with the laptop-controlled moisture meter. Comparisons of measurements taken with the moisture meter with embedded system and with a laboratory vector network analyzer were not significantly different at the 1% probability level. The integration of the embedded system with the microwave moisture meter provided a cost-effective, portable, and robust solution for microwave moisture sensing. The completed microwave moisture meter is shown in Figure 14.11.

14.2 PEANUT DRIER CONTROL BY MONITORING KERNEL MOISTURE CONTENT

14.2.1 BACKGROUND INFORMATION

Peanut drying is an essential operation that takes place at peanut buying stations and shelling plants preceding the grading process and sale of farmers' peanuts. Although peanuts are left in the field in windrows to dry naturally before being harvested with combines, they arrive at buying stations considerably high in moisture content, usually 20–30% (wet basis). Therefore, they must be dried to the mandatory kernel moisture content of 10.49% or lower before the grading and sale can be completed.

To facilitate drying, peanuts are loaded into drying wagons or trailers equipped with perforated floors above drying air plenum chambers into which heated air is blown for the required drying of the peanuts. Driers fueled with propane or natural gas are connected to the wagons or trailers through canvas ducts, and heated air is blown into the airspace below the bed of peanuts. The plenum is pressurized and air is forced up through the peanuts to decrease the moisture content of the peanut bed (Lewis and Trabelsi, 2012; Butts, 1995; Butts et al., 2004). Drying is continued until the target kernel moisture content is reached. Then, a representative sample of peanuts is extracted from the wagon and taken to be graded. During the grading process, if the kernel moisture content is determined to be greater than 10.49%, the sample is marked "NO SALE," and the corresponding lot of peanuts has to be taken back to the drying facility for further drying.

There are decision support systems and other commercial control systems that are currently used to estimate the drying time based on initial atmospheric conditions and the initial kernel moisture content of the bed of peanuts (Butts, 1995; Butts et al., 2003, 2004; Butts and Williams, 2004; Butts and Hall, 1994; Microtherm, 2000). Such systems are based on peanut drying models and either assume certain atmospheric conditions or require frequent updates for any changes. However, because of the current methods for checking kernel moisture content (collecting pod peanut samples, shelling and separating the kernels from the pods, and testing the kernels for moisture in electrical moisture meters), these systems demand heavy user interaction and updating for atmospheric conditions as well. Atmospheric conditions affect drying dynamics greatly (Beasley and Dickens, 1963; Troeger, 1989). If these parameters are not updated appropriately, the peanuts are likely to be overdried or underdried. In many instances, peanut buying points are understaffed, and an operator may or may not be able to manage the drying process with the frequency of moisture testing needed. Systems capable of monitoring the atmospheric conditions in real-time still lack such capabilities for sensing kernel moisture content.

The main parameter of interest during the peanut drying process is peanut kernel moisture content. However, the inability to obtain the kernel moisture content in real-time makes control of the drying process indirect and less efficient. Therefore, provision of a simple, in-process solution for real-time determination of kernel moisture content during drying would improve the efficiency of the drying process and diminish dependence on the operator. For this reason, a feedback-controlled drier system that monitored kernel moisture content and atmospheric conditions in real-time was designed and tested (Lewis et al., 2013). Kernel moisture content was determined in-shell with a microwave moisture meter that uses dielectric properties to determine moisture content (Nelson et al., 1998; Trabelsi et al., 2001, 2010; Trabelsi and Nelson, 1998, 2008). Atmospheric conditions were monitored in real-time to facilitate automated control of drying parameters. Such a control system showed promise in eliminating underdrying and overdrying, reapportioning labor, and minimizing energy consumption while providing efficient, automated drying control.

14.2.2 PEANUT DRIER CONTROL SYSTEM

The automated drying system consisted of six major components that work together to automate the peanut drying process. These components are the sensor network, data acquisition unit, microcontroller unit, data storage, drier unit, and display (Lewis et al., 2013). The sensor network had three temperature sensors, a relative humidity sensor, and the microwave in-shell kernel moisture sensor. The atmospheric conditions of interest, relative humidity, and ambient air temperature were measured with the relative humidity sensor and one of the temperature sensors. The other two temperature sensors were used to measure the temperature of the air blown from the drier into the inlet of the peanut trailer air plenum and the temperature of the peanut bed. The data from these sensors were assessed every 12 s within the microcontroller unit, and the appropriate control actions were relayed to the drier unit (Lewis et al., 2013). Real-time data were shown on the LCD display and stored by the microcontroller on a compact flash card, providing easy data retrieval. The microcontroller unit with display was a modified version of the embedded solution that controlled the microwave moisture meter shown in Figure 14.11 (Lewis and Trabelsi, 2012).

The six components of the automated drying system work together to govern the drying process and enforce the following conditions without human interaction:

1. Terminate aeration when kernel moisture content equals the desired level.
2. Suspend aeration when the relative humidity of ambient air is greater than 85%.
3. Limit temperature of drying air to 15°F **above** the ambient air temperature to ensure optimal drying rate (Beasley and Dickens, 1963; Blankenship and Davidson, 1984; Butts et al., 2008).
4. Limit temperature of drying air to no greater than 95°F to prevent off-flavor development and degradation of milling quality (Whitaker et al., 1974; Beasley and Dickens, 1963; Butts et al., 2008).

These four conditions comprise the control criteria that are used nationwide to facilitate the peanut drying process. While these conditions are normally monitored and enforced manually by an operator, this control system provided an automated solution. The automated drying system provided feedback control, data acquisition, data storage, and real-time monitoring (Lewis et al., 2013).

14.2.3 DRIER CONTROL SYSTEM TESTS

A quarter-scale peanut drying system, based on commercial peanut wagons, was constructed and equipped with the drier control system utilizing the microwave in-shell moisture content sensor (Lewis et al., 2013). This system was tested under varying atmospheric conditions and initial kernel moisture content to evaluate its efficiency and autonomy (Lewis et al., 2014). The initial temperature of the peanuts ranged from 18.5°C to 23.5°C, and initial kernel moisture content was between 12.1% and 16.7% wet basis. The drying process was fully automated, and pod and kernel moisture content were determined in real-time. Established drying parameters were maintained to provide optimal drying rates. Peanuts were dried until the target kernel moisture content, 10.49%, was reached; which required 5.7–11.2 h, depending mainly on ambient relative humidity during the drying period.

Kernel moisture content was determined in-shell throughout the drying process with a standard error of performance of 0.42–0.61% overall for the different trials. These values were within standard error ranges reported for the microwave moisture meter in measuring kernel moisture content of static samples. Samples taken from the peanut bed at intervals during the drying process for shelling and measurements of kernel moisture content, as they are usually monitored in current practice, agreed closely with values provided by in-shell monitoring. Statistical analyses revealed no significant differences between the kernel moisture contents determined within the drying system and the kernel moisture content determined by the standard oven method.

Assessment of the performance of the drying system revealed appropriate behavior and automated control. Implementation of such a system would improve the efficiency of reaching the target kernel moisture content and minimize energy consumption by reducing overdrying—minimizing bottlenecks within the buying point by reducing underdrying, and minimizing the human interaction required for drying peanuts.

REFERENCES

ASAE, ASAE S401.1, 2002. Moisture Measurement—Peanuts. ASAE Standards American Society of Agricultural Engineers, St. Joseph, MI, pp. 600–601.

Beasley E.O., Dickens J.W., 1963. Engineering Research in Peanut Curing. North Carolina Agricultural Experiment Station Tech. Bull. No. 155: North Carolina State University, Raleigh, NC.

Blankenship, P.D., Davidson, J.I., 1984. Automatic monitoring and cutoff of peanut driers. Peanut Sci. 11 (2), 58–60.

Butts, C.L., 1995. Incremental cost of overdrying farmers' stock peanuts. Appl. Eng. Agric. 11 (5), 671–675.

Butts, C.L., Hall, J. III, 1994. Development of a Graphical Decision Support System for a Peanut Drying Operation. ASAE Paper No. 94-6515. American Society of Agricultural Engineeers, St. Joseph, MI.

Butts, C.L., Williams, E.J., 2004. Measuring airflow distribution in peanut drying trailers. Appl. Eng. Agric. 20 (3), 335–339.

Butts, C.L., Davidson, M.I., Lamb, M.C., Kandala, C.V., Troeger, J.M., 2003. A Decision Support System for Curing Farmers' Stock Peanuts. ASAE Paper No. 03-6047. American Society of Agricultural Engineers, St. Joseph, MI.

Butts, C.L., Davidson, J.I., Lamb, M.C., Kandala, C.V., Troeger, J.M., 2004. Estimating drying time for a stock peanut curing decision support system. Trans. ASAE 47 (3), 925–932.

Butts, C.L., Tuggle, J., Williams, E.J., 2008. Good agricultural practices for peanut buying operations. Available at: <http://admin.peanutsusa.com/documents/Documents_Library/2008%20Final%20Chapter%202%20%20Buying%20Point%20Operations.pdf> (accessed 13.11.11.).

Lewis, M.A., Trabelsi, S., 2012. Integrating an embedded system in a microwave moisture meter. Appl. Eng. Agric. 28 (6), 923–931.

Lewis, M.A., Trabelsi, S., Nelson, S.O., Tollner, E.W., Haidekker, M.A., 2013. An automated approach to peanut drying with real-time microwave monitoring of in-shell kernel moisture content. Appl. Eng. Agric. 29 (4), 1–11.

Lewis, M.A., Trabelsi, S., Nelson, S.O., 2014 Assessment of real-time, in-shell kernel moisture content monitoring with a microwave moisture meter during peanut drying. Applied Engineering in Agriculture 30(4): 649-656. American society of Agricultural and Bioloogical Engineers, St. Joseph, Michigan.

Microtherm, I., 2000. Opticure 2000 dryer control system. Available at: <http://www.microtherm.com/dryctrl. htm> (accessed 13.01.10.).

Nelson, S.O., 1990. Use of dielectric mixture equations for estimating permittivities of solids from data on pulverized samples. In: Cody, G.D., Geballe, T.H., Sheng, P. (Eds.), Physical Phenomena in Granular Materials, vol. 195. Materials Research Society, Pittsburgh, PA, pp. 295−300.

Nelson, S.O., 1992. Correlating dielectric properties of solids and particulate samples through mixture relationships. Trans. ASAE 35 (2), 625−629.

Nelson, S.O., Kraszewski, A.W., Trabelsi, S., 1998. Advances in sensing grain moisture content by microwave measurements. Trans. ASAE 41 (2), 483−487.

Trabelsi, S., Nelson, S.O., 1998. Density-independent functions for on-line microwave moisture meters: a general discussion. Meas. Sci. Technol. 9, 570−578.

Trabelsi, S., Nelson, S.O., 2004. Microwave dielectric properties of shelled and unshelled peanuts. Trans. ASAE 47 (4), 1215−1222.

Trabelsi, S., Nelson, S.O., 2006. Microwave sensing technique for nondestructive determination of bulk density and moisture content in unshelled and shelled peanuts. Trans. ASAE 49 (5), 1563−1568.

Trabelsi, S., Nelson, S.O., 2007. A Low-Cost Microwave Sensor for Simultaneous and Independent Determination of Bulk Density and Moisture Content in Grain and Seed. ASABE Paper No. 076240, American Society of Agricultural and Biological Engineers, St. Joseph, MI.

Trabelsi, S., Nelson, S.O., 2008. Microwave moisture sensor for grain and seed. Biol. Eng. 1 (2), 195−202.

Trabelsi, S., Nelson, S.O., 2010a. Microwave moisture sensor for rapid and nondestructive grading of peanuts. In: IEEE SoutheastCon CD: Institute of Electrical and Electronics Engineers (IEEE), Piscataway, New Jersey, pp. 57−59.

Trabelsi, S., Nelson, S.O., 2010b. Microwave moisture meter for granular and particulate materials. In: Proceedings of the 2010 IEEE Instrumentation and Measurement Technology Conference, CD: Institute of Electrical and Electronics Engineers (IEEE), Piscataway, New Jersey, pp. 1304−1308.

Trabelsi, S., Kraszewski, A., Nelson, S.O., 1998. New density-independent calibration function for microwave sensing of moisture content in particulate materials. IEEE Trans. Instrum. Meas. 47 (3), 613−622.

Trabelsi, S., Kraszewski, A.W., Nelson, S.O., 1999. Determining physical properties of grain by microwave permittivity measurements. Trans. ASAE 42 (2), 531−536.

Trabelsi, S., Kraszewski, A.W., Nelson, S.O., 2000. Phase-shift ambiguity in microwave dielectric properties measurements. IEEE Trans. Instrum. Meas. 49 (1), 56−60.

Trabelsi, S., Kraszewski, A.W., Nelson, S.O., 2001. New calibration technique for microwave moisture sensors. IEEE Trans. Instrum. Meas. 50, 877−881.

Trabelsi S., Nelson S.O., Lewis M.A., 2008. Practical microwave meter for sensing moisture and density of granular materials. In: IEEE International Instrumentation and Measurement Technology Conference Proceedings: Institute of Electrical and Electronics Engineers (IEEE), Piscataway, New Jersey, pp. 1021−1025.

Trabelsi, S., Lewis, M.A., Nelson, S.O., 2009. Rapid and Nondestructive Determination of Moisture Content in Peanut Kernels from Microwave Measurement of Dielectric Properties of Pods. ASABE Paper No.097043, American Society of Agricultural and Biological Engineers, St. Joseph, MI.

Trabelsi, S., Lewis, J.E., Nelson, S.O., 2010. Microwave Moisture Meter for Rapid and Nondestructive Grading of Peanuts. ASABE Paper No. 1009183. American society of Agricultural and Biological Engineers, St. Joseph, MI.

Trabelsi, S., Lewis, M.A., Nelson, S.O., 2011. Microwave moisture meter for nondestructive and instantaneous peanut grading application. In: Proceedings of the 9th International Working Conference on Electromagnetic Wave Interaction with Water and Moist Substances (ISEMA 2011), Kansas City, Missouri, pp. 249−255.

Troeger, J.M., 1989. Modeling quality in bulk peanut curing. Peanut Sci. 16 (2), 105—108.

Volgyi, F., Burrows, J., Shepherdson, R., 2002. Calibration methods for non density-sensitive microwave-based moisture sensors. In: 4th International Symposium on Humidity and Moisture, ISHM'02, Taipei, Taiwan, R.O.C. 16-19 Sept. 2002, pp. 371—378.

Whitaker, T.B., Dickens, J.W., Bowen, H.D., 1974. Effects of curing on the internal oxygen concentration of peanuts. Trans. ASAE 17 (3), 557—569.

DIELECTRIC PROPERTIES DATA 15

An attempt is made in this chapter to accumulate, for reference purposes, data on the permittivities, or dielectric properties, of agricultural materials, and to identify other sources of such data that have already been tabulated. Included are references to such data presented in other chapters of this book, where those data provide useful information on dielectric properties.

15.1 EARLIER TABULATIONS

During and following the Second World War, extensive Tables of Dielectric Materials were prepared to aid "government agencies, engineers, and manufacturers in the proper application of dielectrics and in the development of better products" by the Laboratory for Insulation Research, Massachusetts Institute of Technology (von Hippel, 1954). They included many measurements taken in that Laboratory. The dielectric properties were organized for inorganic and organic solids and liquids and tabulated at frequencies generally between 100 Hz and 15 GHz at given temperatures. Graphical data were included for the dielectric constant and loss tangent as functions of temperature. These data, however, have little application to agricultural materials, except possibly as components.

In the early 1970s, the available data on the dielectric properties of agriculturally related materials were compiled into a reference tabulation published with a critical review of information relating to electrical properties of agricultural products (Nelson, 1973a). Major categories included animal tissues, foods, plant materials, fruits and vegetables, grain and seed, wood, and textiles. Values were tabulated for the dielectric constant, loss factor, loss tangent, and conductivity at given frequencies, temperatures, moisture contents, and specific gravities. These dielectric constant and loss-factor data were included in a broader tabulation of dielectric properties of materials for microwave processing (Tinga and Nelson, 1973). Both of these tabulations are useful references for agricultural materials.

The most complete reference for the dielectric properties of grain and seed is available in Engineering Data, ASAE Data D293.4 January 2012, Dielectric Properties of Grain and Seed, ASABE Standards, published annually by the American Society of Agricultural and Biological Engineers, St. Joseph, MI. It includes graphical data for many kinds of grain and seed as functions of frequency at different moisture contents, tabular data for many additional kinds of grain and seed at frequencies from 1 to 50 MHz, audio-frequency dielectric properties from 25 Hz to 20 kHz, microwave dielectric properties of cereal grains and oilseeds, and the models discussed in Chapter 13.

Dielectric Properties of Agricultural Materials and Their Applications. DOI: http://dx.doi.org/10.1016/B978-0-12-802305-1.00015-4

A significant bibliography and tabulation of the dielectric constants and loss factors of food materials was compiled and published in 1987 (Kent, 1987). Dielectric constants and loss factors of food materials, arranged in alphabetical order, were tabulated for frequencies between 10 Hz and 10 GHz at given temperatures. Data for several agricultural products, such as grain and seed, dairy products, fish, fruit, meat, and vegetables were included. Dielectric constants and loss factors for many food materials at microwave frequencies were also compiled from the literature and tabulated as functions of temperature, and composition (Datta et al., 1995). Dielectric properties of a number of foods and agricultural products have also been presented graphically as functions of frequency, moisture content, and temperature (Mudgett, 1995; Nelson and Datta, 2001).

Dielectric properties of animal tissues have been studied extensively because of their usefulness in studying electrophysiologic phenomena and evaluating the effects of electromagnetic energy absorption on living systems (Schwan, 1957; Foster and Schwan, 1989). Tabular and graphical data have been presented for the dielectric constant and conductivity, from 10 Hz to 35 GHz, of different excised human body tissues and animal tissues (Schwan, 1957; Foster and Schwan, 1989; Pethig and Kell, 1987). Many dielectric properties (real permittivity and conductivity values) have been collected and presented graphically for human and animal tissues at body temperatures over the general frequency range from 10 Hz to 20 GHz (Gabriel et al., 1996a,b). While these data were obtained mainly for medical applications, some may have useful agricultural or food applications.

15.2 GRAIN AND SEED DATA

Some data on dielectric properties that are not available in references identified in Section 15.1 are included here for reference. Because dielectric properties of grain and seed vary not only with frequency but also with moisture content, temperature, and bulk density, their values and relationships can be presented more efficiently in graphical format. Therefore, graphical data for the dielectric constant, ε', and loss factor, ε'', are included here for reference and to illustrate relationships and ranges of dielectric properties data. Values for the loss tangent, $\tan \delta = \varepsilon''/\varepsilon'$ and the ac conductivity, $\sigma = \omega\varepsilon_0\varepsilon'' = 2\pi f \varepsilon_0 \varepsilon''$, can be calculated, if desired (see Section 1.1).

15.2.1 HARD RED WINTER WHEAT DATA

Mean values for the dielectric constants and loss factors of hard red winter wheat, *Triticum aestivum* L., are shown in Figures 15.1 and 15.2 for the frequency range from 250 Hz to 12 GHz at moisture contents between 2.7% and 23.8% at 24°C (Nelson and Stetson, 1976), where logarithmic scales are used to show values better over the wide range of variables.

The dependence of the same properties on moisture content at a selection of frequencies is shown in Figures 15.3 and 15.4 at 24°C (Nelson and Stetson, 1976).

In Figures 15.5 and 15.6, dielectric properties of wheat are shown for three specific microwave frequencies (Nelson, 1973b), because scales are compressed for the higher frequencies in Figures 15.3 and 15.4.

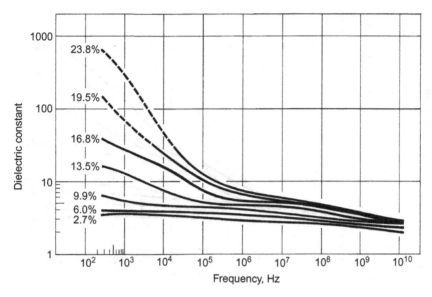

FIGURE 15.1

Frequency dependence of the dielectric constant of hard red winter wheat at indicated moisture contents and 24°C (Nelson and Stetson, 1976).

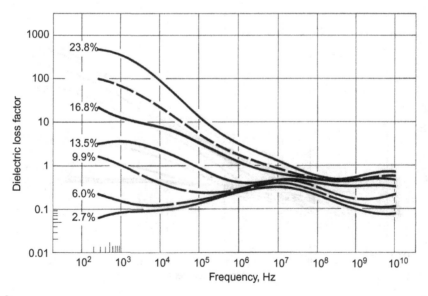

FIGURE 15.2

Frequency dependence of the dielectric loss factor of hard red winter wheat at indicated moisture contents and 24°C (Nelson and Stetson, 1976).

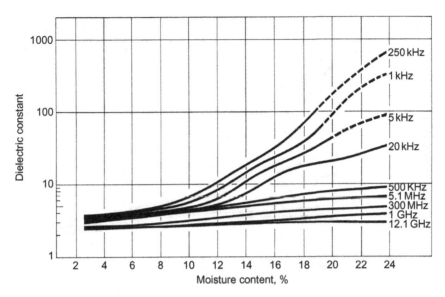

FIGURE 15.3

Variation of the dielectric constant of hard red winter wheat with moisture content, wet basis, at selected frequencies at 24°C (Nelson and Stetson, 1976).

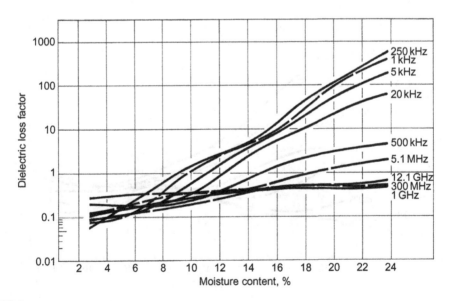

FIGURE 15.4

Variation of the dielectric loss factor of hard red winter wheat with moisture content, wet basis, at selected frequencies at 24°C (Nelson and Stetson, 1976).

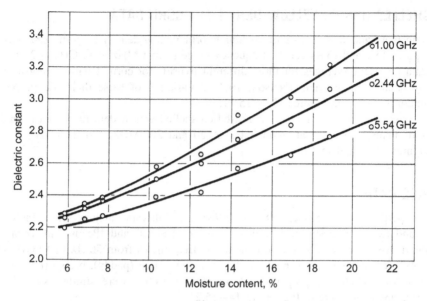

FIGURE 15.5

Variation of the dielectric constant of 'Scout 66' hard red winter wheat at indicated microwave frequencies at 24°C (Nelson, 1973b).

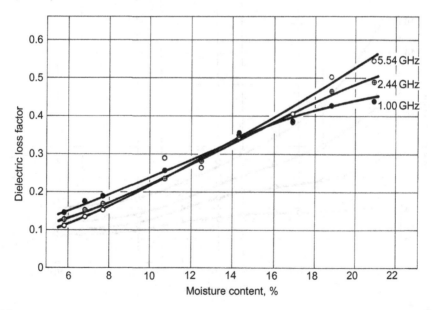

FIGURE 15.6

Variation of the dielectric loss factor of 'Scout 66' hard red winter wheat at indicated microwave frequencies at 24°C (Nelson, 1973b).

15.2.2 SHELLED HYBRID YELLOW-DENT FIELD CORN DATA

Dielectric constants and loss factors of shelled hybrid yellow-dent field corn, *Zea mays* L., are shown in Figures 15.7 and 15.8 over the frequency range from 1 MHz to 11 GHz at 24°C.

These data represent averages for three different hybrids for corn at natural moisture contents harvested by hand from the field (Nelson, 1978a). Variation of these dielectric properties with moisture content is shown in Figures 15.9–15.11.

Dielectric constants and loss factors, measured on shelled yellow-dent field corn from 21 different hybrids of natural moisture at frequencies of 20 and 300 MHz, and 2.45 GHz at 24°C, are shown in Figures 15.12 and 15.13.

15.2.3 RICE DATA

Dielectric properties of three rice, *Oryza sativa* L., cultivars, a long-grain Indica variety, 'Lebonnet', and two medium-grain Japonica varieties, 'Brazos' and 'Pecos', all Texas releases, were measured over a range of moisture contents at frequencies from 50 Hz to 12 GHz (Noh and Nelson, 1989). Measurements were made at 24°C on rough rice (paddy), brown (hulled) rice, and white (milled) rice. Dielectric properties of all three varieties were similar, so only data for 'Lebonnet' rice are shown here in Figures 15.14–15.21.

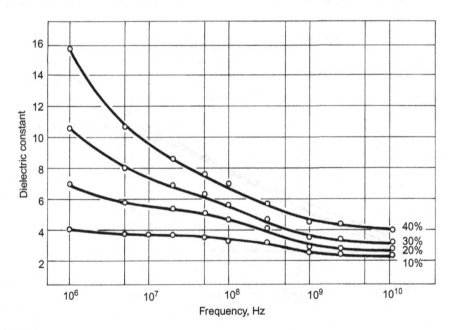

FIGURE 15.7

Frequency dependence of the dielectric constant of shelled yellow-dent field corn at indicated moisture contents, wet basis, at 24°C (Nelson, 1978a).

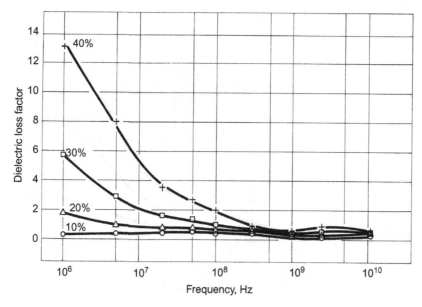

FIGURE 15.8

Frequency dependence of the dielectric loss factor of shelled yellow-dent field corn at indicated moisture contents, wet basis, at 24°C (Nelson, 1978a).

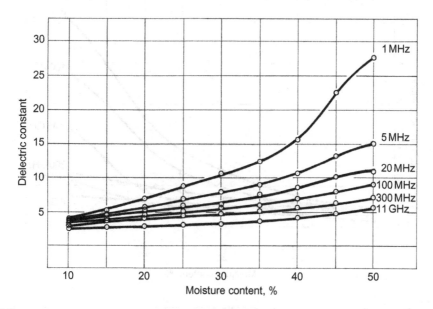

FIGURE 15.9

Variation of the dielectric constant of shelled yellow-dent field corn at indicated frequencies at 24°C (Nelson, 1978a).

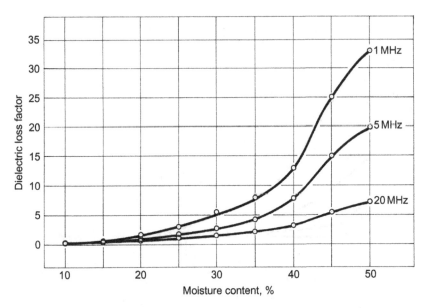

FIGURE 15.10

Variation of the dielectric loss factor of shelled yellow-dent field corn at indicated frequencies at 24°C (Nelson, 1978a).

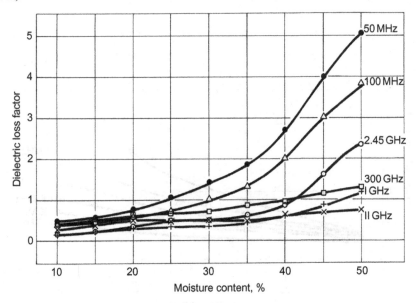

FIGURE 15.11

Variation of the dielectric loss factor of shelled yellow-dent field corn at indicated frequencies at 24°C (Nelson, 1978a).

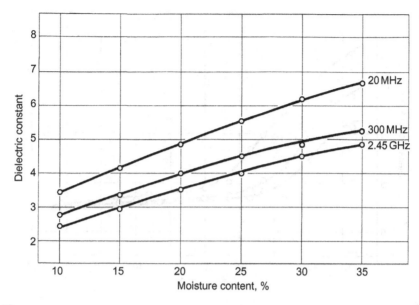

FIGURE 15.12

Variation of the dielectric constant of shelled yellow-dent field corn at indicated microwave frequencies at 24°C (Nelson, 1979, 1978b).

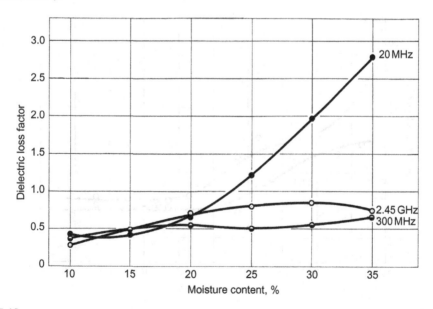

FIGURE 15.13

Variation of the dielectric loss factor of shelled yellow-dent field corn at indicated microwave frequencies at 24°C (Nelson, 1979, 1978b).

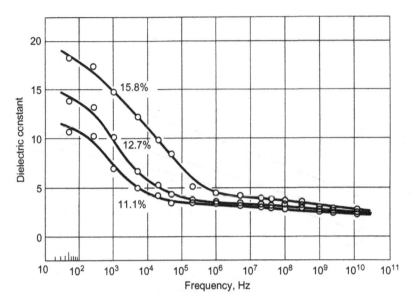

FIGURE 15.14

Frequency dependence of the dielectric constant of rough rice at indicated moisture contents, wet basis, and 24°C (Noh and Nelson, 1989).

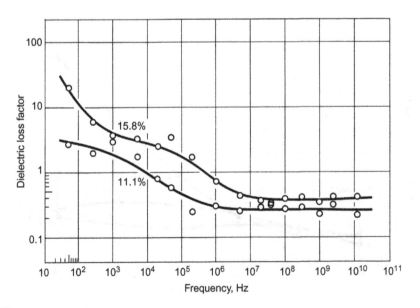

FIGURE 15.15

Frequency dependence of the dielectric loss factor of rough rice at indicated moisture contents, wet basis, and 24°C (Noh and Nelson, 1989).

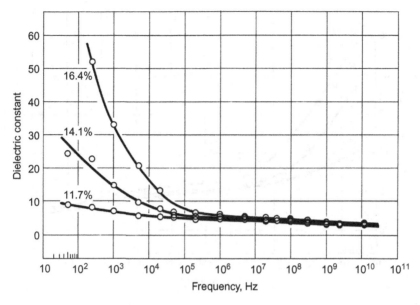

FIGURE 15.16

Frequency dependence of the dielectric constant of brown rice at indicated moisture contents, wet basis, and 24°C (Noh and Nelson, 1989).

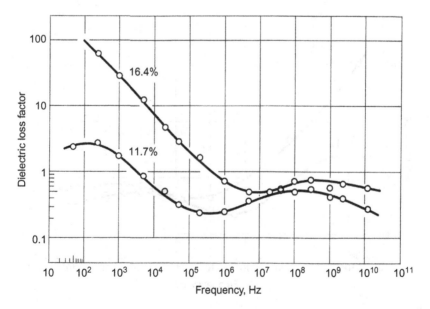

FIGURE 15.17

Frequency dependence of the dielectric loss factor of brown rice at indicated moisture contents, wet basis, and 24°C (Noh and Nelson, 1989).

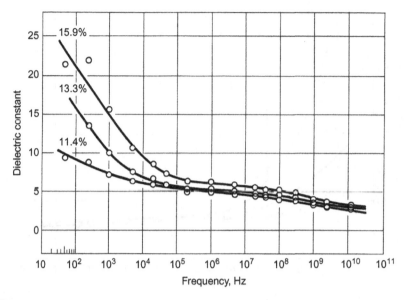

FIGURE 15.18

Frequency dependence of the dielectric constant of white rice at indicated moisture contents, wet basis, and 24°C (Noh and Nelson, 1989).

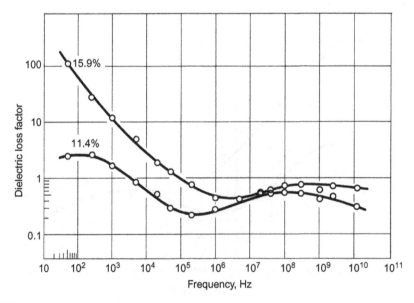

FIGURE 15.19

Frequency dependence of the dielectric loss factor of white rice at indicated moisture contents, wet basis, and 24°C (Noh and Nelson, 1989).

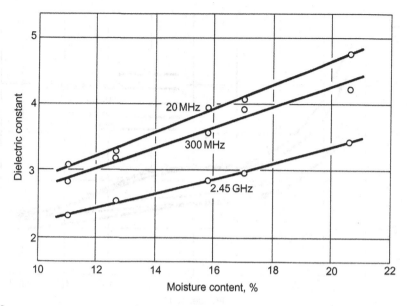

FIGURE 15.20

Variation of the dielectric constant of rough rice with moisture content, wet basis, at indicated frequencies (Noh and Nelson, 1989).

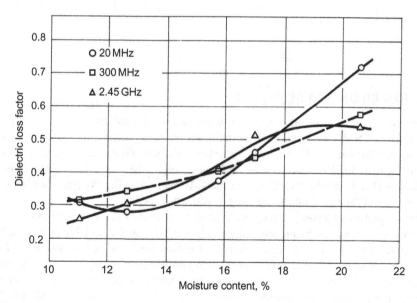

FIGURE 15.21

Variation of the dielectric loss factor of rough rice with moisture content, wet basis, at indicated frequencies (Noh and Nelson, 1989).

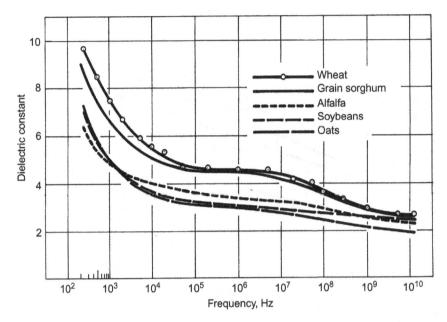

FIGURE 15.22

Frequency dependence of the dielectric constant of indicated kinds of grain and seed (Nelson and Stetson, 1975) at the following moisture contents: wet basis: wheat, 12.5%; grain sorghum, 11.4%; oats, 10.7%; soybeans, 8.5%, and alfalfa, 7.5% (Nelson and Stetson, 1975).

15.2.4 OTHER GRAIN AND SEED DATA

Dielectric properties of several other types of grain and seed were measured over the broad frequency range from 250 Hz to 12 GHz at normally encountered moisture contents and densities at 24°C (Nelson and Stetson, 1975). Dielectric constants and loss factors are shown in Figures 15.22 and 15.23 for "Gage" hard red winter wheat, *T. aestivum* L., NC + Hybrid T700 grain sorghum, *Sorghum bicolor* (L.) Moench, "Neal" spring oats, *Avena sativa* L., "Wayne" soybeans. *Glycene max* (L.) Merrill, and "Ranger" alfalfa, *Medicago sativa* L., seed.

Dielectric properties of several other small grains and soybeans were measured at 24°C, over ranges of moisture content at frequencies of 20 and 300 MHz and 2.45 GHz in sample holders of identical cross-section, so that grain and seed bulk densities would be comparable at all frequencies (Nelson, 1987). These included spring barley, *Hordeum vulgare* L., spring oats, *A. sativa* L., a hardy winter barley, *H. vulgare* L., a semihardy winter barley of the same species, soft red winter wheat, *T. aestivum* L., winter rye, *Secale cereale* L., and soybeans, *G. max* (L.) Merrill. The range of natural bulk densities for each of the seed lots over the range of moisture contents was provided in the original report along with details of the measurements (Nelson, 1987). Dielectric constants and loss factors of these grain and seed lots are shown in Figures 15.24–15.37.

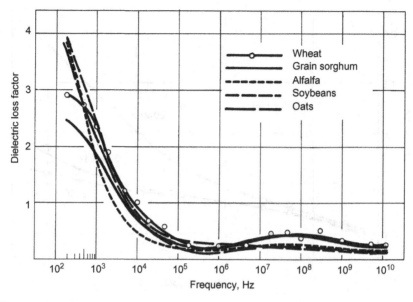

FIGURE 15.23

Frequency dependence of the dielectric loss factor of indicated kinds of grain and seed (Nelson and Stetson, 1975) at the following moisture contents: wet basis: wheat, 12.5%; grain sorghum, 11.4%; oats, 10.7%; soybeans, 8.5%, and alfalfa, 7.5% (Nelson and Stetson, 1975).

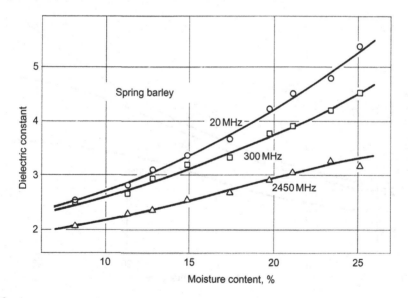

FIGURE 15.24

Variation of the dielectric constant of spring barley with moisture content at indicated frequencies (Nelson, 1987).

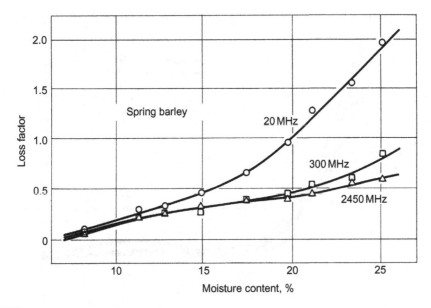

FIGURE 15.25

Variation of the dielectric loss factor of spring barley with moisture content at indicated frequencies (Nelson, 1987).

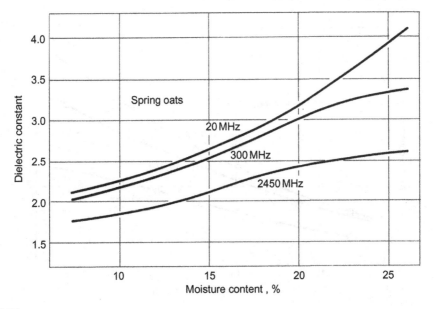

FIGURE 15.26

Variation of the dielectric constant of spring oats with moisture content at indicated frequencies (Nelson, 1987).

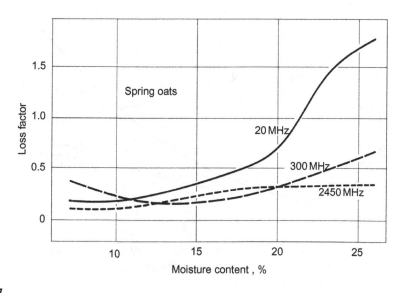

FIGURE 15.27

Variation of the dielectric loss factor of spring oats with moisture content at indicated frequencies (Nelson, 1987).

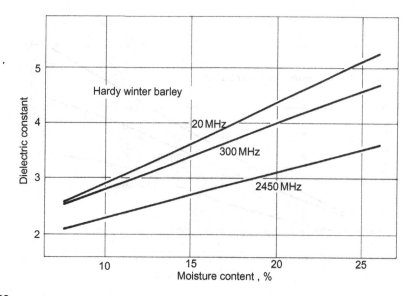

FIGURE 15.28

Variation of the dielectric constant of hardy winter barley with moisture content at indicated frequencies (Nelson, 1987).

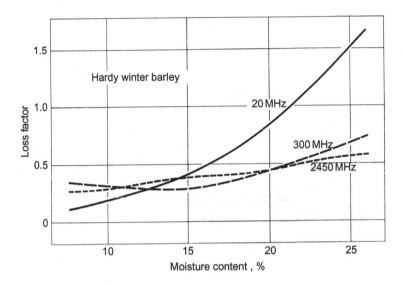

FIGURE 15.29

Variation of the dielectric loss factor of hardy winter barley with moisture content at indicated frequencies (Nelson, 1987).

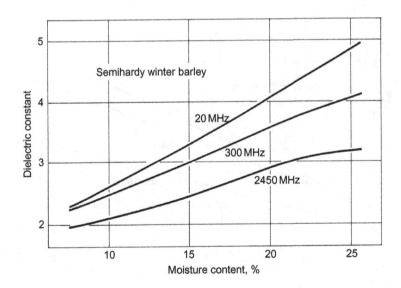

FIGURE 15.30

Variation of the dielectric constant of semihardy winter barley with moisture content at indicated frequencies (Nelson, 1987).

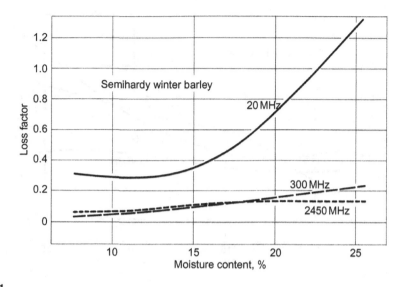

FIGURE 15.31

Variation of the dielectric loss factor of semihardy winter barley with moisture content at indicated frequencies (Nelson, 1987).

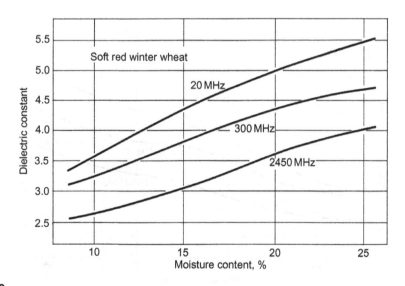

FIGURE 15.32

Variation of the dielectric constant of soft red winter wheat with moisture content at indicated frequencies (Nelson, 1987).

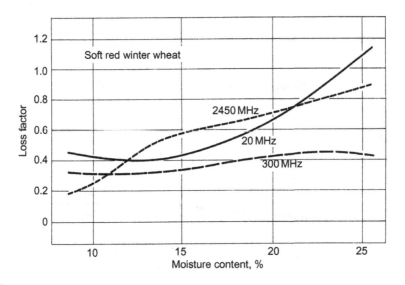

FIGURE 15.33

Variation of the dielectric loss factor of soft red winter wheat with moisture content at indicated frequencies (Nelson, 1987).

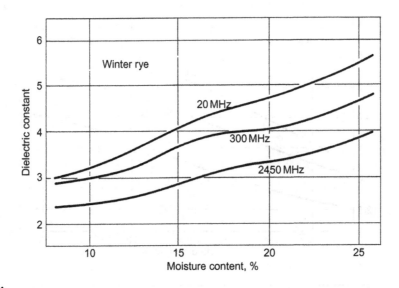

FIGURE 15.34

Variation of the dielectric constant of winter rye with moisture content at indicated frequencies (Nelson, 1987).

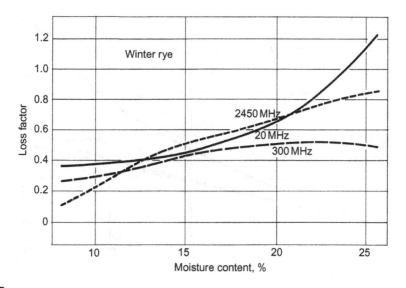

FIGURE 15.35

Variation of the dielectric loss factor of winter rye with moisture content at indicated frequencies (Nelson, 1987).

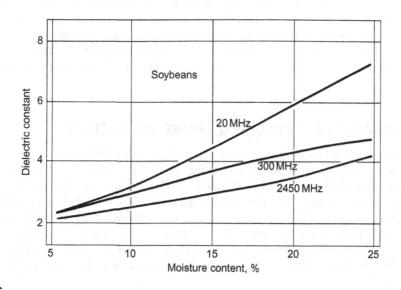

FIGURE 15.36

Variation of the dielectric constant of soybeans with moisture content at indicated frequencies (Nelson, 1987).

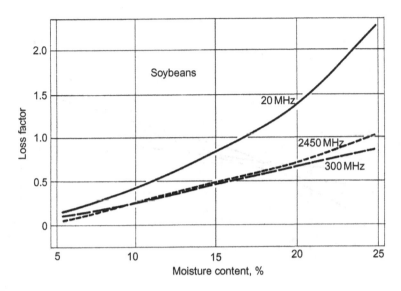

FIGURE 15.37

Variation of the dielectric loss factor of soybeans with moisture content at indicated frequencies (Nelson, 1987).

15.2.5 INDIVIDUAL GRAIN KERNELS AND SEEDS DATA

The preceding data on dielectric properties of grain and seed are properties of the granular materials, including the air space between the kernels and seeds. In work with dielectric mixture equations and other applications, the dielectric properties of the individual grain kernels and seeds are required. Some of these data have been acquired at microwave frequencies (Nelson, 1976, 1983, 1984; You and Nelson, 1988; Nelson and You, 1989), and some data are listed in Table 15.1.

15.2.6 MICROWAVE DATA FOR CEREAL GRAINS AND OILSEEDS

Dielectric properties of some cereal grains and oilseeds have been measured at microwave frequencies from 5 to 15 GHz by free-space transmission measurement techniques with precautions to obtain reliable data (Trabelsi and Nelson, 2005, 2012). Because of the advantages noted in Chapter 7 for moisture sensing with microwave frequencies, these data are included here for wheat ("Arapahoe" hard red winter wheat, *T. aestivum* L.), corn (Pioneer 3335 hybrid yellow-dent field corn, *Z. mays* L.), barley ("Canlon" spring barley, *H. vulgare* L.), oats ("Chapman" winter oats, *A. sativa* L.), grain sorghum (*S. bicolor* L.), soybeans (*G. max* (L.) Merrill), and canola (*Brassica napus* L.). The resulting data for the dielectric constants and loss factors are shown in Tables 15.2–15.7.

Table 15.1 Dielectric Properties of Individual Grain Kernels and Seeds

Kind of Grain or Seed	Moisture Content, %	Kernel or Seed Density, g/cm³	Dielectric Constant		Dielectric Loss Factor	
			11 GHz	22 GHz	11 GHz	22 GHz
Wheat, hard red winter	10.5	1.386	4.52	4.28	0.46	0.45
	12.0	1.382	4.76	4.42	0.52	0.69
	14.4	1.375	5.04	–	0.85	–
	16.8	1.372	5.39	4.55	0.84	0.88
Corn, yellow-dent field	12.9	1.201	4.31	3.70	0.51	0.26
	14.9	1.195	4.53	4.22	0.78	0.61
	16.6	1.195	5.81	4.51	1.04	0.95
	20.4	1.185	6.25	4.82	1.07	0.94
Soybean	7.5	1.233	3.91	3.49	0.22	0.22
	10.4	1.230	4.20	3.62	0.42	0.32
	12.3	1.225	4.61	3.77	0.52	0.48
	15.0	1.222	5.12	3.94	0.68	0.58
Rough Rice, 'Lebonnet'	11.2	1.406	4.71	4.38	0.77	0.65
	20.6	1.388	6.25	5.78	0.84	1.33
'Brazos'	11.5	1.409	5.09	4.54	0.73	0.80
	21.8	1.387	6.58	6.01	0.84	1.13
White Rice, 'Lebonnet'	12.2	1.476	5.15	5.05	0.72	1.16
	15.7	1.454	5.83	5.48	1.26	1.10
'Brazos'	12.7	1.463	5.18	5.17	0.88	1.00
	16.7	1.455	5.90	5.42	1.28	1.13

Table 15.2 Dielectric Constant (ε') and Loss Factor (ε'') of Wheat at 23°C and Indicated Moisture Content (M) and Density (ρ)

Frequency, GHz	$M = 10.2\%$	11.3	12.4	13.4	14.7	15.9	18.0
	$\rho = 0.822$ g/cm³	0.813	0.774	0.794	0.786	0.767	0.746
5	$\varepsilon' = 2.62$	2.68	2.65	2.74	2.87	2.86	3.07
	$\varepsilon'' = 0.23$	0.26	0.30	0.34	0.37	0.43	0.48
6	$\varepsilon' = 2.59$	2.65	2.61	2.70	2.81	2.81	2.98
	$\varepsilon'' = 0.23$	0.26	0.30	0.34	0.37	0.43	0.49
7	$\varepsilon' = 2.57$	2.62	2.59	2.67	2.77	2.78	2.92
	$\varepsilon'' = 0.23$	0.26	0.30	0.33	0.37	0.43	0.50
8	$\varepsilon' = 2.55$	2.61	2.58	2.66	2.75	2.75	2.89
	$\varepsilon'' = 0.23$	0.26	0.29	0.33	0.37	0.43	0.50
9	$\varepsilon' = 2.53$	2.59	2.56	2.64	2.71	2.73	2.86
	$\varepsilon'' = 0.24$	0.26	0.28	0.33	0.37	0.42	0.52
10	$\varepsilon' = 2.51$	2.57	2.54	2.62	2.69	2.71	2.82
	$\varepsilon'' = 0.24$	0.26	0.27	0.32	0.37	0.41	0.53

(Continued)

Table 15.2 Dielectric Constant (ε') and Loss Factor (ε'') of Wheat at 23°C and Indicated Moisture Content (M) and Density (ρ) *Continued*

Frequency, GHz	$M = 10.2\%$ $\rho = 0.822$ g/cm^3	11.3 0.813	12.4 0.774	13.4 0.794	14.7 0.786	15.9 0.767	18.0 0.746
11	$\varepsilon' = 2.50$	2.56	2.53	2.59	2.67	2.69	2.78
	$\varepsilon'' = 0.24$	0.26	0.27	0.31	0.37	0.40	0.53
12	$\varepsilon' = 2.49$	2.55	2.52	2.58	2.65	2.67	2.77
	$\varepsilon'' = 0.23$	0.26	0.26	0.31	0.37	0.39	0.53
13	$\varepsilon' = 2.49$	2.55	2.51	2.57	2.64	2.65	2.76
	$\varepsilon'' = 0.23$	0.25	0.26	0.30	0.36	0.38	0.52
14	$\varepsilon' = 2.49$	2.54	2.50	2.56	2.64	2.64	2.75
	$\varepsilon'' = 0.22$	0.25	0.26	0.30	0.36	0.38	0.52
15	$\varepsilon' = 2.48$	2.53	2.49	2.55	2.64	2.63	2.74
	$\varepsilon'' = 0.22$	0.25	0.26	0.30	0.36	0.37	0.53

Table 15.3 Dielectric Constant (ε') and Loss Factor (ε'') of Corn at 23°C and Indicated Moisture Content (M) and Density (ρ)

Frequency, GHz	$M = 10.8\%$ $\rho = 0.799$ g/cm^3	11.8 0.805	13.7 0.791	15.7 0.767	17.8 0.745	19.6 0.710
5	$\varepsilon' = 2.68$	2.81	2.94	3.06	3.36	3.43
	$\varepsilon'' = 0.30$	0.33	0.42	0.49	0.66	0.72
6	$\varepsilon' = 2.65$	2.76	2.87	3.00	3.22	3.32
	$\varepsilon'' = 0.31$	0.34	0.44	0.51	0.70	0.76
7	$\varepsilon' = 2.64$	2.73	2.81	2.95	3.16	3.29
	$\varepsilon'' = 0.31$	0.35	0.44	0.53	0.67	0.78
8	$\varepsilon' = 2.64$	2.72	2.80	2.92	3.14	3.18
	$\varepsilon'' = 0.31$	0.36	0.45	0.56	0.70	0.79
9	$\varepsilon' = 2.62$	2.69	2.77	2.88	3.08	3.13
	$\varepsilon'' = 0.31$	0.37	0.47	0.58	0.72	0.81
10	$\varepsilon' = 2.60$	2.67	2.74	2.86	3.05	3.08
	$\varepsilon'' = 0.31$	0.39	0.50	0.61	0.72	0.83
11	$\varepsilon' = 2.59$	2.65	2.70	2.83	3.01	3.03
	$\varepsilon'' = 0.32$	0.40	0.51	0.64	0.73	0.84
12	$\varepsilon' = 2.58$	2.65	2.70	2.82	2.98	3.00
	$\varepsilon'' = 0.33$	0.41	0.51	0.66	0.75	0.86
13	$\varepsilon' = 2.56$	2.64	2.69	2.80	2.97	2.97
	$\varepsilon'' = 0.34$	0.42	0.51	0.67	0.75	0.88
14	$\varepsilon' = 2.56$	2.64	2.69	2.80	2.97	2.95
	$\varepsilon'' = 0.35$	0.43	0.52	0.69	0.79	0.92
15	$\varepsilon' = 2.57$	2.64	2.70	2.80	2.91	2.91
	$\varepsilon'' = 0.38$	0.44	0.53	0.70	0.84	0.98

Table 15.4 Dielectric Constant (ε') and Loss Factor (ε'') of Barley at 23°C and Indicated Moisture Content (M) and Density (ρ)

Frequency, GHz	$M = 10.8\%$	13.0	15.8	17.6	20.8	24.8
	$\rho = 0.686$ g/cm³	0.660	0.661	0.650	0.588	0.567
5	$\varepsilon' = 2.23$	2.28	2.47	2.53	2.63	2.82
	$\varepsilon'' = 0.15$	0.19	0.27	0.32	0.41	0.47
6	$\varepsilon' = 2.21$	2.26	2.44	2.49	2.58	2.76
	$\varepsilon'' = 0.16$	0.20	0.28	0.32	0.42	0.49
7	$\varepsilon' = 2.20$	2.24	2.41	2.47	2.54	2.72
	$\varepsilon'' = 0.16$	0.20	0.29	0.33	0.43	0.51
8	$\varepsilon' = 2.19$	2.23	2.38	2.44	2.52	2.68
	$\varepsilon'' = 0.17$	0.21	0.30	0.34	0.44	0.53
9	$\varepsilon' = 2.17$	2.20	2.35	2.41	2.48	2.64
	$\varepsilon'' = 0.18$	0.21	0.30	0.34	0.45	0.55
10	$\varepsilon' = 2.15$	2.18	2.33	2.37	2.45	2.59
	$\varepsilon'' = 0.17$	0.21	0.30	0.34	0.46	0.57
11	$\varepsilon' = 2.14$	2.17	2.31	2.35	2.42	2.55
	$\varepsilon'' = 0.17$	0.20	0.30	0.33	0.47	0.57
12	$\varepsilon' = 2.14$	2.17	2.31	2.35	2.40	2.52
	$\varepsilon'' = 0.16$	0.20	0.29	0.33	0.47	0.58
13	$\varepsilon' = 2.14$	2.17	2.31	2.34	2.37	2.49
	$\varepsilon'' = 0.16$	0.20	0.29	0.34	0.47	0.58
14	$\varepsilon' = 2.14$	2.17	2.31	2.34	2.35	2.46
	$\varepsilon'' = 0.16$	0.20	0.29	0.35	0.47	0.59
15	$\varepsilon' = 2.13$	2.17	2.29	2.34	2.34	2.44
	$\varepsilon'' = 0.17$	0.21	0.29	0.36	0.48	0.61

Table 15.5 Dielectric Constant (ε') and loss Factor (ε'') of Oats at 23°C and Indicated Moisture Content (M) and Density (ρ)

Frequency, GHz	$M = 9.1\%$	11.0	13.3	15.1	17.5	18.3
	$\rho = 0.479$ g/cm³	0.467	0.479	0.469	0.456	0.456
5	$\varepsilon' = 1.71$	1.77	1.87	1.95	2.02	2.04
	$\varepsilon'' = 0.07$	0.09	0.14	0.17	0.19	0.21
6	$\varepsilon' = 1.70$	1.76	1.85	1.93	1.99	2.00
	$\varepsilon'' = 0.06$	0.10	0.15	0.17	0.20	0.22
7	$\varepsilon' = 1.69$	1.75	1.83	1.91	1.97	1.97
	$\varepsilon'' = 0.06$	0.10	0.15	0.17	0.21	0.22
8	$\varepsilon' = 1.69$	1.74	1.81	1.90	1.95	1.96
	$\varepsilon'' = 0.07$	0.10	0.16	0.18	0.22	0.23

(Continued)

Table 15.5 Dielectric Constant (ε') and loss Factor (ε'') of Oats at 23°C and Indicated Moisture Content (M) and Density (ρ) Continued

Frequency, GHz	$M = 9.1\%$ $\rho = 0.479$ g/cm³	11.0 0.467	13.3 0.479	15.1 0.469	17.5 0.456	18.3 0.456
9	$\varepsilon' = 1.69$	1.73	1.79	1.88	1.93	1.93
	$\varepsilon'' = 0.07$	0.10	0.16	0.18	0.22	0.23
10	$\varepsilon' = 1.68$	1.72	1.78	1.87	1.91	1.91
	$\varepsilon'' = 0.08$	0.10	0.15	0.18	0.22	0.24
11	$\varepsilon' = 1.67$	1.72	1.77	1.86	1.89	1.89
	$\varepsilon'' = 0.07$	0.11	0.15	0.18	0.22	0.23
12	$\varepsilon' = 1.67$	1.72	1.77	1.85	1.88	1.88
	$\varepsilon'' = 0.07$	0.11	0.15	0.18	0.22	0.23
13	$\varepsilon' = 1.68$	1.71	1.77	1.85	1.88	1.87
	$\varepsilon'' = 0.07$	0.10	0.15	0.18	0.21	0.22
14	$\varepsilon' = 1.68$	1.71	1.77	1.85	1.87	1.87
	$\varepsilon'' = 0.07$	0.10	0.15	0.18	0.22	0.23
15	$\varepsilon' = 1.68$	1.71	1.77	1.85	1.87	1.86
	$\varepsilon'' = 0.08$	0.10	0.15	0.18	0.22	0.23

Table 15.6 Dielectric Constant (ε') and Loss Factor (ε'') of Soybeans at 23°C and Indicated Moisture Content (M) and Density (ρ)

Frequency, GHz	$M = 8.48\%$ $\rho = 0.734$ g/cm³	9.8 0.733	11.9 0.728	13.3 0.732	16.1 0.734	18.7 0.737	20.3 0.741
5	$\varepsilon' = 2.40$	2.50	2.68	2.77	3.12	3.58	3.74
	$\varepsilon'' = 0.13$	0.19	0.25	0.30	0.49	0.74	0.81
6	$\varepsilon' = 2.38$	2.47	2.64	2.73	3.04	3.45	3.64
	$\varepsilon'' = 0.14$	0.19	0.31	0.36	0.52	0.78	0.85
7	$\varepsilon' = 2.37$	2.45	2.61	2.70	2.98	3.35	3.55
	$\varepsilon'' = 0.14$	0.19	0.30	0.36	0.53	0.78	0.87
8	$\varepsilon' = 2.37$	2.45	2.61	2.69	2.95	3.31	3.49
	$\varepsilon'' = 0.14$	0.19	0.30	0.36	0.53	0.77	0.89
9	$\varepsilon' = 2.36$	2.44	2.60	2.67	2.93	3.27	3.42
	$\varepsilon'' = 0.14$	0.19	0.29	0.35	0.55	0.77	0.89
10	$\varepsilon' = 2.35$	2.43	2.59	2.65	2.91	3.22	3.36
	$\varepsilon'' = 0.14$	0.19	0.29	0.34	0.58	0.79	0.89
11	$\varepsilon' = 2.35$	2.42	2.57	2.64	2.89	3.18	3.31
	$\varepsilon'' = 0.15$	0.19	0.29	0.34	0.60	0.80	0.90
12	$\varepsilon' = 2.35$	2.43	2.57	2.63	2.89	3.16	3.28
	$\varepsilon'' = 0.15$	0.19	0.30	0.35	0.61	0.81	0.92

Table 15.6 Dielectric Constant (ε') and Loss Factor (ε'') of Soybeans at 23°C and Indicated Moisture Content (M) and Density (ρ) *Continued*

Frequency, GHz	$M = 8.48\%$	9.8	11. 9	13.3	16.1	18.7	20.3
	$\rho = 0.734$ g/cm³	0.733	0.728	0.732	0.734	0.737	0.741
13	$\varepsilon' = 2.36$	2.43	2.56	2.62	2.79	3.14	3.25
	$\varepsilon'' = 0.15$	0.20	0.30	0.37	0.58	0.82	0.94
14	$\varepsilon' = 2.36$	2.44	2.56	2.63	2.79	3.13	3.22
	$\varepsilon'' = 0.16$	0.20	0.30	0.39	0.59	0.84	0.96
15	$\varepsilon' = 2.37$	2.45	2.56	2.65	2.79	3.12	3.19
	$\varepsilon'' = 0.16$	0.21	0.32	0.4	0.60	0.88	1.00

Table 15.7 Dielectric Constant (ε') and Loss Factor (ε'') of Canola at 23°C and Indicated Moisture Content (M) and Density (ρ)

Frequency, GHz	$M = 6.6\%$	8.5	10.4	12.4	14.4	16.7	18.8
	$\rho = 0.678$ g/cm³	0.675	0.662	0.678	0.665	0.668	0.676
5	$\varepsilon' = 2.18$	2.29	2.47	2.65	2.85	3.15	3.57
	$\varepsilon'' = 0.09$	0.16	0.26	0.36	0.47	0.56	–
6	$\varepsilon' = 2.16$	2.26	2.42	2.61	2.76	3.04	3.44
	$\varepsilon'' = 0.10$	0.16	0.26	0.36	0.47	0.59	0.73
7	$\varepsilon' = 2.14$	2.24	2.39	2.57	2.71	2.95	3.32
	$\varepsilon'' = 0.10$	0.15	0.25	0.36	0.46	0.58	0.73
8	$\varepsilon' = 2.14$	2.23	2.37	2.55	2.68	2.90	3.26
	$\varepsilon'' = 0.10$	0.15	0.24	0.35	0.45	0.57	0.71
9	$\varepsilon' = 2.13$	2.21	2.35	2.52	2.65	2.86	3.20
	$\varepsilon'' = 0.10$	0.15	0.24	0.34	0.45	0.58	0.70
10	$\varepsilon' = 2.12$	2.20	2.34	2.50	2.62	2.81	3.15
	$\varepsilon'' = 0.10$	0.15	0.24	0.33	0.44	0.59	0.69
11	$\varepsilon' = 2.09$	2.19	2.32	2.47	2.59	2.76	3.10
	$\varepsilon'' = 0.10$	0.15	0.23	0.31	0.43	0.59	0.68
12	$\varepsilon' = 2.11$	2.19	2.31	2.46	2.57	2.73	3.06
	$\varepsilon'' = 0.10$	0.14	0.22	0.30	0.41	0.57	0.67
13	$\varepsilon' = 2.10$	2.18	2.30	2.43	2.55	2.71	3.03
	$\varepsilon'' = 0.09$	0.13	0.21	0.29	0.40	0.56	0.67
14	$\varepsilon' = 2.10$	2.18	2.29	2.42	2.54	2.70	2.99
	$\varepsilon'' = 0.09$	0.13	0.21	0.29	0.39	0.55	0.67
15	$\varepsilon' = 2.10$	2.17	2.28	2.41	2.52	2.69	2.96
	$\varepsilon'' = 0.09$	0.13	0.21	0.29	0.38	0.54	0.67

Table 15.8 Dielectric Constants (ε') and Loss Factors (ε'') of Fresh Fruits and Vegetables at Indicated Frequencies at 24°C

Fruit or Vegetable	10 MHz		100 MHz		1 GHz	
	ε'	ε''	ε'	ε''	ε'	ε''
Apple	109	281	71	33	64	10
Avocado	245	759	66	89	56	14
Banana	166	834	76	91	65	18
Cantaloupe	260	629	70	72	63	14
Carrot	598	1291	87	157	72	23
Cucumber	123	361	80	39	77	9
Grape	122	570	78	60	73	13
Honeydew melon	110	540	86	57	72	10
Orange	197	617	78	69	72	14
Potato	183	679	73	77	62	16
Watermelon	120	580	75	60	70	10

15.2.7 FRESH FRUIT AND VEGETABLE DATA

Dielectric properties of some fruits and vegetables have been measured over wide frequency ranges (Nelson et al., 1994, 2006, 2007; Nelson, 2005; Guo et al., 2007), and values are included here for reference. The dielectric constant and loss factor values for several are listed for frequencies of 10 and 100 MHz and 1 GHz in Table 15.8.

They are also shown graphically for frequencies between 10 MHz and 1.8 GHz in Chapter 11, Figure 11.10 for fresh apple tissue, Figure 11.9 for banana, Figure 11.11 for cantaloupe, Figure 11.12 for navel orange, Figure 11.14 for honeydew melon, and Figure 11.15 for watermelon.

Data on dielectric properties were also obtained over the frequency range from 200 MHz to 20 GHz for a larger number of fresh fruits and vegetables, and these data are tabulated for six frequencies in this range in Table 15.9.

Graphical data for the dielectric constant and loss factor are shown for this frequency range in Chapter 11, Figure 11.14 for honeydew melon and Figure 11.15 for watermelon.

15.2.8 PECAN NUT DATA

Dielectric properties of chopped pecan kernels were measured over wide ranges of frequency and moisture content (Nelson, 1981). Values for the dielectric constant and loss factor are shown as functions of moisture content at different frequencies between 50 kHz and 10 GHz in Figures 15.38 and 15.39.

In Figures 15.40 and 15.41 the dielectric properties are displayed as contour plots, providing another option for reference and interpolation.

These data are for chopped pecans of normal densities. Bulk densities of chopped pecans and kernel densities as functions of moisture content are shown in Figures 15.42 and 15.43, respectively.

Table 15.9 Dielectric Constants (ε') and Loss Factors (ε'') of Fresh Fruits and Vegetables at Indicated Frequencies Between 200 MHz and 20 GHz at 24°C, Along with Other Physical Characteristics

Fruit or Vegetable	Cultivar or Other Description	Moisture Content, %	Density, g/cm³	Soluble Solids, %		Frequency, Hz					
						2×10^8	5×10^8	1.3×10^9	3.2×10^9	8×10^9	2×10^9
Apple, *Malus domestica* Borkh.	'Golden Delicious'	86.5	0.76	12.8	ε'	63.4	61.9	60.2	57.0	46.1	28.4
					ε''	16.0	9.8	8.3	13.2	21.8	26.0
	'Granny Smith'	88.4	0.76	10.9		58.3	57.8	56.3	53.3	43.5	26.9
						17.2	9.8	8.0	12.3	20.4	24.7
	'Red Delicious'	87.3	0.80	11.7		58.1	57.3	55.6	52.8	42.9	24.8
						16.7	9.7	7.9	12.2	20.6	25.4
Avocado, *Persea americana* Miller, var. americana		71.3	0.99	12.6		50.7	48.5	46.2	44.9	36.1	22.1
						56.9	25.8	13.5	12.3	18.5	19.8
Banana, *Musa x paradisiaca*	'Cavendish'	78.1	0.94	25.2		69.3	66.3	63.0	57.5	43.7	24.9
						56.9	27.8	17.6	19.0	25.9	23.1
Cantaloupe, *Cucumis melo* L.	Muskmelon	92.0	0.93	9.8	ε'	70.1	69.2	67.8	65.0	55.2	31.7
					ε''	37.0	17.9	11.6	14.6	24.9	31.2
Carrot, *Daucus carota* subsp. sativus (Hoffm.) Arcang.		86.9	0.99	10.5		62.7	61.6	58.6	55.2	44.0	24.0
						56.5	27.3	15.7	16.1	24.8	25.4
Cucumber, *Cucumis sativus* L.		96.6	0.85	9.0		72.4	71.5	70.4	68.4	60.1	33.5
						29.2	14.4	9.7	13.3	25.6	34.3
Grape, *Vitis amurensis* Rupr.	Chilean 'Thompson' seedless	82.0	1.10	20.9		72.0	70.8	68.4	63.5	49.6	26.1
						36.4	19.1	13.7	18.0	28.3	28.6
Grapefruit, *Citrus x paradise* Macfad.	Pink	91.5	0.83	8.5		77.2	76.6	73.9	72.3	57.9	33.8
						38.0	18.6	12.3	16.4	27.8	30.0
Honeydew, *Cucumis melo* L.		89.1	0.95	12.7		74.7	73.9	70.7	68.0	55.0	31.6
						58.3	27.4	16.3	17.4	27.7	28.1
Kiwi fruit, *Actinidia chinensis* Planchon		81.7	0.99	16.1		73.4	71.3	68.1	65.6	51.5	29.1
						59.9	28.2	16.4	17.5	27.9	27.2
Lemon, *Citrus limon* (L.) Burman f.		91.2	0.88	7.2		74.8	74.5	71.7	70.3	57.8	34.3
						44.9	21.2	12.9	15.1	27.3	29.6
Lime, *Citrus aurantiifolio* (Christm.) Swingle		90.2	0.97	8.2		76.1	74.3	70.7	68.9	56.9	33.0
						50.1	24.1	14.4	15.7	27.5	29.5

(*Continued*)

Table 15.9 Dielectric Constants (ε') and Loss Factors (ε'') of Fresh Fruits and Vegetables at Indicated Frequencies Between 200 MHz and 20 GHz at 24°C, Along with Other Physical Characteristics *Continued*

Fruit or Vegetable	Cultivar or Other Description	Moisture Content, %	Density, g/cm³	Soluble Solids, %	Frequency, Hz					
					2×10^8	5×10^8	1.3×10^9	3.2×10^9	8×10^9	2×10^9
Mango, *Mangifera indica* L.		85.5	0.96	12.1	67.3	65.9	63.7	60.0	49.9	28.6
					34.7	17.7	11.9	15.0	24.3	29.8
Onion, *Alium cepa* L. var. cepa	Red	90.3	0.85	9.0	71.1	69.5	67.6	64.1	53.2	29.6
					42.2	20.6	13.2	15.8	26.3	30.2
	White	92.0	0.97	9.0	68.0	66.7	65.5	62.8	51.6	28.7
					33.2	16.4	11.0	14.5	25.7	29.6
	Yellow	92.9	0.91	7.4	72.1	70.8	69.1	65.9	55.8	31.2
					28.4	15.0	10.8	14.6	26.0	31.2
Orange, *Citrus aurantium* L. Subsp. bergamia	Navel	87.5	0.92	14.8	75.9	74.8	71.2	68.0	51.7	29.2
					33.5	17.1	13.2	17.7	28.4	27.8
Papaya, *Carica papaya* L.		88.1	0.96	12.9	71.1	70.4	68.9	65.6	52.9	29.5
					26.2	13.9	10.6	15.6	27.4	30.0
Pear, *Pyrus communis* L.	'Danjou'	83.9	0.94	14.5	69.3	68.7	66.8	63.1	50.7	28.0
					23.5	12.9	10.3	15.4	26.4	29.3
Potato, *Lolanum tuberosum* L.	'Rusett Burbank'	79.3	1.03	7.9	67.9	64.1	60.3	55.6	45.5	25.8
					65.7	31.5	18.5	17.1	23.2	28.7
Radish, *Raphanus sativus* L.	Red	95.8	0.76	2.7	70.2	69.7	67.3	66.3	57.3	34.2
					71.9	32.0	16.3	15.1	24.2	28.7
Squash, *Cucurbita mixta* Pang.	Yellow crookneck	95.1	0.70	3.5	66.8	65.1	62.5	61.4	52.8	31.3
					50.0	22.9	12.6	13.0	22.1	28.1
Strawberry, *Fragaria chiloensis* (L.) Duchesne		92.1	0.76	8.1	75.1	74.4	71.7	70.1	57.3	33.2
					36.7	18.2	11.8	15.0	27.3	28.7
Sweet potato, *Ipomea batatus* (L.) Lam.		80.5	0.95	8.0	59.1	57.0	54.2	50.7	41.2	24.8
					45.8	22.4	13.5	14.3	20.9	24.7
Turnip, *Brassica rapa* L. var. rapa	'Purple Top'	92.3	0.89	6.3	65.1	64.3	62.8	59.9	51.1	29.8
					37.4	18.0	11.3	13.9	23.3	31.0

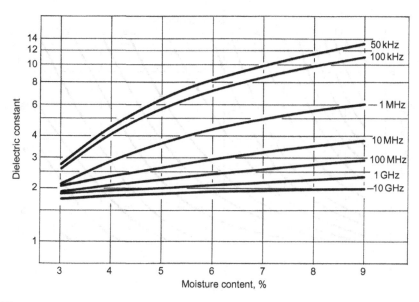

FIGURE 15.38

Moisture dependence of the mean dielectric constant for chopped pecan kernels of six cultivars at 22°C at indicated frequencies (Nelson, 1981).

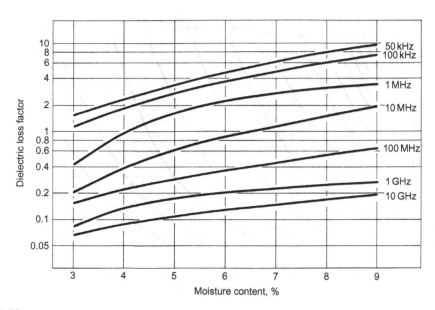

FIGURE 15.39

Moisture dependence of the mean dielectric loss factor for chopped pecan kernels of six cultivars at 22°C at indicated frequencies (Nelson, 1981).

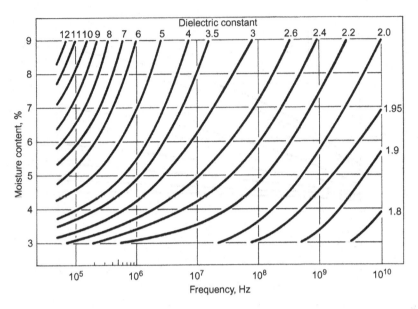

FIGURE 15.40

Mean values of the dielectric constant of chopped pecan kernels of six cultivars at 22°C as a function of frequency and moisture content, wet basis (Nelson, 1981).

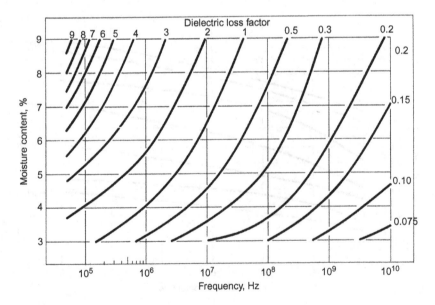

FIGURE 15.41

Mean values of the dielectric loss factor of chopped pecan kernels of six cultivars at 22°C as a function of frequency and moisture content, wet basis (Nelson, 1981).

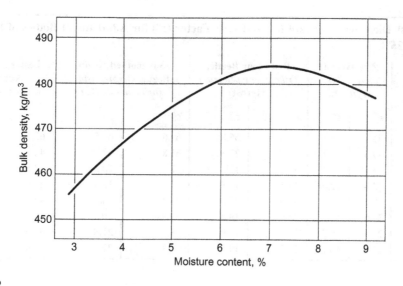

FIGURE 15.42

Variation of mean bulk density of chopped pecan kernels of six cultivars at 22°C with moisture content, wet basis (Nelson, 1981).

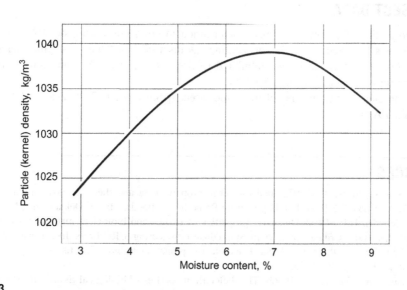

FIGURE 15.43

Variation of mean kernel density for pecans of six cultivars at 22°C with moisture content, wet basis (Nelson, 1981).

Table 15.10 Dielectric Constant (ε') and Loss Factor (ε'') for Adult Insect Bodies of Indicated Species at 25°C

Frequency, GHz	Rice Weevil, *Sitophilus oryzae* L.		Red Flour Beetle, *Triboleum castaneum* (Herbst)		Sawtoothed Grain Beetle, *Oryzaephylus surinamensis* (L.)		Lesser Grain Borer, *Rhyzopertha dominica* (F.)	
	ε'	ε''	ε'	ε''	ε'	ε''	ε'	ε''
0.20	55.1	48.3	61.4	55.8	69.6	67.5	62.9	55.3
0.50	49.0	23.7	54.6	27.3	61.8	32.7	42.1	17.8
1.08	45.4	15.3	50.7	17.5	57.7	20.8	52.5	18.1
2.47	42.0	12.6	47.1	14.8	53.3	16.9	42.8	15.2
5.02	38.2	14.3	42.9	16.3	48.7	18.6	43.2	17.5
9.40	30.6	16.0	34.0	19.4	39.8	20.8	34.3	18.6
10.76	28.9	15.7	31.0	18.9	38.1	21.8	32.5	18.5
20.00	22.7	16.1	25.1	19.3	27.8	22.2	24.9	18.1

Information on the variation of the dielectric properties of chopped pecan kernels with temperatures from 0°C to 40°C at frequencies between 100 kHz and 110 MHz is available elsewhere (Lawrence et al., 1992).

15.3 INSECT DATA

Dielectric properties of insects have been determined in connection with studies on the potential control of insects by selective dielectric heating, as discussed in Chapter 4. Such data on a few stored-grain insect species have been tabulated (Nelson et al., 1998; Nelson, 2001), and they are included here for reference in Table 15.10.

Data on the dielectric properties of the same insect species at 10−70°C are available elsewhere (Nelson et al., 1998).

REFERENCES

Datta, A.K., Sun, E., Solis, A., 1995. Food dielectric property data and their composition-based prediction. In: Rao, M.A., Rizvi, S.S.H. (Eds.), Engineering Properties of Foods. Marcel Dekker, Inc., New York, NY.

Foster, K.R., Schwan, H.P., 1989. Dielectric properties of tissues and biological materials: a critical review. In: Bourne, J.R. (Ed.), Critical Reviews in Biomedical Engineering. CRC Press, Inc., Boca Raton, FL.

Gabriel, C., Gabriel, S., Corthout, E., 1996a. The dielectric properties of biological tissues: I. Literature survey. Phys. Med. Biol. 41, 2231−2249.

Gabriel, S., Lau, R.W., Gabriel, C., 1996b. The dielectric properties of biological tissues: II. Measurements in the frequency range 10 Hz to 20 GHz. Phys. Med. Biol. 41, 2251−2269.

Guo, W.-C., Nelson, S.O., Trabelsi, S., Kays, S.J., 2007. Dielectric properties of honeydew melons and correlation with quality. J. Microw. Power Electromagn. Energy 41 (2), 44−54.

Kent, M., 1987. Electrical and Dielectric Properties of Food Materials. Science and Technology Publishers, Hornchurch, Essex, England.

Lawrence, K.C., Nelson, S.O., Kraszewski, A., 1992. Temperature dependence of the dielectric properties of pecans. Trans. ASAE 35 (1), 251–255.

Mudgett, R.E., 1995. Electrical properties of foods. In: Rao, M.A., Rizvi, S.S.H. (Eds.), Engineereing Properties of Foods. Marcel Dekker, Inc., New York, NY.

Nelson, S.O., 1973a. Electrical properties of agricultural products—a critical review. Trans. ASAE 16 (2), 384–400.

Nelson, S.O., 1973b. Microwave dielectric properties of grain and seed. Trans. ASAE 16 (5), 902–905.

Nelson, S.O., 1976. Microwave dielectric properties of insects and grain kernels. J. Microw. Power 11 (4), 299–303.

Nelson, S.O., 1978a. Frequency and moisture dependence of the dielectric properties of high-moisture corn. J. Microw. Power 13 (2), 213–218.

Nelson, S.O., 1978b. Radiofrequency and Microwave Dielectric Properties of Shelled Field Corn. Agricultural Research Service, USDA, ARS-S-184.

Nelson, S.O., 1979. RF and microwave dielectric properties of shelled, yellow-dent field corn. Trans. ASAE 22 (6), 1451–1457.

Nelson, S.O., 1981. Frequency and moisture dependence of the dielectric properties of chopped pecans. Trans. ASAE 24 (6), 1573–1576.

Nelson, S.O., 1983. Observations on the density dependence of the dielectric properties of particulate materials. J. Microw. Power 18 (2), 143–152.

Nelson, S.O., 1984. Density dependence of the dielectric properties of wheat and whole-wheat flour. J. Microw. Power 19 (1), 55–64.

Nelson, S.O., 1987. Frequency, moisture, and density dependence of the dielectric properties of small grains and soybeans. Trans. ASAE 30 (5), 1538–1541.

Nelson, S.O., 2001. Radio-frequency and microwave dielectric properties of insects. J. Microw. Power Electromagn. Energy 36 (1), 47–56.

Nelson, S.O., 2005. Dielectric spectroscopy of fresh fruit and vegetable tissues from 10 to 1800 MHz. J. Microw. Power Electromagn. Energy 40 (1), 31–47.

Nelson, S.O., Datta, A.K., 2001. Dielectric properties of food materials and electric field interactions. In: Datta, A.K., Anantheswaran, R.C. (Eds.), Handbook of Microwave Technology for Food Applications. Marcel Dekker, Inc., New York, NY.

Nelson, S.O., Stetson, L.E., 1975. 250-Hz to 12-GHz dielectric properties of grain and seed. Trans. ASAE 18 (4), 714, 715, 718.

Nelson, S.O., Stetson, L.E., 1976. Frequency and moisture dependence of the dielectric properties of hard red winter wheat. J. Agric. Eng. Res. 21, 181–192.

Nelson, S.O., You, T.-S., 1989. Microwave dielectric properties of corn and wheat kernels and soybeans. Trans. ASAE 32 (1), 242–249.

Nelson, S.O., Forbus Jr., W.R., Lawrence, K.C., 1994. Microwave permittivities of fresh fruits and vegetables from 0.2 to 20 GHz. Trans. ASAE 37 (1), 181–189.

Nelson, S.O., Bartley Jr., P.G., Lawrence, K.C., 1998. RF and microwave dielectric properties of stored-grain insects and their implications for potential insect control. Trans. ASAE 41 (3), 685–692.

Nelson, S.O., Trabelsi, S., Kays, S.J., 2006. Dielectric spectroscopy of honeydew melons from 10 MHz to 1.8 GHz for quality sensing. Trans. ASABE 49 (6), 1977–1981.

Nelson, S.O., Guo, W., Trabelsi, S., Kays, S.J., 2007. Dielectric spectroscopy of watermelons for quality sensing. Meas. Sci. Technol. 18, 1887–1892.

Noh, S.H., Nelson, S.O., 1989. Dielectric properties of rice at frequencies from 50 Hz to 12 GHz. Trans. ASAE 32 (3), 991–998.

Pethig, R., Kell, D.B., 1987. The passive electrical properties of biological systems: their significance in physiology, biophysics and biotechnology. Phys. Med. Biol. 32 (8), 933–970.

Schwan, H.P., 1957. Electrical properties of tissue and cell suspensions. Adv. Biol. Med. Phys. 5, 147–209.

Tinga, W.R., Nelson, S.O., 1973. Dielectric properties of materials for microwave processing—tabulated. J. Microw. Power 8 (1), 23–65.

Trabelsi, S., Nelson, S.O., 2005. Microwave Dielectric Properties of Cereal Grain and Oilseed. ASAE Paper No. 056165, American Society of Agricultural and Biological Engineers, St. Joseph, MI.

Trabelsi, S., Nelson, S.O., 2012. Microwave dielectric properties of ceral grains. Trans. ASABE 55 (5), 1989–1996.

von Hippel, A.R., 1954. Dielectric Materials and Applications. The Technology Press of M.I.T. and John Wiley & Sons, New York, NY.

You, T.-S., Nelson, S.O., 1988. Microwave dielectric properties of rice kernels. J. Microw. Power Electromagn. Energy 23 (3), 150–159.

CLOSELY RELATED PHYSICAL PROPERTIES DATA FOR GRAIN AND SEED

16

Some physical properties of grain have been studied in some detail because of their influence on the dielectric properties of those materials (Nelson, 1980). The moisture dependence of density, both kernel density and bulk density, is prominent among those. Also, kernel dimensions and kernel and seed densities are of interest, and they have been determined for a range of agricultural grain and crop seeds (Nelson, 2002a,b).

16.1 MOISTURE DEPENDENCE OF KERNEL AND BULK DENSITIES FOR WHEAT AND CORN

Information on bulk densities (test weights) of grain is important because the test weight is an important factor in determining the grade, and, consequently, the selling price. Because moisture content is also an important factor that affects the grade and test weight of grain, relationships between bulk density and moisture content are of interest. Such data are also of value in working with problems related to grain storage, drying, and aeration.

Earlier work on relationships between bulk density and moisture content of grain has been reported (Chung and Converse, 1971; Hall, 1972; Gustafson and Hall, 1972; Hall and Hill, 1974; Brusewitz, 1975). Newer data on the moisture dependence of bulk and kernel densities of hard red winter wheat and shelled, yellow-dent field corn were obtained incidental to studies of the dielectric properties of these kinds of grain (Nelson and Stetson, 1976; Nelson, 1978). Detailed information on seven cultivars of hard red winter wheat, *Triticum aestivum* L., and 21 lots of hybrid yellow-dent field corn, *Zea mays* L., and their history and preparation for these studies, were reported along with the methods and procedures (Nelson, 1980).

16.1.1 WHEAT LOTS

Wheat lots were obtained soon after harvest and stored at 5°C and 50% relative humidity (RH) until the measurements were taken. Equilibrium moisture contents under those storage conditions ranged between 11.3% and 11.9%, wet basis (w.b.).

Dielectric Properties of Agricultural Materials and Their Applications. DOI: http://dx.doi.org/10.1016/B978-0-12-802305-1.00016-6

The moisture content of wheat lots was adjusted in increments of 1% or 2% moisture by adding distilled water or drying in a hot-air oven. After water was added, the lots were sealed in glass jars, conditioned at 5°C, and agitated frequently to aid the uniform distribution of moisture. For reduction of moisture content below the equilibrium level of about 11%, part of the lot was dried at temperatures of 55−60°C for 3−24 h; however, for reduction below 4% moisture content, 2−3 days drying time at 80°C was necessary. The time between tempering or drying and measurement was 1 week or longer. Chemical compositions of wheat lots were determined at the completion of the study in accordance with established procedures for crude protein, fat, and ash determinations (AOAC, 1970).

16.1.2 CORN LOTS

Twenty-one lots of hybrid, yellow-dent field corn, of known genetic background, were picked by hand at relatively high moisture contents from fields and plots. They were stored in closed plastic bags at 4°C for a few days before they were shelled by hand. Then they were sealed in glass jars and returned to cold storage. Kernels near the ends of the ears were excluded to limit kernel size and shape variation. Lots were thoroughly mixed during storage and when samples were drawn for measurements. Lots were permitted to dry in controlled increments in shallow pans at 24°C and 40% RH, and then they were returned to cold storage for at least 7 days for equilibration of moisture distribution before the next measurements were obtained. Crude protein determinations and fat acidity tests were made on each lot at the completion of the study (Nelson, 1978). All fat acidity levels were well below levels that indicate any quality deterioration in corn.

16.1.3 MOISTURE DETERMINATIONS

Moisture contents of wheat samples were determined by grinding and drying triplicate, 2-g samples in a forced-air oven, for 1 h at 130°C (AOAC, 1970). Two-stage procedures were used whenever moisture contents exceeded 16%, w.b. Two oven-moisture-determination methods were used for corn moisture measurements. One was the method specified in the Official Grain Standards of the United States (USDA, 1970), that is, drying whole-kernel, 15-g samples for 72 h at 103°C. The second oven method involved drying 2-g, ground samples for 3 h at 130°C. The latter method is believed to give results in closer agreement with the Karl Fischer titration method[1], which is specific for water, and was therefore used as the basis for data presented here. Two-stage procedures were followed for the latter method when moisture contents exceeded 13%. The 3-h method gave moisture contents about 1% point greater than the 72-h method for the 10−20% moisture range, and 0.6−0.8% point greater for the 20−35% moisture range (Nelson, 1978).

16.1.4 DENSITY MEASUREMENTS

The bulk densities for wheat lots at all moisture levels were determined by using standard test-weight apparatus and procedures (USDA, 1953). Resulting test weights, which were averages of at least three replicated measurements, were converted from pound per bushel to specific gravities, or densities in

---

[1]Personal communication, Frank Jones, Institute of Basic Standards, National Bureau of Standards, US Department of Commerce, Washington, DC, 1978.

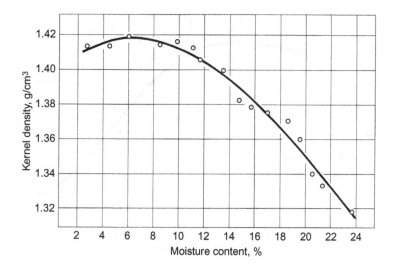

FIGURE 16.1

Moisture dependence of the kernel density of hard red winter wheat at 24°C.

grams per cubic centimeter, by multiplying the test weights by 0.012872. Bulk densities of corn lots were also determined by standard test-weight measurements; however, such measurements are quite variable on lots with moisture contents as high as 30% or 40%. Also, since the principal interest was in the density of the samples when their dielectric properties were determined, bulk-density values were obtained from the weights of the samples in the coaxial-line sample holder used for the electrical measurement and the volume occupied by the sample in that sample holder. That volume was 116.485 cm^3, which is the volume between two concentric, right circular cylinders 5.377 and 2.336 cm in diameter and 6.32 cm long. Consistent procedures were used in filling the coaxial sample holder (Nelson, 1978), and resulting bulk densities are averages for three replicated measurements. Kernel densities for both wheat and corn were calculated from kernel-volume measurements obtained with a Beckman[2] model 930 air comparison pycnometer on samples of at least 25 g and from the exact weights of those samples. Kernel-density values are averages of at least three replicated measurements. The 1−2 atmosphere mode of operation was used, and calibration of the pycnometer was checked frequently to maintain the calibration. The kernels in the corn samples measured with the pycnometer were counted so that mean kernel weights and volumes might also be obtained at each moisture level. Length, width, and thickness of 50 kernels from each corn lot were also measured with a dial-type caliper at the final moisture level, which was about 10.5%.

16.1.5 RESULTS FOR HARD RED WINTER WHEAT

Relationships between densities and moisture content, averaged over all seven wheat lots, are summarized in Figures 16.1 and 16.2.

---

[2]Mention of trade names or commercial products in this publication is solely for the purpose of providing specific information and does not imply recommendation or endorsement by the US Department of Agriculture.

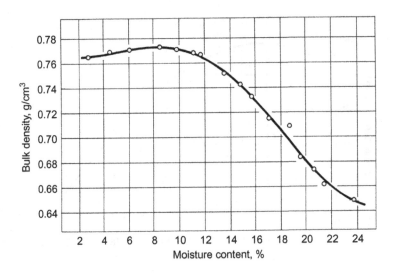

FIGURE 16.2

Moisture dependence of the bulk density of hard red winter wheat at 24°C.

Fifteen moisture levels between 2% and 24% were chosen, and the measured moisture contents of each of the seven lots closest to these levels were averaged, as were the corresponding measured values of bulk and kernel densities. The curves of Figures 16.1 and 16.2 represent polynomial regression curves of best fit for the experimental data. The polynomial equations and corresponding correlation coefficients for kernel density, ρ_k, and bulk density, ρ_b are as follows (M represents moisture content in percent):

$$\rho_b = 0.7744 - 0.00703M + 0.001851M^2 - 0.00014896M^3 \quad r = 0.997 \tag{16.1}$$

$$\rho_k = 1.3982 + 0.00680M - 0.0006086M^2 + 0.00000747M^3 \quad r = 0.991 \tag{16.2}$$

where r is the correlation coefficient. Since the data presented in Figures 16.1 and 16.2 represent several cultivars grown at different locations over a 4-year period, and exhibiting some range in test weight, they should be useful for providing reasonable estimates for density–moisture relationships for hard red winter wheat.

From the curves of Figures 16.1 and 16.2, it is obvious that the kernel density and bulk density of wheat are highly correlated. A plot of ρ_b versus ρ_k for this wide range of moisture contents revealed a linear relationship, and the linear regression of ρ_b on ρ_k for these data had a correlation coefficient of 0.975. Given either the bulk density or kernel density in grams per cubic centimeter, the other can be reasonably well estimated, independent of moisture content, by the following equations:

$$\rho_b = 1.311\rho_k - 1.084 \tag{16.3}$$

$$\rho_k = 0.763\rho_b + 0.827 \tag{16.4}$$

16.1.6 RESULTS FOR YELLOW-DENT FIELD CORN

Since the corn lots were harvested at high moisture levels and permitted to dry in several steps, exact moisture levels could not be controlled as accurately as they were for the wheat that was raised in moisture content by addition of water. Therefore bulk and kernel density values measured for each moisture level were plotted against moisture content for each lot, and smooth curves were drawn through the points. Density values were then taken from the curves at 5% moisture intervals and averaged over all 21 corn lots to provide the data presented in Figures 16.3 and 16.4.

The curves shown for kernel density ρ_k and the bulk density as measured in the coaxial-line sample holder ρ_b are the curves of best fit for third-order polynomials. The polynomial regression equations and corresponding correlation coefficients are as follows (M represents moisture content in percent):

$$\rho_k = 1.2519 + 0.00714M - 0.0005971M^2 + 0.00001088M^3 \quad r = 0.998 \tag{16.5}$$

$$\rho_b = 0.6829 + 0.01422M - 0.0009843M^2 + 0.00001548M^3 \quad r = 0.996 \tag{16.6}$$

Because of the small size of the coaxial sample holder used for the electrical measurements, the bulk density of corn in the sample holder was somewhat less than that determined by standard test-weight procedures in which a 1-qt kettle is used. Standard test-weight measurements were also obtained on all corn lots for most moisture levels. It was observed that test weights measured immediately after drying corn lots in open pans were slightly greater than test weights measured after the same corn lots had been conditioned for 1 week in sealed jars at 4°C and permitted to come to 24°C in the sealed jars for test-weight measurements. Therefore test weights were taken after the conditioning period during which moisture distribution in the kernels had come to equilibrium. For correction of the bulk density values from the sample holder, ρ_b, so they would be

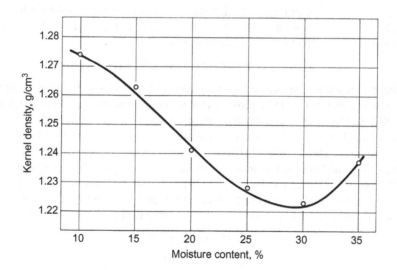

FIGURE 16.3

Moisture dependence of the kernel density of shelled, yellow-dent field corn at 24°C.

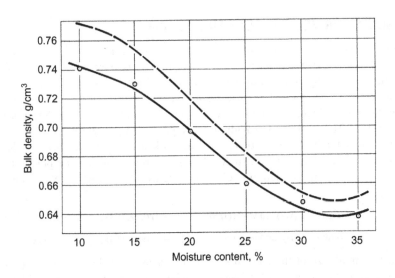

FIGURE 16.4

Moisture dependence of the bulk density of shelled, yellow-dent field corn at 24°C. Solid curve: density in coaxial-line sample holder, ρ_b. Dashed curve: bulk density corrected to correspond with standard test weight, ρ_{bt}.

equivalent to those from standard test-weight measurements, ρ_{bt}, a linear regression analysis was run for all moisture levels and over all corn lots. The following equation was obtained:

$$\rho_{bt} = 1.1783\rho_b - 0.1028 \quad r = 0.907 \tag{16.7}$$

The curve, as corrected to provide bulk densities equivalent to those obtained by standard test-weight procedures, is shown as the dashed curve for ρ_{bt}, in Figure 16.4. Its equation can be obtained by combining Eqs (16.6) and (16.7) as follows:

$$\rho_{bt} = 0.7019 + 0.01676M - 0.0011598M^2 + 0.00001824M^3 \tag{16.8}$$

Because the data reported here are based on an average for 21 lots with a reasonable range of test weights, Eqs (16.5) and (16.8) should provide characteristic estimates for kernel density and bulk density of shelled, yellow-dent field corn. In addition to normal variation among lots of corn, differences in relationships between test weight and moisture content have been found between lots dried at different temperatures (Hall, 1972), between lots with different amounts of mechanical damage (Hall and Hill, 1974), and between lots in which moisture changes were by absorption and desorption (Chung and Converse, 1971). As was indicated for wheat up to 24% moisture, the kernel density and bulk density of corn were highly correlated between 10% and 30% moisture. In that moisture range, linear regression analysis yielded the following relationships, with a correlation coefficient of 0.970:

$$\rho_b = 1.849\rho_k - 1.608 \tag{16.9}$$

$$\rho_k = 0.541\rho_b + 0.870 \tag{16.10}$$

Thus, if either the bulk density or kernel density is known, the other can be estimated, independent of moisture content, in terms of the other.

A few additional observations on these data are noteworthy. When mean kernel weights were plotted against moisture content, linear relationships were obtained for each lot. If W represents the weight of a kernel and M represents the moisture content in percent, these lines can be expressed as $W = a_0 + a_1 M$, where a_0 represents the zero-moisture intercept and a_1 is the slope of the W-versus-M line. The true relationship is hyperbolic, but it does not deviate greatly from linearity over the moisture range from 10% to 50%. The value of a_0 will be the estimated dry weight, and the value of a_1 is known approximately, because at 50% moisture, w.b., $W = 2a_0$. Therefore the slope of the line is $a_1 = (2a_0 - a_0)/50 = a_0/50$. For the regression lines of the mean kernel weights versus moisture content, the ratio a_0/a_1 should, therefore, have values of about 50. The mean a_0/a_1 value for the 21 corn lots in this study was 57.1. The discrepancy was no doubt attributable to the nonlinear true relationship between kernel weight and moisture content and to the narrow range of moisture contents for which data were available on several lots. Kernel volumes were also plotted against moisture content, and a linear relationship was observed. Kernel sizes varied considerably between different lots. Therefore, to combine kernel volume data for all 21 lots, the kernel volumes at each moisture content were 'normalized' by dividing them by the kernel volume at the equilibrium moisture content near 10.5% moisture. A linear regression analysis on all normalized data for all lots resulted in the following equation for the normalized kernel volume, V_{kn}:

$$V_{kn} = 0.811 + 0.0173M \quad r = 0.947 \tag{16.11}$$

Equation (16.11) gives $V_{kn} = 1$ for a moisture content, M, of 10.9% and can be used to estimate kernel volume for yellow-dent field corn in the range from about 10% to 35% moisture if the kernel volume and moisture content are known at any point in that range.

16.1.7 POROSITY OF WHEAT AND CORN

The kernel volume-fraction (V_{fk}), that fraction of the space in bulk grain occupied by the kernels, and the fractional porosity (P_f), that fraction of the space in the bulk grain not occupied by the kernels, can also be obtained from the data presented here for hard red winter wheat and yellow-dent field corn. The kernel volume-fraction is ρ_b/ρ_k, and the fractional porosity is $1 - \rho_b/\rho_k$. For calculation of these values, Eqs (16.1) and (16.2) provide the information needed for wheat, and Eqs (16.5) and (16.8) should be appropriate for corn. Resulting values for kernel volume fraction and fractional porosity are given in Figure 16.5.

At moisture contents above 10%, porosity increases with moisture content. For porosities calculated from bulk and kernel densities of individual lots, coefficients of variation were 1.6% and 2.5% for wheat and corn, respectively. The wide moisture ranges over which these data were collected and the fact that the data represent averages over several lots of different descriptions should make them useful for predictive purposes. Although Chung and Converse (1971) observed differences in test weights for separated kernel shape and size classes, moisture dependent relationships for the different classes were similar. The exclusion of round kernels (end-of-the-ear regions) from lots on which data for work reported here were obtained would not be expected to materially alter the results, because each lot still had substantial size variations.

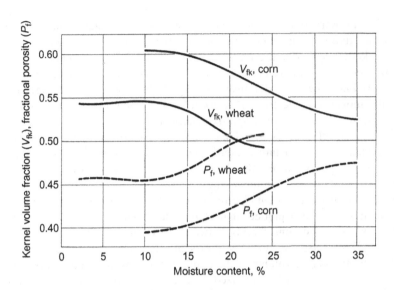

FIGURE 16.5

Moisture dependence of kernel volume fraction V_{fk} and fractional porosity P_f for hard red winter wheat and shelled, yellow-dent field corn at 24°C.

16.1.8 SUMMARY

The moisture dependence of kernel density and bulk density (test weight) of hard red winter wheat and shelled, yellow-dent field corn over moisture ranges of 3–24% and 10–35%, respectively, can be accurately represented by third- and fourth-order polynomial equations. The kernel and bulk densities of wheat are linearly and highly correlated, both increasing slightly as moisture content increases from 3% to about 8% and then decreasing markedly and continuously as moisture content increases to 24%.

Bulk and kernel densities of corn are linearly and highly correlated in the range from 10% to 30% moisture, both decreasing continuously with increasing moisture content. Above 30% moisture, kernel density increases markedly, whereas bulk density continues to decrease and does not show much tendency to increase as moisture content rises up to 35%. Both the weight and volume of corn kernels are positively and almost linearly correlated with moisture content.

The porosity of hard red winter wheat is somewhat greater than that of shelled corn. For both kinds of grain, porosity increases with increasing moisture content at levels above 10% moisture.

16.2 GRAIN KERNEL AND SEED DIMENSIONS AND DENSITIES FOR AGRICULTURAL CROPS

Because there can be confusion concerning the use of certain terms relating to cereal grains and seed of other crops, these terms are defined, as used here, as follows: "Grain" is the collective name for harvested seed of cereal grasses, or the so-called cereal grains. "Grain" can also refer to a single seed, but it is not used in that sense here. "Seed" or "seeds" can refer to either the collective name for the fertilized and ripened ovules of flowering plants or to the individual seeds of such plants. "Kernel" is an agronomic term for the whole seed of a cereal plant or some other agricultural crops.

Certain physical properties of grain and seed have long been used as indicators of product quality. Moisture content is important in determining storability and in maintenance of quality. Test weight, the bulk density determined under prescribed conditions (USDA, 1953), can indicate the degree of kernel filling during growth or the presence of shriveled kernels and is therefore an indicator of quality. Grain yields are generally determined by use of standard test weights established for each commodity, so adjustments must be made for low test weights, and prices are usually adjusted both for test weights and moisture contents outside of desired ranges. The porosity of the bulk material is important in connection with grain and seed aeration and drying.

The porosity is determined as a function of the bulk density and kernel density of the grain. The volume fraction (V_f) occupied by the kernels is given by the ratio of the bulk density (ρ_b) to the kernel density (ρ_k) of the grain ($V_f = \rho_b/\rho_k$), and the fractional porosity is $1 - V_f$. Thus, interest developed in the determination of kernel density (Zink, 1935; Thompson and Isaacs, 1967). Kernel densities of wheat have long been of interest in relation to milling and baking qualities (Bailey, 1916).

Early determinations of specific gravity, or kernel density, were made by Bailey and Thomas (1912) with a specific gravity bottle and toluene as the measurement liquid. Liquid density gradient columns were used to determine kernel densities of wheat (Peters and Katz, 1962), popcorn (Haugh et al., 1976), and other grain and seed (Shelef and Mohsenin, 1968; Mohsenin, 1986). Liquid toluene displacement was also used for kernel density determination for wheat and canola (Sokhansanj and Lang, 1996). Gas pycnometers have been used in studies of kernel densities and their relationship with moisture content (Chung and Converse, 1971; Brusewitz, 1975; Nelson and Stetson, 1976; Nelson, 1976, 1978, 1979; Fortes and Okos, 1980; Stroshine, and Crane, 1987).

Seed densities have been determined by water displacement in studies on swelling of seeds following imbibition of water (Leopold, 1983). Density separation of imbibed seeds to separate low-density, poor-quality seeds in aqueous solutions of different specific gravities has also been studied for upgrading seed lots (Hill et al., 1989).

Cereal grain kernel densities have been of interest in breakage susceptibility and hardness studies (Martin et al., 1987; Chang, 1988). Chang (1988) distinguished differences in kernel density and kernel tissue density, which he termed 'true density,' accounting for void spaces within the kernel. Variation in test weight and kernel densities with changes in moisture content during drying have been studied (Hall, 1972; Gustafson and Hall, 1972; Fortes and Okos, 1980). Moisture-dependent kernel- and bulk-density relationships and effects on porosity for wheat and shelled corn have also been reported (Nelson, 1980).

Kernel and bulk density data have been used in research on determining the dielectric properties of cereal grain kernels (Nelson, 1983a,b, 1984; You and Nelson, 1988; Sokhansanj and Nelson, 1988; Noh and Nelson, 1989; Nelson and You, 1989) and for determining volume fractions for use in dielectric mixture equations (Nelson, 1992). Basic data on dimensional, density, and moisture content characteristics of several different kinds of seed are summarized here. This information may be used in further studies and may serve as a useful reference. A simple method is also presented for estimating the volume and density of a particular kind of seed.

16.2.1 GRAIN AND SEED LOTS

Most of the grain and seed lots had been collected earlier for dielectric properties measurements and were stored at 4°C and 40–50% RH for maintenance of condition and viability. All were clean grain and seed lots, and many, including wheat, oats, barley, and soybeans, were obtained as Certified seed that had not been treated with any chemical protectants. With other seed lots, the cultivar was determined whenever possible as part of the seed lot description. When obtained through commercial channels, the cultivar was generally unknown. Moisture contents were determined by oven drying in accordance with standard methods (ASAE, 2000b; ISTA, 1966; USDA, 1986). The method used for each kind of seed was specified in the original article (Nelson, 2002a).

16.2.2 DIMENSIONAL AND WEIGHT MEASUREMENTS

Fifty kernels or seeds were randomly selected from grain or seed lots for dimensional measurements. A dial caliper was used to obtain three orthogonally oriented dimensions on each kernel or seed. The largest of the three dimensions was designated as length, the second largest as width, and the smallest as thickness. For canola, which is a small spherical seed, only two dimensions were measured, the major and minor diameters. Kernel or seed weights were also determined with an analytical balance, reading weights in grams to four decimal places with an accuracy of ± 0.2 mg.

16.2.3 DENSITY MEASUREMENTS

For most grain and seed lots, standard test weight determinations were made with a Fairbanks Morse grain tester, code 11192, weight-per-bushel apparatus equipped with a one-quart measure. This provides a bulk density determination under prescribed conditions. Kernel or seed densities were determined by measuring the volume occupied by the kernels or seeds in a known sample weight, randomly taken from each grain or seed lot. Volume measurements were made on samples of about 20–25 g, or somewhat less for low-density seeds, with a Beckman model 930 air comparison pycnometer. Measurements were repeated at least four times, and mean values were used for calculation of kernel and seed densities by dividing the weighed mass by the measured volume. The number of seeds in the sample weighed for pycnometer measurements was determined by manual count for determination of mean kernel or seed weight and volume. The volume measurement accuracy of the pycnometer is within 0.05 cm^3. For a 20-cm^3 sample, this represents, at most, an error of 0.25%, so the retention of the third decimal place for kernel and seed densities is justified.

16.2.4 RESULTING DATA

Results of measurements are summarized in Table 16.1 for 20 different kinds of grain and seed.

Although the moisture contents for most grain and seed lots were the equilibrium moisture contents for the storage conditions, some with higher or lower levels, including some wheat, rice, popcorn, and soybean samples, had been conditioned earlier and stored in sealed containers. An examination of kernel or seed density data reveals a range from 1.00 to 1.46 g/cm^3.

There is a correlation between kernel or seed density and bulk density, or test weight, but it is not perfect, because variations in kernel and seed shape, surface characteristics, and packing susceptibility also affect the test weight.

To provide further information on the variation of these data among different cultivars and growing regions, additional data are presented for barley in Table 16.2 and for soybeans in Table 16.3.

Table 16.1 Physical Properties of Grain and Seed, Including Individual Kernel or Seed Characteristics

Description of Grain or Seed	Moisture Content, %	Test Weight		Seed Weight, mg	Seed Volume, mm³	Seed Density, gcm³	Seed Dimensions		
		lb/bu	g/cm³				Length, mm	Width, mm	Thickness, mm
Alfalfa, *Medicago sativa* L., cv. Vernal	7.2	63.2	0.814	1.95	1.5	1.301	2.2	1.3	0.9
Barley, *Hordeum vulgare* L.									
Spring, cv. Custer	11.2	44.0	0.566	35.1	25.9	1.356	9.5	3.1	2.4
Winter, cv. Hitchcock	11.2	47.8	0.615	26.6	19.7	1.352	7.9	2.9	2.2
Canola, *Brassica napus* L.	6.2	52.1	0.671	2.9	2.7	1.111	1.6	1.4	–
Corn, *Zea mays* L., yellow-dent hybrid	10.6	62.9	0.810	348.8	274.0	1.273	12.6	8.3	4.5
Cotton, *Gossypium hirsutum* L., acid-delinted	9.8	–	–	114.3	92.9	1.231	10.0	5.3	4.7
Flax, *Linum ussitatissimum* L.	7.9	50.3	0.647	5.9	5.2	1.135	4.3	2.3	0.9
Grain sorghum *Sorghum bicolor* (L.), Moench	11.2	60.2	0.775	33.2	24.7	1.344	4.5	4.1	3.4
Kidney bean, *Phaseolus vulgaris* L.									
Light red	12.6	57.7	0.743	605.1	420.9	1.438	17.2	8.9	6.1
Dark red	13.2	56.4	0.726	653.1	460.9	1.417	17.4	9.0	6.6
Lentil, *Lens culinaris* Medik.	10.5	58.2	0.748	55.3	40.1	1.380	6.3	6.1	2.2
Oats, *Avena sativa* L.									
Winter, cv. Chapman	10.7	32.6	0.419	34.8	26.8	1.295	10.9	2.8	2.1
Spring, cv. Larry	10.6	35.3	0.454	28.1	21.4	1.314	10.2	2.8	2.2
Peanut, *Arachis hypogaea* L., cv. Florunner	6.0	49.5	0.637	59.5	54.0	1.102	14.5	9.6	8.0
Pea, garden, *Pisum sativum* L.	11.5	64.7	0.833	273.4	202	1.353	8.3	7.0	6.2
Popcorn, *Zea mays* L.									
White	11.1	–	–	108.7	78.0	1.393	8.1	5.1	4.0
Yellow	11.0	–	–	116.2	84.1	1.382	7.4	5.5	4.3

(Continued)

Table 16.1 Physical Properties of Grain and Seed, Including Individual Kernel or Seed Characteristics *Continued*

Description of Grain or Seed	Moisture Content, %	Test Weight lb/bu	Test Weight g/cm³	Seed Weight, mg	Seed Volume, mm³	Seed Density, gcm³	Length, mm	Width, mm	Thickness, mm
Rice, *Oryza sativa* L.									
Long-grain, cv. Lebonnet Rough	15.7	51.3	0.660	23.6	17.0	1.388	8.9	2.3	2.0
Brown	11.5	55.6	0.716	20.9	14.6	1.432	7.4	2.1	1.7
White	11.5	60.1	0.773	18.9	12.7	1.460	6.5	2.1	1.7
Medium-grain, cv. Pecos Rough	15.4	49.8	0.641	24.9	18.0	1.382	7.8	2.9	2.0
Brown	12.0	62.3	0.802	21.5	14.9	1.434	5.6	2.5	1.8
White	12.0	66.1	0.851	17.5	12.0	1.462	5.3	2.5	1.7
Rye, *Secale cereale* L., cv. Wrens Abruzzi	11.5	51.8	0.667	16.6	11.9	1.394	6.4	2.3	2.0
Safflower, *Carthamus tinctorius* L.	9.6	40.9	0.526	34.2	34.1	1.002	7.3	3.6	3.2
Soybean, *Glycene max* (L.), Merrill, cv. Benning	16.7	54.8	0.705	189.5	152.8	1.243	7.3	6.7	5.9
	13.4	55.3	0.712	173.9	139.1	1.250	7.1	6.6	5.7
	10.0	56.2	0.723	167.6	134.1	1.250	7.0	6.6	5.7
Sunflower, *Helianthus annuus* L.									
Oil type	7.6	30.0	0.386	59.5	58.2	1.023	10.7	5.2	3.1
Confectionery	8.7	26.3	0.339	115.8	105.4	1.099	14.4	8.1	4.6
Sweetclover, *Melilotus officinalis* Lam.	8.6	52.3	0.673	1.80	1.40	1.297	1.9	1.3	0.9
Wheat, *Triticum aestivum* L.									
Hard red winter, cv. Arapahoe	16.9	56.1	0.722	27.7	20.2	1.373	5.9	2.8	2.6
	8.6	60.0	0.772	26.0	18.5	1.409	5.8	2.6	2.4
Hard red winter, cv. Keene	12.1	59.3	0.763	29.2	21.0	1.388	5.5	2.9	2.6
Soft red winter, cv. Gore	11.8	58.7	0.756	35.7	26.4	1.345	5.6	3.2	2.9
Soft white winter, cv. Madsen	13.8	–	–	39.7	28.6	1.385	6.4	3.4	2.9
Durum, *Triticum durum* Desf, cv. Ben	10.9	61.2	0.788	36.8	26.1	1.411	6.9	2.8	2.8

Seed Dimensions

Table 16.2 Physical Properties of Barley, *Hordeum vulgare*, L., Seed Lots and Kernels of Indicated Cultivars

Barley Type, (Origin), and Cultivar	Moisture Content, %	Test Weight		Kernel Weight, mg	Kernel Volume, mm³	Kernel Density, g/cm³	Kernel Dimensions		
		lb/bu	g/cm³				Length, mm	Width, mm	Thickness, mm
Spring Barley (Nebraska)									
Beacon	11.0	42.6	0.548	32.9	24.6	1.337	8.7	3.2	2.5
Bowers	11.1	47.4	0.610	31.9	23.5	1.362	7.6	3.1	2.4
Custer	11.2	44.0	0.566	35.1	25.9	1.356	9.4	3.1	2.4
Steptoe	10.9	42.7	0.549	32.9	24.2	1.350	9.5	3.0	2.3
Winter Barley (Georgia)									
Dundy	11.1	48.5	0.624	28.5	20.6	1.381	8.2	2.9	2.3
Hitchcock	11.2	47.8	0.615	26.6	19.7	1.352	7.9	2.9	2.2
Kline	10.8	39.1	0.503	27.8	2102	1.312	8.8	3.1	2.4
Volbar	11.0	43.0	0.554	32.7	24.8	1.317	8.9	3.2	2.5

Table 16.3 Physical Properties of Soybean, *Glycene max* (L.) Merrill, Seed Lots of Different Origins and Cultivars

Description	Moisture Content, %	Test Weight		Seed Weight, mg	Seed Volume, mm³	Seed Density, g/cm³	Seed Dimensions		
		lb/bu	g/cm³				Length, mm	Width, mm	Thickness, mm
Illinois Origin									
Burlson	8.3	58.0	0.747	198.7	158.9	1.250	7.3	6.9	6.0
Edison	8.3	59.1	0.761	170.7	136.	1.248	6.8	6.5	5.8
Resnik	8.4	58.7	0.756	153.0	122.4	1.250	6.8	6.5	5.8
Williams	8.2	59.8	0.769	180.8	144.9	1.248	7.3	6.7	5.7
Georgia Origin									
Bryan	8.2	58.5	0.753	145.0	117.7	1.232	7.2	6.0	5.1
Colquit	7.9	57.0	0.764	152.4	123.6	1.232	7.1	6.5	5.4
Gaysoy 17	8.1	57.4	0.739	129.9	105.0	1.237	6.5	6.3	5.2
Kirby	8.5	60.1	0.774	142.1	113.5	0.252	7.1	6.3	5.2
Nebraska Origin									
Hamilton	8.0	57.4	0.739	142.5	114.9	1.241	6.9	6.4	5.2
Kenwood	7.9	55.8	0.718	105.6	86.7	1.219	6.4	5.7	4.9
Pella 86	8.0	57.5	0.735	157.6	128.1	1.231	7.4	6.5	5.2
Resnik	8.0	57.7	0.743	037.4	110.6	1.243	6.9	6.2	5.3

For both of these commodities, the moisture contents were in equilibrium with the storage chamber, about 40°C and 45% RH. The range of kernel and seed densities was about 5% for barley and only about 2.5% for soybeans relative to mean values.

Dependence of kernel and seed densities on moisture content was not well characterized by these measurements because moisture ranges were not selected for that purpose. However, those relationships have been well studied and presented previously for hard red winter wheat (Nelson and Stetson, 1976), shelled yellow-dent field corn (Nelson, 1979, 1980), and rice (Noh and Nelson, 1989). In those studies, the kernel density for hard red winter wheat decreased from approximately 1.41 to about 1.32 g/cm^3 as moisture content increased from 8% to 24%, which agrees well with the data from Table 16.1. For shelled yellow-dent field corn, kernel densities decreased from above 1.27 to 1.23 g/cm^3 over a similar moisture range, agreeing well with the 1.27 g/cm^3 value for 10.6% moisture corn in Table 16.1. The earlier reported kernel densities for rough, brown, and white rice are also similar to those shown in Table 16.1. The kernel density for canola is very close to that reported earlier (Sokhansanj and Lang, 1996), which varied little at moisture contents from 5% to 19%. Seed densities for other kinds of seed are also in reasonable agreement with those tabulated by Mohsenin (1986).

The kernel densities measured for nine lots of hybrid yellow-dent field corn from Illinois, Iowa, Nebraska, and Georgia (not shown in Table 16.1), all of about 10−11% moisture content, ranged from 1.24 to 1.33 g/cm^3, agreeing well with values obtained earlier by helium pycnometer determinations (Gustafson and Hall, 1972; Martin et al., 1987; Chang, 1988) and air comparison pycnometer determinations for 21 hybrids reported previously (Nelson, 1978, 1979).

Values of kernel and seed densities tabulated as specific gravity values in the ASAE Standards (2000a), which are mainly those reported by Zink (1935), are generally lower than those reported here. Since Zink's data were obtained by mercury displacement measurements, the high surface tension of the mercury may have prevented it from displacing all air on the surface of some kernels or seeds (Chang, 1988). Zink stated that his kernel specific gravities for wheat were about 0.1 lower than those obtained by the more precise method of Bailey and Thomas (1912), in which toluene was used as the liquid and trapped air was aspirated. The kernel or seed densities reported here in Tables 16.1−16.3 are about 0.1 g/cm^3 greater than Zink's earlier data for barley, grain sorghum, rice, wheat, and soybeans, and about 0.2 g/cm^3 greater for oats, rye, and rough rice. Whether modern varieties have greater density than older varieties might be questioned, but values for corn and flaxseed agree well, and the differences for other seeds most likely resulted from differences in the volume measurement technique.

Kernel and seed weights and volumes in Tables 16.1−16.3 are based on the sample weight for the pycnometer measurements and the kernel or seed count and volume determination for that sample. The kernel and seed dimensions listed in Tables 16.1−16.3 are mean values for 50 kernels or seeds measured for each seed lot. Kernel or seed weights were also determined for each of the 50 seeds for which dimensions were measured from each of the 85 lots of different kinds, types, and cultivars. These data were subjected to further analysis. The product of the three orthogonally oriented (mutually perpendicular) dimensions for each seed provides information about the volume of that seed, but it does not give the true volume because seeds are not right rectangular parallelepipeds. Shape classifications, such as roundness and sphericity, have been developed for objects like seeds (Mohsenin, 1986), but the product of the three orthogonal dimensions seems more useful

Table 16.4 Volume Coefficients for Determining Seed Volume from Dimensional Measurements

Kind of Grain or Seed	Number of Lots	Mean Seed Weight, mg	Mean Seed Density, g/cm³	Calculated Seed Volume, mm³	Volume Coefficient		
						Sample Standard Deviation	
					Value	Among Lots	Within Lots
Alfalfa	1	1.9	1.301	1.5	0.562	–	0.075
Barley	8	33.8	1.346	25.2	0.397	0.018	0.041
Corn	9	354.4	1.292	274.1	0.574	0.024	0.065
Cotton	1	118.4	1.231	96.1	0.384	–	0.038
Flax	2	5.6	1.136	4.9	0.528	0.018	0.015
Grain sorghum	1	33.8	1.344	25.2	0.486	–	0.024
Kidney bean	4	581.6	1.405	413.9	0.485	0.015	0.050
Lentil	2	50.6	1.396	36.3	0.460	0.005	0.023
Oats	4	28.9	1.296	22.3	0.372	0.017	0.063
Peanut	1	621.7	1.102	564.2	0.505	–	0.028
Peas, garden	2	246.3	1.342	183.3	0.579	0.018	0.037
Popcorn	2	122.2	1.388	88.1	0.510	0.003	0.047
Rice, Rough	3	25.3	1.385	18.2	0.412	0.032	0.058
Brown	3	21.3	1.429	14.9	0.552	0.010	0.031
White	3	19.0	1.458	13.0	0.535	0.018	0.030
Rye	1	19.1	1.394	13.7	0.465	–	0.051
Safflower	2	37.0	1.016	36.4	0.416	0.017	0.046
Sunflower, oil type	2	56.0	1.028	54.5	0.222	0.015	0.051
Confectionery	2	134.4	1.109	121.2	0.365	0.046	0.061
Soybean	19	162.9	1.243	130.9	0.514	0.007	0.021
Sweetclover	1	1.8	1.297	1.4	0.588	–	0.054
Wheat	12	33.4	1.369	24.4	0.525	0.019	0.038

for the various shapes encountered in this study. Since the mean seed densities were determined from pycnometer measurements, the volume of each seed can be closely estimated by dividing the seed weight by the density. Then, by dividing the estimated seed volume by the product of the three dimensions for that particular seed, a volume coefficient is obtained. This volume coefficient can be used to estimate seed volumes of other kernels or seeds for a particular kind of crop from the dimensions of those seeds. The product of the three seed dimensions, when multiplied by the volume coefficient for that kind of seed, provides an estimate of the seed volume. In addition, dividing the seed weight by the volume of the seed provides an estimate of its density.

The results of this analysis, applied to all of the kernels and seeds for which three dimensions were measured, are presented in Table 16.4.

The volume coefficients listed are mean values for the number of lots measured. The volume-coefficient sample standard deviations, among lots, for those means of the 50-kernel averages provide a measure of the variation among the lots measured for a given type. The standard deviations among the means of the lots for a given type of grain or seed are rather small, in relation to the volume coefficients, for most of the grain and seed types, indicating that shapes within types are quite uniform.

The sample standard deviations, within lots, listed in Table 16.4, are mean values of the average standard deviations for 50-kernel samples of each lot. These standard deviations provide a measure of the variation in shape among the individual kernels or seeds in a given lot, and they are consistently larger than the standard deviations of volume coefficients among lots. These standard deviations include not only variations in kernel or seed shape but also variations in kernel or seed densities within a given lot. Some information is available on the density distributions among kernels of given grain lots (Peters and Katz, 1962; Shelef and Mohsenin, 1968). Field corn is notorious for its variation in kernel shape and size, and it has relatively high standard deviations for volume coefficients in both categories. Soybean volume coefficients, on the other hand, have low standard deviations both within lots and among lots.

Lentil and soybean have the lowest standard deviations within lots of those grain and seed types represented in Table 16.4. Soybeans are nearly spherical, and they have a volume coefficient of 0.514. It is interesting that a sphere of unit diameter in a cube of unit dimensions on each side occupies 0.524 of the space in the cubic parallelepiped. Thus, the soybean and the sphere have about the same volume coefficients, as defined here. Volume coefficients developed, as shown in Table 16.4, range from 0.222 for the oil-type sunflower to 0.588 for sweetclover seed. Volume coefficients for the confectionery sunflower and oats are also relatively low, whereas the volume coefficients for garden peas, field corn, and the other small-seeded legume, alfalfa, are very high.

16.2.5 SUMMARY

Data are presented for reference on kernel and seed dimensions, weights, volumes, and densities for a selection of different grains and seeds of other crops. Comparisons of these data are presented for several cultivars of barley and soybeans that were grown in different regions of the USA. Within grain and crop types, kernel and seed densities are relatively consistent, but they do vary within given seed lots depending on moisture content. Generally, oil seeds have lower seed densities, and cereal grains such as rice, corn, wheat, and barley have higher seed densities. A volume coefficient was developed for seeds of the various crops that is useful for estimating the volume of kernels or seeds of those crops by multiplying it times the product of three orthogonally-oriented dimensional measurements (length, width, and thickness) of the seeds. The volume coefficient varied from 0.222 for oil-type sunflower seeds to 0.588 for sweetclover seeds and had relatively low values for other oil seeds, except flax, peanut, and soybean, and relatively high values for cereal grains such as rice, corn, and wheat. Kernel or seed densities can also be estimated from seed dimensions and weight by using the volume coefficients presented for the various grains and seeds of other crops.

REFERENCES

AOAC, 1970. Sec. 14.004 Air Oven Method. In: Horowitz, W. (Ed.), Official Methods of Analysis of the Association of Official Analytical Chemists, eleventh ed. Association of Analytical Chemists, Washington, DC.

ASAE, D241.4, 2000a. Density, Specific Gravity, and Mass–Moisture Relationships for Grain for Storage, forty seventh ed. ASAE Standards, American Society of Agricultural Engineers, St. Joseph, MI.

ASAE S352.2., 2000b. Moisture measurement — Unground grain and seeds, ASAE Standards 2002, forty ninth ed. American Society of Agricultural Engineers, p. 589. Published by the American Society of Agricultural Engineers, St. Joseph, Michigan.

Bailey, C.H., 1916. The relation of certain physical characteristics of the wheat kernels to milling quality. J. Agric. Sci. 7 (Part IV), 432–442.

Bailey, C.H. and Thomas, L.M., 1912. A method for the determination of the specific gravity of wheat and other cereals, USDA Bureau of Plant Industry. Circular No. 99, US Department of Agriculture, Washington, DC.

Brusewitz, G.H., 1975. Variation of the bulk density of cereals with moisture content. Trans. ASAE 18 (5), 935–938.

Chang, C.S., 1988. Measuring density and porosity of grain kernels using a gas pycnometer. Cereal Chem. 65 (1), 13–15.

Chung, D.S., Converse, H.H., 1971. Effects of moisture content on some physical properties of grains. Trans. ASAE 14 (4), 612–614.

Fortes, M., Okos, M.R., 1980. Changes in physical properties of grain during drying. Trans. ASAE 23 (4), 1004–1008.

Gustafson, R.J., Hall, G.E., 1972. Density and porosity changes of shelled corn during drying. Trans. ASAE 15 (3), 523–525.

Hall, G.E., 1972. Test weight changes of shelled corn during drying. Trans. ASAE 15 (2), 320–323.

Hall, G.E., Hill, L.D., 1974. Test weight adjustment based on moisture content and mechanical damage of corn kernels. Trans. ASAE 17 (3), 578–579.

Haugh, C.G., Lein, R.M., Haynes, R.E., Ashman, R.B., 1976. Physical properties of popcorn. Trans. ASAE 19 (1), 168–171.

Hill, H.J., Taylor, A.G., Min, T.-G., 1989. Density separation of imbibed and primed vegetable seeds. J. Soc. Hortic. Sci. 114 (4), 661–665.

ISTA, 1966. International rules for seed testing. Proc. Int. Seed Test. Assoc. 31 (1), 128–134.

Leopold, C.A., 1983. Volumetric components of seed imbibition. Plant Physiol. 73, 677–680.

Martin, C.R., Czuchajowska, Z., Pomeranz, Y., 1987. Evaluation of digitally filtered aquagram signals of wet and dry corn mixtures. Cereal Chem. 64 (5), 356–358.

Mohsenin, N.N., 1986. Physical Properties of Plant and Animal Materials. Gordon and Breach Science Publishers, New York, NY.

Nelson, S.O., 1976. Microwave dielectric properties of insects and grain kernels. J. Microw. Power 11 (4), 299–303.

Nelson, S.O., 1978. Radiofrequency and Microwave Dielectric Properties of Shelled Field Corn. Agricultural Research Service, USDA, ARS-S-184, Washington, DC.

Nelson, S.O., 1979. RF and microwave dielectric properties of shelled, yellow-dent field corn. Trans. ASAE 22 (6), 1451–1457.

Nelson, S.O., 1980. Moisture-dependent kernel- and bulk-density relationships for wheat and corn. Trans. ASAE 23 (1), 139–143.

Nelson, S.O., 1983a. Observations on the density dependence of the dielectric properties of particulate materials. J. Microw. Power 18 (2), 143–152.

Nelson, S.O., 1983b. Density dependence of the dielectric properties of particulate materials. Trans. ASAE 26 (6), 1823–1825, 1829.

Nelson, S.O., 1984. Density dependence of the dielectric properties of wheat and whole-wheat flour. J. Microw. Power 19 (1), 55–64.

Nelson, S.O., 1992. Correlating dielectric properties of solids and particulate samples through mixture relationships. Trans. ASAE 35 (2), 625–629.

Nelson, S.O., 2002a. Dimensional and density data and relationships for seeds of agricultural crops. Seed Technol. 24 (1), 76–88.

Nelson, S.O., 2002b. Dimensional and density data for seeds of cereal grain and other crops. Trans. ASAE 45 (1), 165–170.

Nelson, S.O., Stetson, L.E., 1976. Frequency and moisture dependence of the dielectric properties of hard red winter wheat. J. Agric. Eng. Res. 21, 181–192.

Nelson, S.O., You, T.-S., 1989. Microwave dielectric properties of corn and wheat kernels and soybeans. Trans. ASAE 32 (1), 242–249.

Noh, S.H., Nelson, S.O., 1989. Dielectric properties of rice at frequencies from 50 Hz to 12 GHz. Trans. ASAE 32 (3), 991–998.

Peters, W.R., Katz, R., 1962. Using a density gradient column to determine wheat density. Cereal Chem. 39 (6), 487–494.

Shelef, L., Mohsenin, N.N., 1968. To determine the density spectrum of individual grains. Agric. Eng. 49 (1), 28.

Sokhansanj, S., Lang, W., 1996. Prediction of kernel and bulk volume of wheat and canola during adsorption and desorption. J. Agric. Eng. Res. 63, 129–136.

Sokhansanj, S., Nelson, S.O., 1988. Dependence of dielectric properties of whole-grain wheat on bulk density. J. Agric. Eng. Res. 39, 173–179.

Stroshine, R.J., Crane, P., 1987. Effect of Kernel Physical Properties on Shelled-Corn Thin-Layer Drying Rates. ASAE Paper No. 87-6557, American Society of Agricultural Engineers, St. Joseph, MI.

Thompson, R.A., Isaacs, G.W., 1967. Porosity determinations of grains and seeds with an air-comparison pycnometer. Trans. ASAE 10 (5), 693–696.

USDA, 1953. The Test Weight per Bushel of Grain: Methods of Use and Calibration of Apparatus. Circular No. 921, US Department of Agriculture, Washington, DC.

USDA, 1970. Official Grain Standards of the United States. US Department of Agriculture, Consumer and Marketing Service, Grain Division, Washington, DC.

USDA, 1986. Air-oven methods. In: F.G.I. Service Moisture Handbook. US Department of Agriculture, Washington, DC (Chapter 4).

You, T.-S., Nelson, S.O., 1988. Microwave dielectric properties of rice kernels. J. Microw. Power Electromagn. Energy 23 (3), 150–159.

Zink, F.J., 1935. Specific gravity and air space of grains and seeds. Agric. Eng. 16 (11), 439–440.

Index

Note: Page numbers followed by '*f*' and '*t*' refer to figures and tables, respectively.

A

Admittance Meter, 11–12, 21
Alfalfa seed, 68, 224, 224*f*
  bacteria on, 38–39
  chopped alfalfa, drying of, 73
Alfalfa seed studies, 58–63
  basic factors, 59–60
  experimental findings, 60–63
    field intensity, 61
    frequency, 60–61
    moisture content, 61
    temperature effects, 62
    variations, 62
  hard-seed problem, 58
  practical application, aspects of, 63
Alsike clover, 65
Amphibole (richterite), 135
Animal tissues, dielectric properties of, 212
Apple, 153, 154*f*, 155
Apple juice, 150*f*, 151, 158
Apple studies, 125–126
Arachis hypogaea L. Peanuts, 94–95, 188
Attenuation, 13, 34–35, 113–115
Attenuation constant, 34
Audio frequencies, 14
Avena sativa L., 65, 111–112, 188, 224

B

Banana, 152–153, 153*f*
Barley. *See also* Spring barley; Winter barley
  loose smut in, 37–38
  microwave data for, 232, 235*t*
  models for, 182–183
  physical properties of, 259*t*
Battery-operated moisture meter, 78
Beckman Model 930 air comparison pycnometer, 135, 137, 256
Biofuels, 167
Birdsfoot trefoil, 65
'Brazos', 183, 216, 233*t*
Broadband measurements, 26–27
  impedance and network analyzer, 27
  open-ended coaxial-line, 27
  time-domain, 26–27
Broad-frequency-range measurements, 142
Brown rice, 216, 221*f*

Bulk densities, 7, 82, 84–86, 89, 115–116, 135, 171, 175–176, 183, 185, 247–249
  determination, for pine pellets, 169–170

C

Calibration function, 84–86, 170
  moisture calibration function, 170, 170*f*
Canola, 188, 232, 237*t*, 256, 260
Cantaloupe, 125, 154–155, 154*f*, 156*f*
Carageenan, 159
Carotene retention, 73
Carya illinoensis (Wangenh.) K. Koch, 97
Cereal grain, composite model for, 185–187
  density dependence, 185–186
  frequency dependence, 186
  microwave data for, 232–237
  model development, 187
  moisture dependence, 186–187
Cheese, 147, 148*f*, 150
Chicken breast meat, 152, 153*f*, 157, 160*f*, 161, 161*f*, 162*f*, 163, 163*f*
Chicken meat, 159–160, 163
  for quality sensing, 159–163
Chlorite (clinochlore), 135
Chopped alfalfa, drying of, 73
Closed-structure methods, 12
Coal, 131
  dielectric properties measurements on, 131–132
  and limestone measurements, 136–142
Coal mines, rock dusting in, 142
Coal–pyrite mixtures, dielectric heating of, 132–134
Coaxial sample holder, 11–16, 18, 19*f*, 20–21, 27
  lumped-circuit parameter model of, 22*f*
  open-circuit, 12
  sectional view of, 15*f*
  sectional view with Q-meter of, 17*f*
Cole–Cole diagram, 3, 4*f*
Commonly used grain moisture meters, 89
Complex Refractive Index, 7–8
Conductance-type moisture meter, 78
Confused flour beetles, 43–47
Corn, 13, 84, 89–90
  models for, 181–182
  single-kernel grain moisture measurements on, 90–93
  single-kernel moisture sensing techniques for, 95–97

Corn lots, 89–90, 92–93, 248
Cryptolestes ferrugineus (Stephens), 50
Curculio caryae (Horn), 51–52
Curing process, 126

D

Data, 211
    earlier tabulations, 211–212
    fresh fruit and vegetable data, 238
    hard red winter wheat data, 212–215, 213*f*, 214*f*, 233*t*
    individual grain kernels and seeds data, 232
    insect data, 244
    microwave data for cereal grains and oilseeds, 232–237
    pecan nut data, 238–244
    rice data, 216–223, 220*f*, 221*f*, 222*f*, 223*f*
    shelled hybrid yellow-dent field corn data, 216, 216*f*, 217*f*, 218*f*, 219*f*, 233*t*
Dates, sensing the moisture content of, 127–129
Dc conductance instrument, 94, 97
Dc conductance meters, 83
Dc conductance of peanut kernels, 95
Debye equation, 3–4, 158*f*
Density dependence, 6–8, 185–186
Density measurements, 248–249, 256
Density ratio, 117
Density-independent moisture calibration function, 84, 86–89, 86*f*, 87*f*, 127, 170–171
Density-independent nature of moisture sensing, 88–89
Dielectric constant, 2, 4*f*, 23, 80, 82–84, 88, 118, 123–124, 126, 142, 147–151, 155, 157, 159
Dielectric heating, 33–34, 41–42, 112–113
    of coal–pyrite mixtures, 132–134
Dielectric loss factor, 2, 4*f*, 23, 25, 59–60, 126, 160–161, 189*f*, 190*f*
Dielectric properties, 77–82
Dielectric relaxation, 5–6, 150, 155
Dielectric-type capacitance-sensing moisture meter, 78
Dimensional and weight measurements, 256
Dried Fruit Moisture Tester, 127–128
Dry seeds, 112
Dutch clover, 65

E

Eastern white pine, 66–67
Electric field intensity, 34–35, 42, 45
Electromagnetic energy, 2
Electronic moisture meter, 78, 94–95

F

Feldspar (labradorite), 135
Food materials, dielectric properties of, 147
    chicken meat, for quality sensing, 159–163
    hydrocolloid food ingredients, measurements on, 159
    measurement of, 147–159
Free water, influence of, 158
Free-space measurement arrangement, 168*f*
Free-space transmission measurements, 25–26, 168
Frequency dependence, 3–6, 132, 186
Fresh fruit and vegetable data, 238

G

'Gage' hard red winter wheat, 224
General Radio 1602-B U-H-F Admittance Meter, 21
General Radio 1608-A precision impedance bridge, 14
Glycene max (L.) Merrill, 74, 93, 182, 224
    physical properties of, 259*t*
Goethite, 135
GR 874 connector, 21, 21*f*
Grain and seed lots, 255–256
Grain moisture content, 87–88
Grain sorghum, microwave data for, 224*f*, 232
Ground whole wheat, permittivity of, 149*f*, 151
Gum arabic, 159

H

Hand sorters, 128
Hard red winter wheat, 47
    bulk density of, 250*f*
    data, 212–215, 213*f*, 214*f*, 233*t*
    kernel density of, 249*f*
    results for, 249–250
Hard seeds, 38–39, 58–59, 65–66
Hematite, 135
Hewlett-Packard 4191A RF Impedance Analyzer, 27
Hewlett-Packard 4192A LF Impedance Analyzer, 27
Hewlett-Packard 8753C network analyzer, 27
Hewlett-Packard vector network analyzer (VNA), 168
High field intensities, 46
Honeydew melon, 155, 156*f*
Hordeum vulgare, L., 37–38, 182–183, 224
    physical properties of, 259*t*
Horn-lens antennas, 168–169
Hydrocolloids, 159
    measurements on, 159

I

Ilmenite, 135
Impedance analyzers, 14, 27
Insect control applications, 41
    developmental stage
        pecan insects, 51–53
        stored-grain insects, 41–51

entomologic factors, 43—45
experimental findings, 42—43
physical factors, 45—50
practical aspects, 50—51
selective dielectric heating, 41—42
Insect data, 244
Ionic conduction, 147—151, 157—158, 160—161, 163

J

Jeffrey pine, 66—67

K

Karl Fischer titration method, 248
Kernel moisture content, 95, 98—99, 204*f*, 206—208
Kernel moisture equilibrium, 96
Kernel moisture meter. *See* Peanut kernel moisture meter
Kernel moisture variation, 89

L

Ladino clover, 65
Landau and Lifshitz, Looyenga dielectric mixture equation,
    82, 137, 140, 142, 196
Landau and Lifshitz, Looyenga relationships, 7—8
'Lebonnet' rice, 216, 233*t*
Limestone, 136—137
Listeria monocytogenes, 38—39
Loblolly pine, 66—67
Locust bean gum, 159
Lolium parenne L, 65
Loss angle, 91—92
Lotus corniculatus L., 65

M

Macaroni, 147, 148*f*, 150
Manganese oxide (hollandite), 135
Materials, dielectric properties of, 1—2
    density dependence, 6—8
    frequency dependence, 3—6
    temperature dependence, 6
Mathematical models for dielectric constant, 82, 175,
    180—181
Measurement of dielectric properties, 11
    from 1 to 50 MHz, 15—19
    from 50 to 250 MHz, 20
    from 200 to 500 MHz, 21—22
    audio frequencies, 14
    broadband measurements, 26—27
        impedance and network analyzer, 27
        open-ended coaxial-line, 27
        time-domain, 26—27
    microwave frequencies, 22—26

free-space techniques, 25—26
short-circuited line technique, 22—24
Medicago sativa L., 38, 58
Melilotus alba Deer., 64—65
Melon studies, 124—125
Mica (muscovite), 135
Mica (phlogopite), 135
Microcontroller, addition of, 205
Microwave attenuation, 36
Microwave data for cereal grains and oilseeds, 232—237
Microwave dielectric properties, 12—13, 23—24
Microwave energy applications, 109—112, 119
Microwave frequencies
    measuring, 22—26
        free-space techniques, 25—26
        short-circuited line technique, 22—24
    models at, 188—191
Microwave heating, 34—36, 73—74, 112—113, 133
Microwave measurements, 87
Microwave moisture meter, 88
Microwave moisture sensing, 84
Microwave moisture-sensing instrumentation, 195
    microwave moisture sensor prototype, 198*f*
    peanut drier control by monitoring kernel moisture content,
        206—208
        background information, 206—207
        dryer control system tests, 207—208
        peanut dryer control system, 207
    peanut kernel moisture meter, 195—205
        background, 195—196
        dielectric properties, 196
        practical meter development, 196—205
Microwave resonant cavity, 98—99
Microwave sensing, 167
Minerals, 134
    dielectric properties measurements of, 134—135
Minimum energy absorbed (MEA), 74
Mining applications, 131
    coal, 131
        dielectric properties measurements on, 131—132
        and limestone measurements, 136—142
    coal—pyrite mixtures, dielectric heating of, 132—134
    minerals, 134
        dielectric properties measurements of, 134—135
    sensing pulverized material mixture proportions, 142—144
        mixture proportions, measuring, 143—144
        resonant cavity measurement, principles of, 143
Mixture proportions, measuring, 143—144
Models
    for barley, 182—183
    for cereal grain. *See* Cereal grain, composite model for
    for corn, 181—182

Models (*Continued*)
  at microwave frequencies, 188−191
  for rice, 183−185
  for soybeans, 182
  for wheat, 175−181
Moisture content, 151−153, 155, 157, 159, 171
  alfalfa seed studies, 61
  determination, 170−171
  and insect control, 46
  measurement, in single kernels of peanuts, 94−95
  sensing, 83−88
Moisture dependence of kernel and bulk densities, 247−254
  corn lots, 248
  density measurements, 248−249
  hard red winter wheat, results for, 249−250
  moisture determinations, 248
  porosity of wheat and corn, 253
  wheat lots, 247−248
  yellow-dent field corn, results for, 251−253, 251*f*, 252*f*
Moisture determinations, 248
Moisture distribution in the kernel, 96
Moisture meter
  battery-operated, 78
  capacitance-type, 78
  conductance-type, 78
  dielectric-type capacitance-sensing, 78
  electronic, 94−95
  microwave, 88
  single-kernel, 93−94
  tag-Heppenstall grain, 89−90
Moisture sensing applications, 77
  background information, 77
  development of, 88−89
  dielectric properties, 78−82
  early history, 77−78
  four single-kernel moisture sensing techniques for corn, comparison of, 95−97
  moisture content sensing, 83−88
  single-kernel grain moisture measurements on corn, 90−93
    conductance and impedance measurements, 92−93
    parallel-plate high-frequency impedance measurement technique, 90−92
  single-kernel microwave-resonator moisture sensing, 99−103
  single kernels of peanuts, measuring moisture content in, 94−95
  single nut and kernel pecan moisture sensing, 97−99
  single soybean seed moisture measurements, 93−94

N
'Neal' spring oats, 224
Nematodes, 109−110
Network analyzer measurements, 27

O
Oats, microwave data for, 224*f*, 232, 235*t*
Oilseeds, microwave data for, 232−237
Onion curing, 126
Onion studies, 126−127
Open-ended coaxial-line probe, 13, 27, 147, 159
Open-structure techniques, 12
Orange tissue, 155, 155*f*
Oryza sativa L., 183, 216

P
Parallel-plate high-frequency impedance measurement technique, sensor principles for, 90−92
Particle densities, 135, 137
Peanut drier control by monitoring kernel moisture content, 206−208
  background information, 206−207
  dryer control system tests, 207−208
  peanut dryer control system, 207
Peanut kernel moisture content, 206−207
Peanut kernel moisture meter, 195−205
  background, 195−196
  dielectric properties, 196
  practical meter development, 196−205
    addition of microcontroller, 205
    field testing of microwave moisture meter, 203−205
    laboratory calibration, 200−203
    moisture determination, 200
Peanut kernels
  dc conductance of, 95
  moisture content of, 87
Peanut-hull pellets, 171−173
Pecan insects, 51−53
Pecan nut data, 238−244
Pecans, quality maintenance in, 74−75
'Pecos', 216
Pelleted sawdust. *See* Pine pellets
Penetration depth, 35
Perennial ryegrass, 65
Permittivity, 2, 7
  -based calibration functions, 84−85
  -based maturity index, 124
Permittivity, absolute, 2
Pest control, assessment of soil treatment for, 109
  attenuation, 113−115
  basic principles, 112−113
  initial assessment, 111−112
  selective heating, 115−117
  soil insect and nematode treatment, 117−118
  soil microorganisms and nematodes, 109−110
  soil treatment for weed control, 110−111
Phase angle, 91−92

Phase constant, 34
Physical properties data for grain and seed, 247
   grain kernel and seed dimensions and densities for
       agricultural crops, 254−262
      density measurements, 256
      dimensional and weight measurements, 256
      grain and seed lots, 255−256
      resulting data, 256−262
   moisture dependence of kernel and bulk
       densities, 247−254
      corn lots, 248
      density measurements, 248−249
      hard red winter wheat, results for, 249−250
      moisture determinations, 248
      porosity of wheat and corn, 253
      wheat lots, 247−248
      yellow-dent field corn, results for, 251−253, 251f, 252f
Pine pellets, 167−171
   bulk density determination, 169−170
   moisture content determination, 170−171
Pine seed, 66−67
Pinus jeffreyi Grev. and Balf., 66−67
Pinus lambertiana, 66−67
Pinus lambertiana Dougl., 66−67
Pinus monticola Dougl., 68
Pinus ponderosa Laws, 66−67
Pinus sabiniana, 67
Pinus sabiniana Dougl., 66−67
Pinus strobus, 66−67
Pinus strobus L., 66−67
Pinus taeda L., 66−67
Polarization, 3
Ponderosa pine, 66−68
Porosity, 255
   of wheat and corn, 253
Potato starch, dielectric constant of, 159
Precision vernier capacitor, 18, 20f
Product conditioning applications, 73
   chopped alfalfa, drying of, 73
   nutritional value of soybeans, improving, 74
   pecans, quality maintenance in, 74−75
Product quality sensing, 36−37
Prototype instrument, 93
Pseudotsuga menziesii var. menziesii (Mirb.)
      Franco, 67−68
Pulverized material mixture proportions,
      sensing, 142−144
   mixture proportions, measuring, 143−144
   resonant cavity measurement, principles of, 143
Pycnometer measurements on limestone samples, 137
Pyrite, 131−133
Pyroxene (salite), 135

Q
Q-factor, 100−103
Q-Meter, 11, 15−18, 16f, 17f, 19f, 20f
Quality sensing, dielectric properties of chicken meat for,
      159−163
Quality sensing in fruits and vegetables, 123
   apple studies, 125−126
   background information, 123
   melon studies, 124−125
   onion studies, 126−127
   sensing the moisture content of dates, 127−129

R
Radio frequency (RF), 1, 11
   dielectric heating, 33, 35−37
   electric fields, 58
   electric fields, 33, 124
   impedance and microwave cavity measurements, 96
   insect-control treatments, 42−43
   treatments, 67
Rectangular-waveguide systems, 135
Red clover, 59, 65
Red winter wheat, 148f, 151, 152f
Relaxation frequency, 5−6
Resonant cavity measurement, principles of, 143
Rice, models for, 183−185
Rice data, 216−223, 220f, 221f, 222f, 223f
Rice weevils, 46
Rock dusting in coal mines, 142
Rock fragmentation, 134, 136
Rough rice, 216, 220f, 223f
RX Meter, 11−12, 20

S
'Scout 66' hard red winter wheat, 215f
Secale cereale L., 65, 224
Seed densities, 255, 260
Seed treatment applications, 57
   alfalfa seed studies, 58−63
      basic factors, 59−60
      experimental findings, 60−63
      practical application, aspects of, 63
   small-seeded legumes and cereals, 65
   sweetclover seed, 64−65
   tree seed, 66−68
      pine seed, 66−67
   vegetable seed, 65−66
   woody plant seeds, 67−68
Seed volume, 260−261
Seed weight, 256, 260−261

Seed-borne pathogens, treating, 37–39
    bacteria on alfalfa seed, 38–39
    loose smut in barley, 37–38
Selective dielectric heating, 41–42
    of pyrite, 131–132
Selective heating, 115–117
Sensor principles for parallel-plate high-frequency impedance
        measurement technique, 90–92
Shelled hybrid yellow-dent field corn data, 216, 216*f*, 217*f*,
        218*f*, 219*f*, 233*t*
Short-circuited line technique, 22–24
Short-circuited-waveguide permittivity measurements, 124
Single corn kernel conductance and impedance measurements,
        92–93
Single kernels of peanuts, measuring moisture content in,
        94–95
Single nut and kernel pecan moisture sensing, 97–99
Single soybean seed moisture measurements, 93–94
Single-kernel grain moisture measurements on corn, 90–93
    conductance and impedance measurements, 92–93
    parallel-plate high-frequency impedance measurement
        technique, sensor principles for, 90–92
Single-kernel microwave-resonator moisture sensing, 99–103
Single-kernel moisture meter, 93–94
Single-kernel moisture sensing techniques for corn,
        comparison of, 95–97
Sitophilus oryzae L., 117–118
Small-seeded legumes and cereals, 65
Soft red winter wheat, 224, 229*f*, 230*f*
Soil insect and nematode treatment, 117–118
Soil microorganisms and nematodes, 109–110
Soil-borne pathogens, 109
Solid biofuels, sensing moisture and density of, 167
    peanut-hull pellets, 171–173
    pine pellets, 167–171
        bulk density determination, 169–170
        moisture content determination, 170–171
Sorghum bicolor (L.) Moench, 224
Soybeans, 224, 231*f*, 232*f*, 233*t*
    microwave data for, 224*f*, 232, 236*t*
    models for, 182
    nutritional value, 74
    physical properties of, 259*t*
Spring barley, 224, 225*f*, 226*f*
Spring oats, 224, 226*f*
Standard errors of calibration (SEC), 98–99
Standing wave ratios (SWRs), 13, 22–23
Steam heating treatments, 75
Stored-grain insects, 41–51
    entomologic factors, 43–45
    experimental findings, 42–43

physical factors, 45–50
practical aspects, 50–51
selective dielectric heating, 41–42
Studies on the use of dielectric properties, 123–129
    apple studies, 125–126
    melon studies, 124–125
    onion studies, 126–127
    sensing the moisture content of dates, 127–129
Sugar content of fresh fruit tissues, 155
Sugar pine, 66–67
Sweetclover seed, 64–65
Sweetness, 124

T
Tag-Heppenstall grain moisture meter, 89–90
Temperature dependence, 6
Test weight, 247–249, 251–252, 255–256
Time-domain measurements, 26–27
Tree seed, 66–68
    pine seed, 66–67
Trifolium hybridum L., 65
Trifolium pratense L., 65
Trifolium repens L., 65
Triticum aestivum L., 212, 224
Trypsin inhibitor, 74

U
Ultrahigh frequency (UHF) radiation control, 111
Unified grain moisture algorithm, 83–84, 88
Ustilago nuda (Jens.) Rostr., 37

V
Vector network analyzer (VNA), 14, 168
Vegetable seed, 65–66
Velocity of light, 2
Velocity of propagation, 2
Vernier capacitor, 15–18
Vidalia onions, 127

W
Water content, 195. *See also* Moisture content
Watermelon, 155, 157*f*
Watermelon studies, 124–125
Waveguide measurement, 12, 14, 22
'Wayne' soybeans, 224
Weed control, soil treatment for, 110–111
Wheat
    dielectric properties of, 157
    microwave data for, 224*f*, 232, 233*t*
Wheat lots, 247–248

Whey protein gel, 148*f*, 149*f*, 150
White rice, 216, 222*f*
Wide frequency ranges, 238
Winter barley, 224, 227*f*, 228*f*, 229*f*
Winter oats, 65
Winter rye, 65, 224, 230*f*, 231*f*
Woody plant seeds, 67–68

Y
Yellow mealworm larvae, 44–45
Yellow-dent field corn, 251–253, 251*f*, 252*f*

Z
Zea mays L., 181–182

Printed in the United States
By Bookmasters